T0190284

NEUROMETHODS

Series Editor
Wolfgang Walz
University of Saskatchewan
Saskatoon, SK, Canada

For further volumes:
http://www.springer.com/series/7657

Spatial Learning and Attention Guidance

Edited by

Stefan Pollmann

Institute of Psychology, Otto-von-Guericke-University-Magdeburg, Magdeburg, Germany

Editor
Stefan Pollmann
Institute of Psychology
Otto-von-Guericke-University-Magdeburg
Magdeburg, Germany

ISSN 0893-2336 ISSN 1940-6045 (electronic)
Neuromethods
ISBN 978-1-4939-9950-7 ISBN 978-1-4939-9948-4 (eBook)
https://doi.org/10.1007/978-1-4939-9948-4

© Springer Science+Business Media, LLC, part of Springer Nature 2020
This work is subject to copyright. All rights are reserved by the Publisher, whether the whole or part of the material is concerned, specifically the rights of translation, reprinting, reuse of illustrations, recitation, broadcasting, reproduction on microfilms or in any other physical way, and transmission or information storage and retrieval, electronic adaptation, computer software, or by similar or dissimilar methodology now known or hereafter developed.
The use of general descriptive names, registered names, trademarks, service marks, etc. in this publication does not imply, even in the absence of a specific statement, that such names are exempt from the relevant protective laws and regulations and therefore free for general use.
The publisher, the authors, and the editors are safe to assume that the advice and information in this book are believed to be true and accurate at the date of publication. Neither the publisher nor the authors or the editors give a warranty, expressed or implied, with respect to the material contained herein or for any errors or omissions that may have been made. The publisher remains neutral with regard to jurisdictional claims in published maps and institutional affiliations.

This Humana imprint is published by the registered company Springer Science+Business Media, LLC, part of Springer Nature.
The registered company address is: 233 Spring Street, New York, NY 10013, U.S.A.

Preface to the Series

Experimental life sciences have two basic foundations: concepts and tools. The *Neuromethods* series focuses on the tools and techniques unique to the investigation of the nervous system and excitable cells. It will not, however, shortchange the concept side of things as care has been taken to integrate these tools within the context of the concepts and questions under investigation. In this way, the series is unique in that it not only collects protocols but also includes theoretical background information and critiques which led to the methods and their development. Thus it gives the reader a better understanding of the origin of the techniques and their potential future development. The *Neuromethods* publishing program strikes a balance between recent and exciting developments like those concerning new animal models of disease, imaging, in vivo methods, and more established techniques, including, for example, immunocytochemistry and electrophysiological technologies. New trainees in neurosciences still need a sound footing in these older methods in order to apply a critical approach to their results.

Under the guidance of its founders, Alan Boulton and Glen Baker, the *Neuromethods* series has been a success since its first volume published through Humana Press in 1985. The series continues to flourish through many changes over the years. It is now published under the umbrella of Springer Protocols. While methods involving brain research have changed a lot since the series started, the publishing environment and technology have changed even more radically. Neuromethods has the distinct layout and style of the Springer Protocols program, designed specifically for readability and ease of reference in a laboratory setting.

The careful application of methods is potentially the most important step in the process of scientific inquiry. In the past, new methodologies led the way in developing new disciplines in the biological and medical sciences. For example, Physiology emerged out of Anatomy in the nineteenth century by harnessing new methods based on the newly discovered phenomenon of electricity. Nowadays, the relationships between disciplines and methods are more complex. Methods are now widely shared between disciplines and research areas. New developments in electronic publishing make it possible for scientists that encounter new methods to quickly find sources of information electronically. The design of individual volumes and chapters in this series takes this new access technology into account. Springer Protocols makes it possible to download single protocols separately. In addition, Springer makes its print-on-demand technology available globally. A print copy can therefore be acquired quickly and for a competitive price anywhere in the world.

Saskatoon, SK, Canada ***Wolfgang Walz***

Preface

Human behavior depends to a large degree on spatial learning. This is obvious when we search for an object, explore a scene or navigate an environment. Research on these topics has asked how visual search can be guided by both short-term and long-term memory, often gained by implicit learning. Recent experiences influence where we look, in the form of priming. However, repeated spatial configurations can also lead to long-lasting improvements of search. What is learned is often of a statistical, probabilistic nature. The search improvements manifest themselves in reduced search times, and also in more efficient gaze paths. Eye movements are important both for encoding environmental aspects into memory and for making use of learned spatial patterns for efficient search. While these processes are mostly implicit, spatial memory is often explicit, e.g., when we hold information in working memory for an ongoing task. Working memory, in turn, interacts with long-term memory and is increasingly seen as an activated part of long-term memory.

The neural code of spatial learning and memory has been intensively investigated in the recent years. In particular, the storage and manipulation of information in working memory are currently an intense focus of research. Novel EEG and MEG analysis methods yield insights in the processes of memory storage, whereas both structural and functional MR methods enable fine-grained analyses of the functional neuroanatomy of spatial learning and memory.

The chapters in this volume treat both well-established and—in the majority—very recent methods. The former have been included because of their widespread use and the latter because of their potential for future use across the field of spatial learning and attention guidance. While some of the methods have been described in dedicated papers, to my knowledge there is no comparable collection of methods in a single volume.

This book is divided into three parts—behavioral, psychophysiological, and functional neuroimaging methods. The chapters in Part I focus on different aspects of visual search. Kristjansson and colleagues discuss visual foraging methods. While classical visual search paradigms are about finding a single target in a search display, visual foraging is about finding many targets in a display or scene. This allows more dynamic search tasks that capture search processes that are not typically seen in classical search tasks. Visual search involves eye movements. The measurement of eye movements has become more and more tractable over the years, so that eye movement recordings are now an important source of data that not only inform us about the total duration of search, like reaction times, but also enable us to investigate attentional strategies that occur during search. Hollingworth and Bahle give an overview of paradigms and methods of eye tracking in visual search. Visual search is affected by priming. Chetverikov and colleagues present a new method to analyze the effects of feature distributions in the search history on the internal representation of visual ensembles. Contextual cueing allows to investigate the impact of implicitly learned spatial contexts on memory-guided search. The chapter by Jiang and Sisk provides a methodological guide on the experimental paradigm of contextual cueing and reviews key findings. In addition, Marek and Pollmann present a virtual reality environment for contextual cueing experiments.

Part II begins with the chapter by Balaban and Luria on the contralateral delay activity (CDA). They review the evidence that the CDA is an event-related potential marker of visual working memory. However, they also show how the CDA can be used to assess the role of visual working memory in online processes. Fahrenfort contributes a tutorial-style guide to analyze attentional selection processes with multivariate pattern analysis (MVPA) of the EEG-signal. He starts with the classical N2pc-potential and goes on to show how comparable information can be identified with MVPA and goes on to describe the use of a forward encoding model for EEG-signal analysis. Merkel and colleagues present a guide to the design and analysis of multiple-object-tracking experiments, as a tool to investigate the processes that maintain the correspondence between objects and their neural representation over time. Part II finishes with the chapter by Grover and Reinhart on the combined use of transcranial direct-current stimulation and EEG to investigate the influence of memory on attention.

Part III begins with a classical topic. Lesion studies have contributed to our understanding of functional neuroanatomy long before the advent of functional imaging methods. Karnath and colleagues provide a tutorial on how to analyze the relationship between structure and function in the lesioned brain with modern univariate and multivariate analysis methods. Parietal cortex has long been known as an important contributor to both attentional and working memory processes. Sheremata shows how the retinotopic structure of parietal cortex can be delineated with fMRI. In medial temporal cortex, grid cells, first discovered in rodents and characterized by their remarkably regular 60° firing patterns, are critical for navigation. Stangl and colleagues show how grid cell activity can be analyzed in the human brain at the neural population level with fMRI. A general problem in fMRI research is the loss of precision by averaging individual brains with their individual differences in anatomy. Guntupalli describes a solution to this problem by aligning not brain structure but functional activation across individuals, a method called hyperalignment. Last but not least, studies of spatial learning and attention guidance can benefit greatly if eye movements, as an indicator of attentional processes (see the chapter by Hollingworth and Bahle), are acquired simultaneously with the fMRI signal. Hanke et al. give an overview of the do's and don'ts of this still challenging endeavor.

We hope that the reader will find this collection useful, perhaps inspiring creative new experiments that will illuminate the relation between attention and memory.

Magdeburg, Germany *Stefan Pollmann*

Contents

Contributors

BRETT BAHLE • *Department of Psychological and Brain Sciences, The University of Iowa, Iowa City, IA, USA*

HALELY BALABAN • *Sagol School of Neuroscience and The School of Psychological Sciences, Tel Aviv University, Tel Aviv, Israel*

ANDREY CHETVERIKOV • *Donders Institute for Brain, Cognition, and Behavior, Radboud University, Nijmegen, The Netherlands; Cognitive Research Lab, Russian Academy of National Economy and Public Administration, Moscow, Russia; School of Health Sciences, University of Iceland, Reykjavík, Iceland*

BIANCA DE HAAN • *Division of Psychology, Department of Life Sciences, College of Health and Life Sciences, Brunel University London, Uxbridge, UK*

JOHANNES JACOBUS FAHRENFORT • *Department of Psychology, University of Amsterdam, Amsterdam, The Netherlands; Amsterdam Brain and Cognition (ABC), University of Amsterdam, Amsterdam, The Netherlands; Department of Experimental and Applied Psychology, Vrije Universiteit Amsterdam, Amsterdam, The Netherlands*

SHREY GROVER • *Department of Psychological and Brain Sciences, Boston University, Boston, MA, USA*

J. SWAROOP GUNTUPALLI • *Vicarious AI, Union City, CA, USA*

MICHAEL HANKE • *Institute of Neuroscience and Medicine, Brain and Behaviour (INM-7), Research Centre Jülich, Jülich, Germany; Medical Faculty, Institute of Systems Neuroscience, Heinrich Heine University Düsseldorf, Düsseldorf, Germany*

SABRINA HANSMANN-ROTH • *School of Health Sciences, University of Iceland, Reykjavík, Iceland*

ANDREW HOLLINGWORTH • *Department of Psychological and Brain Sciences, The University of Iowa, Iowa City, IA, USA*

JENS-MAX HOPF • *Department of Neurology, Otto-von-Guericke University, Magdeburg, Germany; Department of Behavioral Neurology, Leibniz Institute for Neurobiology, Magdeburg, Germany*

YUHONG V. JIANG • *Department of Psychology, University of Minnesota, Minneapolis, MN, USA*

HANS-OTTO KARNATH • *Division of Neuropsychology, Center of Neurology, Hertie-Institute for Clinical Brain Research, University of Tübingen, Tübingen, Germany*

ÁRNI KRISTJÁNSSON • *Faculty of Psychology, School of Health Sciences, University of Iceland, Reykjavík, Iceland; School of Psychology, National Research University Higher School of Economics, Moscow, Russia*

TÓMAS KRISTJÁNSSON • *Faculty of Psychology, School of Health Sciences, University of Iceland, Reykjavík, Iceland*

ROY LURIA • *Sagol School of Neuroscience and The School of Psychological Sciences, Tel Aviv University, Tel Aviv, Israel*

NICO MAREK • *Department of Psychology, Institut für Psychologie, Otto-von-Guericke Universität, Magdeburg, Germany*

SEBASTIAAN MATHÔT • *Department of Psychology, University of Groningen, Groningen, The Netherlands*

CHRISTIAN MERKEL • *Department of Neurology, Otto-von-Guericke University, Magdeburg, Germany*

INGA M. ÓLAFSDÓTTIR • *Faculty of Psychology, School of Health Sciences, University of Iceland, Reykjavík, Iceland*

EDUARD ORT • *Department of Experimental and Applied Psychology, Institute for Brain and Behaviour, Vrije Universiteit Amsterdam, Amsterdam, The Netherlands*

NORMAN PEITEK • *Leibniz Institute for Neurobiology, Magdeburg, Germany*

STEFAN POLLMANN • *Department of Psychology, Institut für Psychologie, Otto-von-Guericke Universität, Magdeburg, Germany; Center for Brain and Behavioral Sciences, Otto-von-Guericke Universität, Magdeburg, Germany; Beijing Key Laboratory of Learning and Cognition and School of Psychology, Capital Normal University, Beijing, China*

ROBERT M. G. REINHART • *Department of Psychological and Brain Sciences, Boston University, Boston, MA, USA; Center for Systems Neuroscience, Cognitive Neuroimaging Center, Center for Research in Sensory Communication and Emerging Neural Technology, Boston University, Boston, MA, USA*

MIRCEA ARIEL SCHOENFELD • *Department of Neurology, Otto-von-Guericke University, Magdeburg, Germany; Department of Behavioral Neurology, Leibniz Institute for Neurobiology, Magdeburg, Germany; Kliniken Schmieder, Heidelberg, Germany*

SUMMER SHEREMATA • *Department of Psychology, Florida Atlantic University, Boca Raton, FL, USA; Center for Complex Systems and Brain Sciences, Florida Atlantic University, Boca Raton, FL, USA; Behavioral Sciences 101, Florida Atlantic University, Boca Raton, FL, USA*

JONATHAN P. SHINE • *German Center for Neurodegenerative Diseases (DZNE), Magdeburg, Germany; Faculty of Medicine, Otto-von-Guericke-University Magdeburg, Magdeburg, Germany*

CAITLIN A. SISK • *Department of Psychology, University of Minnesota, Minneapolis, MN, USA*

CHRISTOPH SPERBER • *Division of Neuropsychology, Center of Neurology, Hertie-Institute for Clinical Brain Research, University of Tübingen, Tübingen, Germany*

JÖRG STADLER • *Leibniz Institute for Neurobiology, Magdeburg, Germany*

MATTHIAS STANGL • *Semel Institute for Neuroscience and Human Behavior, University of California, Los Angeles, CA, USA*

ÖMER DAĞLAR TANRIKULU • *School of Health Sciences, University of Iceland, Reykjavík, Iceland*

ADINA WAGNER • *Institute of Neuroscience and Medicine, Brain and Behaviour (INM-7), Research Centre Jülich, Jülich, Germany*

DANIEL WIESEN • *Division of Neuropsychology, Center of Neurology, Hertie-Institute for Clinical Brain Research, University of Tübingen, Tübingen, Germany*

THOMAS WOLBERS • *German Center for Neurodegenerative Diseases (DZNE), Magdeburg, Germany; Center for Behavioral Brain Sciences, Magdeburg, Germany*

Part I

Behavioral Methods

Neuromethods (2020) 151: 3–21
DOI 10.1007/7657_2019_21
© Springer Science+Business Media, LLC 2019
Published online: 11 May 2019

Visual Foraging Tasks Provide New Insights into the Orienting of Visual Attention: Methodological Considerations

Árni Kristjánsson, Inga M. Ólafsdóttir, and Tómas Kristjánsson

Abstract

The topic of visual attention has played an increasingly large role in visual perception research in the past half-century or so. This highlights the need for paradigms that allow a thorough understanding of the function of visual attention and that the experimental tasks that are used are varied and dynamic enough to sample the operational characteristics of visual attention. We discuss newly developed foraging tasks that are more dynamic than many tasks used in the literature, such as the visual search task. Our orienting in the visual environment may not be particularly well encapsulated by the analogy of search for a single item, a search that then ends once the single target is found. Multitarget foraging tasks might cast further light upon the orienting of visual attention, especially in dynamic, multitarget environments. During foraging, observers are asked to select a certain number of target types among distractor items. Here we discuss such foraging tasks, the main considerations for efficient design and effective data analysis. We propose that these tasks will be a highly valuable addition to the toolbox of scientists who investigate the operation of visual attention and visual cognition more generally.

Keywords Visual foraging, Visual attention, Optimal foraging, Patch leaving, Visual search

1 Introduction

The topic of visual attention has played a central role in visual perception research in the past half-century. Selective visual attention has been shown to influence more or less all aspects of the visual process, amplifying the processing of the stimuli that are most relevant to behavior at any given moment [1–4]. Attention has been shown to have an influence at processing stages all the way from low-level processing in the lateral geniculate nucleus [5] to higher-level object representations in the ventral visual stream [6].

This highlights the need for paradigms that allow a thorough understanding of the function of visual attention and that the experimental tasks that are used are varied and dynamic enough to sample the operational characteristics of visual attention sufficiently well. One way of studying visual attention involves how we locate or select the visual stimuli that are important to us at a given moment. A popular way of modeling this procedure experimentally involves the well-known visual search task [7, 8].

In the version of the visual search task that has become dominant in the field, observers typically search for a single target among distractors [9, 10]. Older experiments often involved multiple target types [11–13], although there was usually only one target to search for in these paradigms in each instance of the task (each trial). These tasks can be of various difficulties that depend both on stimulus factors, such as whether noise is included to decrease visibility in the display [8, 14] and instructional manipulations such as what rules distinguish the target from the distractors: Is the target the single item that has a certain unique color or the single item that has a certain color *and* a certain shape [15]? Observers are typically asked to indicate whether a predesignated target is present or absent in an array of visual search items and their response times (usually with key press) are recorded. When they have responded, the search trial ends and a new trial follows. Results from this task have been extensively used in theories of the operation of visual attention for decades. A common assumption is that the search process occurs in processing stages [16, 17], and this is the main idea in Anne Treisman's highly influential *feature integration theory* of visual attention [18]. Neisser [19] proposed the idea of a preattentive stage, where individual items are not processed. Observers noted that individual nontarget letters in his search task were "only a blur." The crux of the feature integration theory is that an attentive stage follows this preattentive stage where free-floating features are integrated and objects identified, a process that takes time and effort (e.g., [10]).

The single-target visual search task has undeniably yielded important insights regarding the operational characteristics of visual attention, such as its processing capacity [13], its processing speed [16, 19], what captures attention [20], how attention is guided across a visual scene [21], what items can be processed in parallel [15], and how this is influenced by the context that the stimuli appear in [22, 23]. Feature integration theory is at its core a two-stage theory, where a distinction is made between preattentive processing and attentive processing, and this distinction has been dominant in the literature on visual search and visual attention [24], where the assumption is that some items can be found preattentively while others need processing by visual attention and effortful feature integration. Many results have however cast doubt upon such a stringent distinction [23–27] necessitating modifications of the basic tenets of many theories of visual attention.

2 Visual Foraging

A number of authors have speculated that more dynamic tasks might cast further light upon the orienting of visual attention. Our orienting in the visual environment may not be particularly well encapsulated by the analogy of search for a single item, a search that then ends once the single target is found. One method that researchers have increasingly used in recent years to investigate human visual orienting is the so-called *foraging* task. During foraging, observers are asked to select a certain number of target types among distractor items (see example in Fig. 1). The foraging task has its roots in research on how animals feed [29–35].

Some of the key variables that researchers have investigated with regard to foraging behavior involve things such as the concept of the *search image* (see Table 1). The idea is that during search and foraging, observers use search images whose content reflects the task goals in each case [36–38]. They have a rather clear analogue in the literature on human visual attention, where they are referred to as templates that are assumed to be stored in working memory [2, 39–43]. Working memory is capacity limited, which may be one reason why participants do not always forage optimally. In the words of Tinbergen ([34], pp. 332–333), birds "perform a highly selective sieving operation on the visual stimuli that reach their retina [and] can only use a limited number of different search images at the same time."

Another key concept is that the search image forces particular *run behavior* (Table 1). This refers to in what order participants forage for the different target types in the display. Early findings on run patterns showed that foraging for two different target types is typically nonrandom in animals. Experimental studies of such run-like behavior involve investigations of patterns of free choice [44, 45], serial-detection responses [36, 46] or direct measures of the order in which items are selected [47], and how *crypsis*, which involves how easily targets can be found in the display, affects foraging patterns.

Early foraging studies of humans involved mechanical displays [48–50]. Recent technological progress, such as the availability of touchscreen technology, has enabled easier assessment of human foraging. Note that foraging tasks are by no means confined to touch, however—foraging can also be done by mouse clicks [51] and eye gaze [52] in real-world displays where observers physically interact with the stimuli that they should collect [49, 50] or in virtual reality [52].

The foraging paradigm yields very rich datasets—and perhaps the large amount of multifaceted performance measures and their interaction can be overwhelming. The typical methods of visual search tasks are far more manageable, and perhaps this partly

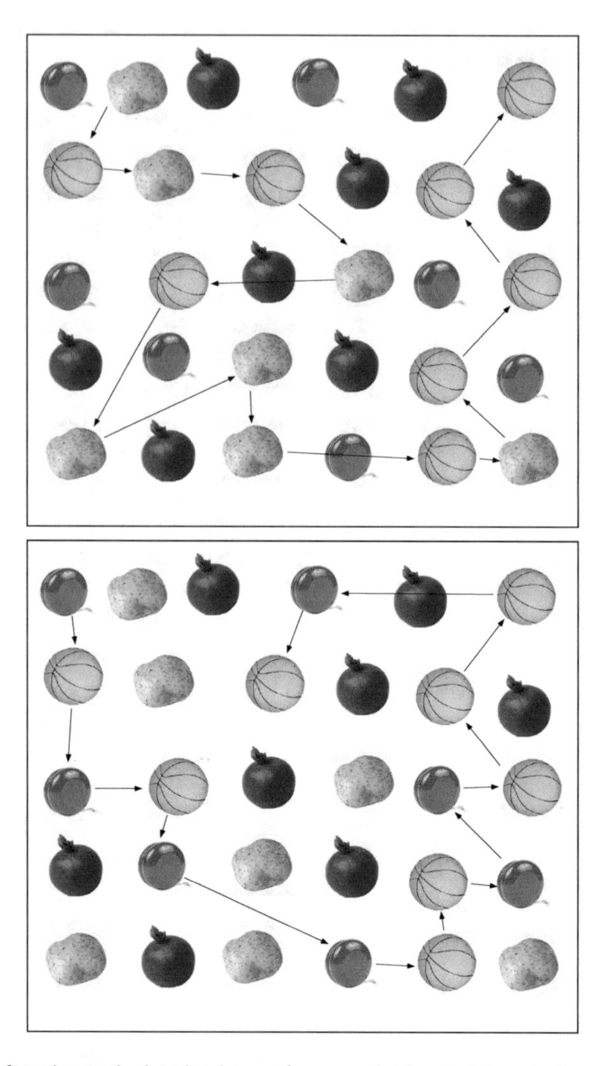

Fig. 1 A foraging task showing how a forager might select targets depending on different instructions or selection rules. In the top panel, the task is to select all the items that have a certain color, while in the lower panel, the observers must forage for all the items connected to sports or games. (The individual images come from Ref. [28])

accounts for the popularity of visual search methodology. For example, slopes that are assumed to assess the involvement of attention allow easy summary interpretations of visual search data, although their blind use as indices of the involvement of visual attention is highly suspect [24].

While it is not our goal here to go over in detail what the findings on human foraging have revealed, foraging results have in many ways deviated from many predictions, such as from theories of visual attention, predictions from priming studies [53], and conceptions of working memory and its capacity. In Kristjánsson

Table 1
Key concepts involved in studies of visual foraging

Search image	The concept that foragers form an internal image that contains the characteristics of the target (or targets) in each case. The concept is roughly similar to templates in visual working memory
Run behavior	In what order do observers select the different target types in the display? Do they select the same target type repeatedly until they are all gone or select the different target types randomly? A run is defined as the repeated consecutive selection of the same type of prey
Patch leaving	In experiments where observers can move to a new foraging display ("patch") before all the targets in their current one have been collected, patch leaving involves how soon they move to the next target source
Optimal foraging theory and the *marginal value theorem*	According to optimal foraging theory, observers will adjust their foraging so that it matches the marginal value theorem which states that foragers will leave a food source when the collection rate drops below the average collection rate within an environment
Collection rate	A measure of how quickly target items are collected. Usually measured in the number of items per second
Intertarget times	The time that passes between each target selection *within* a foraging trial
Cruise phase	The phase during foraging trials where intertarget times are low and constant from one selection to the next. Typically excludes the first and last target selection during the foraging trial
Mid- and end peaks	Mid-peaks are seen in difficult foraging tasks (e.g., conjunction foraging) and reflect when observers switch between target categories. End peaks are seen at the end of foraging trials (in tasks where observers must find all targets)
Switch costs	Involve the increase that appears in ITTs when observers switch between target types
Foraging organization	How observers organize their foraging within a trial. Typically measured in three ways: with *best-R (see below), distance traveled, and the number of intersections*

et al. [53] where observers foraged for multiple instances of two target types simultaneously, interestingly, some observers could forage simultaneously for two conjunction targets, going against the predictions of theoretical conceptions such as feature integration theory, and the findings of Kristjánsson and Kristjánsson [54] argued against influential slot conceptions of working memory [55, 56]. This highlights how foraging tasks can help to decide between theoretical accounts and the additional information that these foraging paradigms can provide over and above the reductionist approach used in many studies of visual search and visual working memory.

In what follows, we discuss a number of issues to consider when foraging studies are designed. Our aim is to provide an overview of methods used in foraging research, including statistical treatment of the large datasets that foraging experiments yield.

2.1 Procedures

2.1.1 Exhaustive Foraging Versus Non-exhaustive Foraging and "Patch Leaving"

One distinction made in the foraging literature is between exhaustive foraging and non-exhaustive foraging. Exhaustive foraging means that every target in the foraging patch must be found, while in non-exhaustive foraging, the forager can freely move to a new "patch" (in computerized foraging tasks a new foraging display), whenever he chooses, called *patch leaving* (Table 1). Patch leaving refers, in other words, to when foraging within a certain patch is terminated and the forager moves to the next patch [51]. Several parameters affect when foragers leave a patch and move to the next one. The availability of targets [51], expectation of the number of targets [57], the time it takes to move to a different patch or the effort that it involves [51], the difficulty of the foraging task [58], and the value of the foraged items [59, 60] can all affect patch leaving. When the targets are abundant, foragers will leave a patch earlier than if targets are scarce, which at first glance might seem counterintuitive. Consider, however, that if you are searching for food and there is little food about, it is likely that you will exhaustively search each area before moving to the next. If, however, there is a lot of food around, you are likely to pick the easily accessible food and leave behind food items that are harder to find. This is also related to the effect of expectation; if the collection rate is lower than expected, foragers will leave that patch earlier than if the collection rate matches or exceeds the expected collection rate. The time or cost of moving to a new patch also affects patch leaving and does so in a very intuitive way, when you have to travel a long distance to a new patch; if moving to the new patch takes a long time or if it puts the forager at risk to move, the forager will stay in the current patch for longer. This is logical as long as the collection rate in the current patch is not zero. If you could have collected several targets in the time it takes to move to a new patch, the collection rate at the new patch must be higher than at the current patch to make leaving the current patch beneficial. In addition, the value of the foraged targets affects the patch leaving behavior. Wolfe et al. [59] showed that when targets have different values, foragers will conduct a roughly exhaustive search of the more valuable targets while leaving more of the less valuable targets behind. This effect seems to override or at least strongly skew the effect of availability, although there were also some individual differences.

Non-exhaustive foraging replicates natural food foraging situations, where humans and animals are generally able to leave a foraging patch. There are however situations where finding every single target is required for the task at hand; in those situations, exhaustive foraging tasks better represent those situations. In addition, studying exhaustive foraging allows for better control over the number of targets selected and therefore a more controlled study of variables such as switch costs and the number of runs. Further, according to *optimal foraging theory* and the *marginal value theorem* [61]; (see Table 1), foragers will leave a food source when the collection rate drops below the average collection rate or the expectancy of such an average. Exhaustive foraging allows the study of foraging after the collection rate drops below average, and the forager would have left the patch in a patch leaving paradigm. Both procedures therefore have their obvious value.

2.1.2 Do the Foraging Targets Disappear or Persist Once They Have Been Tapped?

In most foraging studies, the targets disappear once they have been tapped. This seems to be a natural procedure as it mimics real-life foraging where food items disappear upon being collected and/or eaten. But studies where targets do not disappear upon being tapped, but persist on the screen, can nevertheless be useful for testing memory for visited locations, both in normal populations and patient populations that may be expected to have problems with spatial memory. Those may include those suffering from hemispatial neglect [62, 63], simultagnosia [64], or impaired spatial working memory [65] and can also address questions such as regarding spatial learning throughout the foraging process.

Observers rarely re-fixate distractors in single-target visual search [66, 67] and seem to have good memory for checked items and locations, but as soon as the targets become two or more, and persist on the screen after they are selected, refixations on both distractors and the previously found targets become significantly more common [66, 68]. Moreover, when targets disappear or if they change into salient, easily distinguishable items when they are found, accuracy is significantly higher than if the targets remain on the screen or change into distractor items once they are found [68].

The presence of already selected targets may misdirect attention. The features of these tapped, but remaining targets, match those of unfound target items, and they may therefore still attract attention, interfering with subsequent foraging [68]. Another explanation of why search accuracy is worse when targets remain on the screen is that found targets may take up working memory resources, thus interfering with further foraging [66, 68]. We should note, however, that Thornton and Horowitz [69] had items either vanish or remain once they had been located. This manipulation had surprisingly little effect on search performance, suggesting that old target items could be efficiently ignored. However, in Horowitz and Thornton [70], where the items moved around, memory for previous targets was more or less gone, suggesting that locations, not objects, had been memorized in their previous study.

Patients with right hemisphere brain damage often fail to attend to objects, sounds, and events on their left, a condition called hemispatial neglect [71–73]. Neglect patients also show abnormal visual search behavior, where they repeatedly examine items in their right visual hemifield while ignoring items in their left hemifield [63, 71–74]. On bedside paper and pen cancellation tests, which require foraging for multiple items, these patients usually start searching from the right and fail to find targets on the left side of the paper test. Many patients also repeatedly re-fixate locations on their right while ignoring targets on their left [63, 65, 75].

Using a touchscreen foraging task, Parton et al. [63] studied the foraging behavior of two groups of stroke patients with right hemisphere brain damage, with and without hemispatial neglect, in four different conditions: cancellation test, where found targets were marked; invisible cancellation test, where found targets persisted on the screen and no visible marks were left; eraser task, where the targets disappeared upon being found; and finally a bold task, where target outlines became thicker upon being touched, which increased the salience of found targets. If the perceptual salience of found targets is the main reason for revisits, then they should be most common in the cancellation and bold tasks and least common in the eraser task. If, on the other hand, a deficit of spatial working memory is to blame for the results, the revisits should occur predominantly in the invisible cancellation test [63]. For both patient groups, revisits were most frequent in the invisible cancellation test, indicating that their memory for previously searched locations was impaired but that the influence of the perceptual salience of previously found targets was minimal. Mannan et al. [62] used a foraging task with persistent targets along with eye tracking to assess whether neglect patients misjudged whether they were finding a target for the first time or not. The patients were instructed to press a response button only when they fixated a target for the first time. Patients with frontal lobe or intraparietal sulcus lesions repeatedly judged that they were selecting previously found targets for the first time, but patients with occipito-temporal lesions did not. These results indicate that repeated refixations occur because patients do not remember where they have found targets before, suggesting that a spatial working memory deficit accompanies neglect. Using a foraging task where targets persist on the screen after being found thus helped map out the deficits that explain the pathological refixation on targets in the right hemifield of hemispatial neglect patients.

2.2 Measures of Foraging Performance

2.2.1 Run Patterns

With multiple target types, it is possible to assess the strategies that observers use to finish the foraging task. Do observers finish all targets of one type before switching to the next, or do they tend to switch between target types within a trial? Measuring this aspect of performance can, among other things, provide information about how many target templates observers can simultaneously keep in mind and how hard it is for them to switch between different target types. Studies on animal foraging have shown that animals often forage for prey in nonrandom "runs" [30, 35, 44]. A run is defined as the repeated consecutive selection of the same type of prey. The typical finding is that the harder the prey is to find, the longer the runs become, while when the prey is easy to detect, switching between target types becomes more frequent, with the animal foraging in many short runs [35, 76]. Studies on the foraging patterns of humans have also revealed that attentional load alters foraging behavior [53, 77]. When targets are defined by a single feature (e.g., color), switching between target categories is frequent and random, but when targets are defined by a conjunction of features (e.g., color and shape), observers tend to exhaustively forage for one target type before switching to the next one (Fig. 2), completing most trials in only two runs [53, 78–80]. Those foraging patterns can be affected, however, by altering task parameters. Switches between target

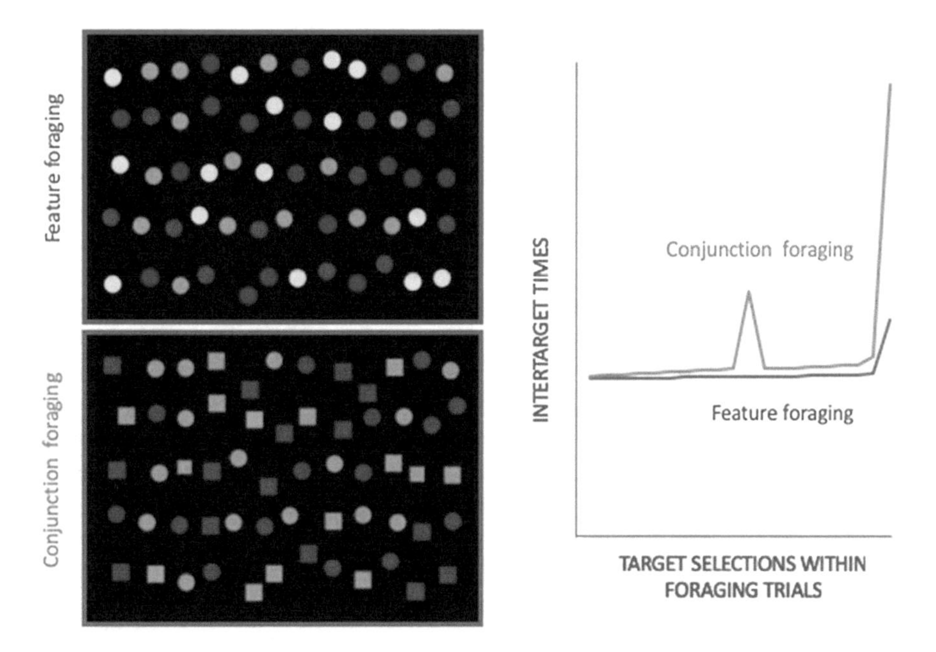

Fig. 2 Feature versus conjunction foraging. The graph shows typical intertrial patterns for the two types of foraging—relatively similar "cruise phases" apart from the mid-peaks for conjunction foraging and the end peaks seen in both conditions. Note that the end peaks are much higher for conjunction foraging

categories during conjunction foraging can be increased by imposing time limits [81], having participants forage using eye gaze and fixation rather than tapping targets [82], and only rewarding switches between target categories [83], indicating that foraging patterns do not only reflect capacity restraints but also a strategy to optimize hit rate with as little mental effort as possible.

2.2.2 Intertarget Times and Collection Rate

Another way to analyze foraging data is to assess how quickly people can forage for a given number of targets. One obvious measure is the total time taken to forage for all targets within a certain time. Another, arguably more interesting measure involves the intertarget times (ITTs, see Fig. 2) that involve the time between each tap on a target. Another way to define foraging speed is with collection rate, which is a measure of how many items are collected per second or in other words 1 divided by the ITTs [59, 84]. So, if the ITTs are 0.25 s, then the collection rate is $1/0.25 = 4$ items per second. Measuring foraging times allows insights into how different foraging conditions affect foraging performance, and whether ITTs are steady throughout a trial, or if they change or fluctuate within trials.

As foraging tasks become more difficult, ITTs become higher, and collection rates therefore drop. ITTs are, for example, higher when the targets are defined by a more complex rule (i.e., by a conjunction of features rather than a single feature [54]), when the number of targets or target types is increased [54], and when the targets are moving [84]. In a series of studies where the researchers used feature versus conjunction foraging to manipulate task complexity (see also discussion above), there was a dramatic difference in total completion times of feature and conjunction-based trials, where observers had to forage for 40 targets of two types among 40 distractors (feature foraging, e.g., green and red targets among blue and yellow distractors; conjunction foraging, e.g., red disks and green squares among red squares and green disks) [53, 77, 82]. But interestingly, the ITTs per tap throughout the majority of the foraging trials were rather similar across conditions. The largest performance differences between ITTs during feature and conjunction foraging involved the mid- and end peaks. The mid-peaks only appeared during conjunction foraging and represented the switch costs that occur when observers switch from one target type to the next. The end peaks appeared in both tasks but were significantly larger during conjunction foraging [78–80]. What is more, there was only a slight upward slope of ITTs throughout the "cruise phase," or, in other words, the taps that did not belong to the mid- or end peaks, meaning that ITTs did not become significantly higher as a function of fewer targets on the screen but remained relatively stable, excluding the mid- and end peaks [78]. Note that, again, this highlights the added information available with foraging tasks over simpler ones, such as visual search tasks.

2.2.3 Switch Costs and Mid-peaks

Switch costs involve the increase in ITTs that occurs when observers switch between target types, instead of continuing a run by selecting a target of the same type as the previously selected one. Switch costs are calculated by deducting the mean ITTs of taps in a run from the mean ITTs when observers switch between target types.

In easy foraging tasks, where there are only two target types, defined by a single feature, and all targets are static on the screen, the switch costs are negligible. As the tasks become more complex, the switch costs grow larger; they are, for example, much larger during conjunction than feature foraging [53] and grow as the number of target types goes up [54, 84]. In accordance with this, when there are more than two target types, observers tend to alternate between two target types instead of switching randomly between targets of all types in the display, and this switching results in lower switch costs than switching between more target types [83]. Notably, however, a number of observers actually were able to rapidly switch between target types even during difficult foraging tasks and were tentatively called "super-foragers" [53, 77, 85] as the results seemed to suggest that their capacity to hold complex target types in memory was higher than for the other participants. But Kristjánsson et al. [81] then found that by introducing time limits during foraging; contrasting conditions where observers had 5, 10, or 15 s to forage; and comparing this with foraging with unlimited time, many more observers could switch rapidly between target types. Kristjánsson et al. [81] suggested that this demonstrated the flexibility of working memory—that observers could perform at high capacity for short bursts that require high levels of concentration and that they choose not to do this during foraging with no time limits because of the effort involved. The switch costs then result in so-called mid-peaks in ITTs during foraging trials for the more difficult tasks, reflecting the point where observers switch between target categories, after having finished all targets from the other target category (see Fig. 2).

2.2.4 End Peaks

No matter how simple or easy a foraging task is, observers seem to have difficulty finding the last target item in the display and are usually much slower at this than for other items in the display. This results in a distinct rise in ITTs for the last target item, a performance pattern that has been named "end peaks" (Kristjánsson et al., under review; see Fig. 2). The same applies to end peaks as to switch costs and ITTs in general: As the task becomes more complex, the end peaks become larger. Moreover, interestingly, Kristjánsson et al. [78] found that set size has a different effect on end peaks during feature and conjunction foraging tasks. During feature foraging, end peaks remain the same size regardless of set size, but during conjunction foraging, the end peaks become larger as set size increases. So, when the size of the end peaks of trials with different set sizes is plotted, set size slopes found in single-target visual search are replicated. What is interesting about these results is that it is only this last selection out of multiple target selections during the foraging task that reflects results from single-target visual search tasks. This highlights the need for varied approaches to studying the complex construct of visual attention, and also the additional information that foraging paradigms provide, since this result suggests that the last target selections in the foraging display simply reduce to one trial in a single-target visual search task.

2.2.5 Measures of Foraging Organization

Another way of assessing foraging performance is to analyze how observers organize their foraging throughout the trial. Do they, for example, start on one side and systematically go through the display in columns or rows, or is their foraging more random than this (see Fig. 3)? As discussed by Woods et al. [86], there are at least three ways to measure foraging organization.

Best-r can be used to test the organization or systematicity of the foraging path that observers take. In a static foraging display, human foragers will typically forage in a systematic way, starting, for example, in the top left quadrant and searching through rows or columns through the foraging display [87–89]. The *best-r* is found by calculating the Pearson correlation coefficient (r) from a linear regression of the x- and y-values independently by the location of each target relative to the order in which they were selected. The higher r-value (either from the x- or y-axis) is then used to assess how systematic or orthogonal the foraging path is. For example, starting in the top left quadrant of the foraging display and selecting targets by columns progressively rightward would result in a higher r-value on the x-axis than the y-axis regression, because the foraging path would be consistently horizontal (left-to-right). A highly organized foraging path would generally result in a high best-r. So, if the r for the regression of the y-axis and selection order was .1, while it was .9 for the same regression relative to the x-axis, we would assume that the search was highly organized by columns. Note, however, that there *are* some systematic paths possible that would result in a low best-r. Best-r is calculated for each trial for each participant, and averages for each participant in each condition can then be used in ANOVAs or t-tests or other statistical treatments.

Total distance traveled is another measure of foraging organization. The most organized foragers should travel the shortest distance on average, going systematically through the display, while more disorganized foragers might be more prone to select targets further away from one another, or forgetting targets in one part of the display, forcing them to go back to collect them and adding unnecessary travel distance. It is also possible to calculate

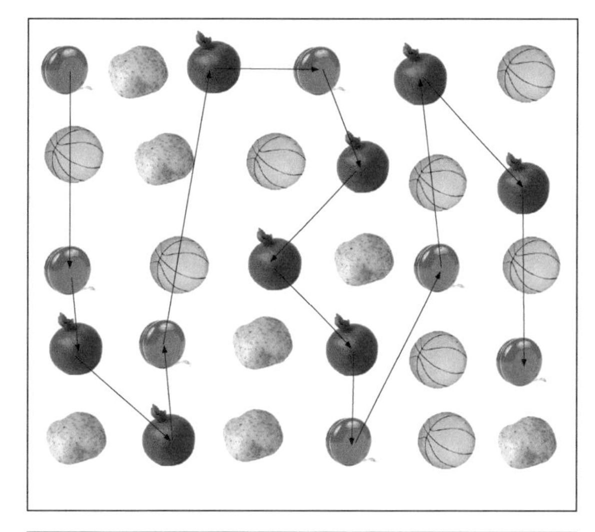

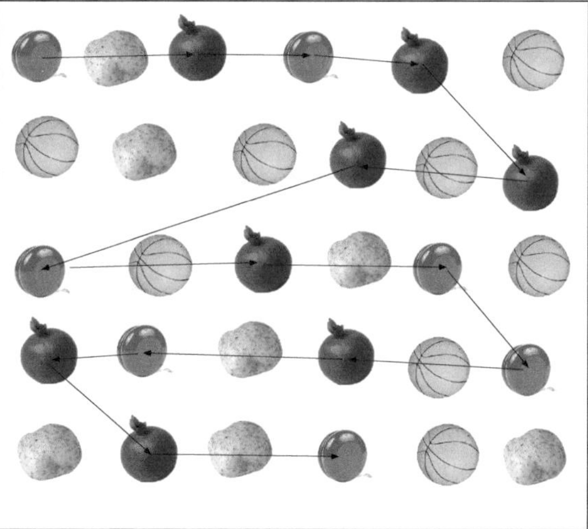

Fig. 3 Different ways of organizing foraging within a given trial, where the targets, in both cases, are the red items

the mean distance between targets [86] with similar results since these measures should be highly correlated.

The number of intersections in the search path is a measure of how often observers revisit parts of the display that they have previously foraged through. If foraging is well organized, there should be no reason to revisit previously searched regions, resulting in few, if any, intersections [86]. The number of intersections should decrease as the foraging becomes more organized. If there is a chance that observers will not collect the same amount of target items (such as in patch leaving tasks or tasks with different set sizes), the total amount of intersections should be divided by the number of items collected, to avoid confounding the assessment with the fact that as more targets are collected, the number of intersections increases on average [86]. Jóhannesson et al. [82] assessed foraging organization during both foraging with fingers and with eye gaze where observers selected the targets by fixating on them. Firstly, they found that foraging was highly organized during feature foraging. This indicates that participants use consistent horizontal or vertical sweeps through the display when attentional load is low. But during conjunction foraging, performance was not as organized, which may indicate that they were less aware of the locations of other targets in the display.

2.2.6 Patch Leaving

Not all foraging tasks involve that observers must exhaustively forage for all target items in the display. Tasks where participants can choose to leave the current display (or "patch") and move to a new one to forage in, even if all targets have not yet been acquired, allow assessment of when participants choose to leave the current patch. Such studies mimic real-life foraging, where animals seldom exhaustively pick all food items from a patch (e.g., all berries from a single bush) before searching for a better one. As the food items become scarce within a given source, animals tend to move to the next available food source [32, 51].

The marginal value theorem (MVT) is a model of patch leaving behavior that has been used extensively in research on foraging [61]. According to MVT, animals will leave a patch when the instantaneous rate of return drops below the average rate of return for the whole environment [61]. In simple foraging tasks, where all items have the same value, the instantaneous rate of return is essentially the collection rate. If the targets are of different types and have different values, the value of each item collected must be considered [59]. The average collection rate is determined by finding the mean collection rate of the whole area (or the whole set of trials within a block). The average collection rate becomes lower as the time and effort involved in traveling between patches increase [51]. This can be modeled in computerized tasks, for example, by manipulating the amount of time that passes before the next patch appears [51, 59, 84].

Some studies indicate that during simple foraging tasks, human foraging is well predicted by MVT, but when the tasks become more complicated, such as when there is large variation in patch quality [51], when the targets have different values [59], or when the targets cannot be seen (e.g., when observers went fishing in a virtual pond where the availability of fish in the pond was not visible [90]), foraging behavior starts to deviate from the predictions of the model, and observers' behavior tends to become increasingly varied. Recently, Kristjánsson et al. [90] tested patch leaving in humans performing an iPad foraging task. They found that observers foraged for much longer within each patch than predicted by MVT. A clear conclusion from this is that many more variables than the average acquisition rate within a patch determine patch leaving in particular or foraging performance more generally.

2.2.7 Summary: Modulation of Foraging Strategies

It seems that all the measures discussed above can be modulated by task demands, such as stimulus factors, temporal constraints, or manipulations of motivation (such as with reward). Human foraging performance seems to be affected by various factors, and how observers orient in these multitarget environments can be assessed in many ways. This multidimensional aspect of foraging performance involves a methodological and statistical challenge (a point that we partly address in Sect. 3), while at the same time, this multidimensionality is also an asset for uncovering the operating characteristics of visual attention, enabling unique insights about how observers orient their attention in multidimensional visual environments.

3 Statistical Treatment of Data from Multitarget Foraging Experiments

We should start by noting that foraging paradigms are highly powerful statistically, since they allow rapid acquisition of a large amount of measurements and multidimensional performance measures as discussed in Sect. 2. Measures such as foraging patterns and intertarget times as a function of group and condition can be addressed with standard statistical methods such as ANOVA, regression, or linear mixed models to name a few potential approaches. The last approach in particular may be very promising in assessing both group characteristics and aspects of individual performance. There are however several less traditional methods that are utilized in foraging studies and some further considerations to highlight.

Studying the foraging process generally involves both within-subject and between-subject comparisons. When those within-subject comparisons are performed, it is important to keep in mind that the measures, even between conditions, are highly interdependent, and therefore, if using ANOVA, repeated measures ANOVAs are generally preferable.

A widely used procedure involves the so-called runs test that involves assessing whether observers choose different target types randomly or have a bias toward choosing one target type or the other [91, 92]. The runs test is widely used in the animal literature to test whether animals (from bees to cattle) behave randomly during their selections or not [94]. The basic idea involves determining whether a sequence of x events deviates from a random series of the same number (x) of events. To determine if run behavior is random, a one-sample runs test can be used (a two-sample runs test is called a Wald-Wolfowitz test [93]). Since it is a one-sample test, it needs to be carried out on the data of each participant separately.

Note that since this gives an estimate of whether the run behavior is random for each trial for every participant, it is vital to use post hoc corrections of *p*-values for multiple tests, such as the Bonferroni correction (although other less conservative options should sometimes be considered; [94]). This measure yields a proportion of trials for each participant, in each condition, where run behavior is random. These proportions can then be compared by *t*-tests or ANOVAs depending on the number of conditions. The one-sample runs test has been criticized for its lack of power and often inappropriate use. However, it remains a good choice in ecological studies and studies that fit the initial purpose of the test.

4 Conclusions

The above discussion shows the multifaceted performance measures that are available in foraging studies. Their sheer magnitude may seem daunting, but the tendency in visual attention research has for the last decades or so gone in the other direction. The visual search task (most often involving single targets) has become the standard for assessing whether attention is involved in particular tasks and slopes of set versus response times have actually taken on the status of a diagnostic tool for the question of whether visual attention is involved in a task, and only two main parameters are often assessed (slope and intercept).

We think that this is a mistake [24]. While the visual search task will most certainly continue to be a part of the toolbox of scientists investigating vision and visual attention, more dynamic tasks that more closely resemble real-life visual orienting are also needed. We believe that the visual foraging task is a good candidate for this. In our laboratory we have started to investigate visual foraging in virtual reality environments in an attempt to increase even more the ecological validity of our paradigms. We hope that the discussion here has convinced the reader that foraging tasks will play an important role in the study of human and animal orienting in the visual environment in future research and that the multidimensionality of the approach should be considered a feature rather than a bug.

References

1. Corbetta M, Shulman GL (2002) Control of goal-directed and stimulus-driven attention in the brain. Nat Rev Neurosci 3(3):201–215
2. Desimone R, Duncan J (1995) Neural mechanisms of selective visual attention. Annu Rev Neurosci 18(1):193–222
3. Driver J (2001) A selective review of selective attention research from the past century. Br J Psychol 92(1):53–78
4. Kristjánsson Á (2006) Rapid learning in attention shifts—a review. Vis Cognit 13:324–362
5. O'Connor DH, Fukui MM, Pinsk MA, Kastner S (2002) Attention modulates responses in the human lateral geniculate nucleus. Nat Neurosci 5(11):1203–1209
6. Wojciulik E, Kanwisher N, Driver J (1998) Covert visual attention modulates face-specific

activity in the human fusiform gyrus: fMRI study. J Neurophysiol 79(3):1574–1578

7. Egeth HE (1966) Parallel versus serial processes in multidimensional stimulus discrimination. Percept Psychophys 1(4):245–252
8. Sternberg S (1967) Two operations in character recognition: some evidence from reaction time measurements. Percept Psychophys 2:45–53
9. Treisman A (1986) Features and objects in visual processing. Sci Am 255(5):114–125
10. Wolfe JM (1998) Visual search. In: Pashler H (ed) Attention. University College London Press, London, pp 13–73
11. Kaplan IT, Carvellas T (1965) Scanning for multiple targets. Percept Mot Skills 21:239–243
12. Metlay W, Sokoloff M, Kaplan IT (1970) Visual search for multiple targets. J Exp Psychol 85(1):148
13. Neisser U, Novick R, Lazar R (1963) Searching for ten targets simultaneously. Percept Mot Skills 17(3):955–961
14. Eriksen CW, Schultz DW (1979) Information processing in visual search: a continuous flow conception and experimental results. Percept Psychophys 25(4):249–263
15. Treisman A, Sykes M, Gelade G (1977) Selective attention and stimulus integration. In: Dornic S (ed) Attention and performance VI. Lawrence Erlbaum, Hillsdale, pp 333–361
16. Neisser U (1963) Decision-time without reaction-time: experiments in visual scanning. Am J Psychol 76(3):376–385
17. Sternberg S (1969) The discovery of processing stages: extensions of Donders' method. Acta Psychol (Amst) 30:276–315
18. Treisman AM, Gelade G (1980) A feature-integration theory of attention. Cogn Psychol 12(1):97–136
19. Neisser U (1964) Visual search. Sci Am 210 (6):94–103
20. Yantis S, Jonides J (1984) Abrupt visual onsets and selective attention: evidence from visual search. J Exp Psychol Hum Percept Perform 10(5):601
21. Wolfe JM, Cave KR, Franzel SL (1989) Guided search: an alternative to the feature integration model for visual search. J Exp Psychol Hum Percept Perform 15(3):419–433
22. Cavanagh JP, Chase WG (1971) The equivalence of target and nontarget processing in visual search. Percept Psychophys 9 (6):493–495
23. Wang D, Kristjánsson Á, Nakayama K (2005) Efficient visual search without top-down or bottom-up guidance. Percept Psychophys 67:239–253
24. Kristjánsson Á (2015) Reconsidering visual search. i-Perception 6(6):2041669515614670
25. Bravo MJ, Nakayama K (1992) The role of attention in different visual-search tasks. Percept Psychophys 51(5):465–472
26. Enns JT, Rensink RA (1990) Sensitivity to three-dimensional orientation in visual search. Psychol Sci 1(5):323–326
27. Joseph JS, Chun MM, Nakayama K (1997) Attentional requirements in a 'preattentive' feature search task. Nature 387 (6635):805–807
28. Brady TF, Konkle T, Alvarez GA, Oliva A (2008) Visual long-term memory has a massive storage capacity for object details. Proc Natl Acad Sci 105(38):14325–14329
29. Bukovinszky T, Rikken I, Evers S (2017) Effects of pollen species composition on the foraging behaviour and offspring performance of the mason bee Osmia bicornis (L.). Basic Appl Ecol 18:21–30
30. Dawkins M (1971) Perceptual changes in chicks: another look at the 'search image' concept. Anim Behav 19(3):566–574
31. Mallott EK, Garber PA, Malhi RS (2017) Integrating feeding behavior, ecological data, and DNA barcoding to identify developmental differences in invertebrate foraging strategies in wild white-faced capuchins (Cebus capucinus). Am J Phys Anthropol 162(2):241–254
32. Pyke GH, Pulliam HR, Charnov EL (1977) Optimal foraging: a selective review of theory and tests. Q Rev Biol 52(2):137–154
33. Schuppli C, Forss SIF, Meulman EJM (2016) Development of foraging skills in two orangutan populations: needing to learn or needing to grow? Front Zool 13(43). https://doi.org/10.1186/s12983-016-0178-5
34. Tinbergen L (1960) The natural control of insects in pinewoods I. Factors influencing the intensity of predation by songbirds. Archives Néerlandaises de Zoologie 13:265–336
35. Dukas R (2002) Behavioural and ecological consequences of limited attention. Philos Trans R Soc Lond B Biol Sci 357(1427):1539–1547
36. Bond AB, Kamil AC (1999) Searching image in blue jays: facilitation and interference in sequential priming. Anim Learn Behav 27 (4):461–471
37. Dukas R, Kamil AC (2001) Limited attention: the constraint underlying search image. Behav Ecol 12(2):192–199
38. Nakayama K, Maljkovic V, Kristjánsson Á (2004) Short term memory for the rapid

deployment of visual attention. In: Gazzaniga MS (ritstj.) The cognitive neurosciences, 3rd edn. MIT Press, Cambridge
39. Awh E, Jonides J (2001) Overlapping mechanisms of attention and spatial working memory. Trends Cogn Sci 5(3):119–126
40. Bundesen C (1990) A theory of visual attention. Psychol Rev 97(4):523–547
41. Carlisle NB, Kristjánsson Á (2018) How visual working memory contents influence priming of visual attention. Psychol Res 82(5):833–839
42. Vickery TJ, King LW, Jiang Y (2005) Setting up the target template in visual search. J Vis 5 (1):81–92
43. Woodman GF, Carlisle NB, Reinhart RM (2013) Where do we store the memory representations that guide attention? J Vis 13 (3):1–17
44. Bond AB (1983) Visual search and selection of natural stimuli in the pigeon: the attention threshold hypothesis. J Exp Psychol Anim Behav Process 9(3):292–306
45. Reid PJ, Shettleworth SJ (1992) Detection of cryptic prey: search image or search rate? J Exp Psychol Anim Behav Process 18(3):273–286
46. Pietrewicz AT, Kamil AC (1979) Search image formation in the blue jay (Cyanocitta cristata). Science 204(4399):1332–1333
47. Punzalan D, Rodd FH, Hughes KA (2005) Perceptual processes and the maintenance of polymorphism through frequency-dependent predation. Evol Ecol 19(3):303–320
48. Ballard DH, Hayhoe MM, Li F, Whitehead SD (1992) Hand-eye coordination during sequential tasks. Philos Trans R Soc Lond B Biol Sci 337(1281):331–339
49. Bond AB (1982) The bead game: response strategies in free assortment. Hum Factors 24 (1):101–110
50. Gilchrist ID, North A, Hood B (2001) Is visual search really like foraging? Perception 30 (12):1459–1464
51. Wolfe JM (2013) When is it time to move to the next raspberry bush? Foraging rules in human visual search. J Vis 13(3):1–17
52. Draschkow D, Kristjánsson (in preparation) Foraging experiments in virtual reality environments. Manuscript in preparation
53. Kristjánsson Á, Jóhannesson ÓI, Thornton IM (2014) Common attentional constraints in visual foraging. PLoS One 9(6):e100752
54. Kristjánsson T, Kristjánsson Á (2018) Foraging through multiple target categories reveals the flexibility of visual working memory. Acta Psychol (Amst) 183:108–115
55. Luck SJ, Vogel EK (1997) The capacity of visual working memory for features and conjunctions. Nature 390(6657):279–281
56. Olivers CN, Peters J, Houtkamp R, Roelfsema PR (2011) Different states in visual working memory: when it guides attention and when it does not. Trends Cogn Sci 15(7):327–334
57. Cain MS, Vul E, Clark K, Mitroff SR (2012) A Bayesian optimal foraging model of human visual search. Psychol Sci 23(9):1047–1054
58. Ehinger KA, Wolfe JM (2016) When is it time to move to the next map? Optimal foraging in guided visual search. Atten Percept Psychophys 78(7):2135–2151
59. Wolfe JM, Cain MS, Alaoui-Soce A (2018) Hybrid value foraging: How the value of targets shapes human foraging behavior. Atten Percept Psychophys 80:609–621
60. Zhang J, Gong X, Fougnie D, Wolfe JM (2017) How humans react to changing rewards during visual foraging. Atten Percept Psychophys 79(8):2299–2309
61. Charnov EL (1976) Optimal foraging: the marginal value theorem. Theor Popul Biol 9 (2):129–135, 110(971), 141–151
62. Mannan SK, Mort DJ, Hodgson TL, Driver J, Kennard C, Husain M (2005) Revisiting previously searched locations in visual neglect: role of right parietal and frontal lesions in misjudging old locations as new. J Cogn Neurosci 17 (2):340–354
63. Parton A, Malhotra P, Husain M (2004) Hemispatial neglect. J Neurol Neurosurg Psychiatry 75(1):13–21
64. Dehaene S, Cohen L (1994) Dissociable mechanisms of subitizing and counting: neuropsychological evidence from simultanagnosic patients. J Exp Psychol Hum Percept Perform 20:958–975
65. Malhotra P, Mannan S, Driver J, Husain M (2004) Impaired spatial working memory: one component of the visual neglect syndrome? Cortex 40(4-5):667–676
66. Cain MS, Adamo SH, Mitroff SR (2013) A taxonomy of errors in multiple-target visual search. Visual Cognition 21(7):899–921
67. Peterson MS, Kramer AF, Wang RF, Irwin DE, McCarley JS (2001) Visual search has memory. Psychol Sci 12(4):287–292
68. Cain MS, Mitroff SR (2012) Memory for found targets interferes with subsequent performance in multiple-target visual search. J Exp Psychol Hum Percept Perform 39 (5):1398–1406
69. Thornton IM, Horowitz TS (2004) The multi-item localization (MILO) task: measuring the

spatiotemporal context of vision for action. Percept Psychophys 66(1):38–50
70. Horowitz TS, Thornton IM (2008) Objects or locations in vision for action? Evidence from the MILO task. Vis Cognit 16(4):486–513
71. Buxbaum LJ, Ferraro MK, Veramonti T, Farne A, Whyte J, Ladavas E, Frassinetti F, Coslett HB (2004) Hemispatial neglect: subtypes, neuroanatomy, and disability. Neurology 62(5):749–756
72. Heilman KM, Valenstein E (1979) Mechanisms underlying hemispatial neglect. Ann Neurol 5(2):166–170
73. Saevarsson S, Halsband U, Kristjánsson Á (2011) Designing rehabilitation programs for neglect: could 2 be more than 1+ 1? Appl Neuropsychol 18(2):95–106
74. Kristjánsson Á, Vuilleumier P (2010) Disruption of spatial memory in visual search in the left visual field in patients with hemispatial neglect. Vision Res 50:1426–1435
75. Husain M, Mannan S, Hodgson T, Wojciulik E, Driver J, Kennard C (2001) Impaired spatial working memory across saccades contributes to abnormal search in parietal neglect. Brain 124 (5):941–952
76. Dukas R, Ellner S (1993) Information processing and prey detection. Ecology 74:1337–1346
77. Jóhannesson ÓI, Kristjánsson Á, Thornton IM (2017) Are foraging patterns in humans related to working memory and inhibitory control? Jpn Psychol Res 59:152–166
78. Kristjánsson T, Thornton IM, Chetverikov A, Kristjánsson Á (Under review) Dynamics of visual attention revealed in foraging tasks. Manuscript under review
79. Ólafsdóttir IM, Kristjánsson T, Gestsdóttir S, Jóhannesson ÓI, Kristjánsson Á (2016) Understanding visual attention in childhood: insights from a new visual foraging task. Cogn Res Princ Implic 1(1):18
80. Ólafsdóttir IM, Gestsdóttir S, Kristjánsson Á (2019) Visual foraging and executive functions: a developmental perspective. Acta Psychol (Amst) 193:203–213
81. Kristjánsson T, Thornton IM, Kristjánsson Á (2018) Time limits during visual foraging reveal flexible working memory templates. J Exp Psychol Hum Percept Perform 44 (6):827–835
82. Jóhannesson ÓI, Thornton IM, Smith IJ, Chetverikov A, Kristjánsson Á (2016) Visual foraging with fingers and eye gaze. i-Perception 7(2):2041669516637279
83. Socé AA, Cain M, Wolfe J (2016) Fitting two target templates into the focus of attention in a hybrid foraging task. J Vis 16(12):1288
84. Wolfe JM, Aizenman AM, Boettcher SEP, Cain MS (2016) Hybrid foraging search: searching for multiple instances of multiple types of target. Vision Res 119:50–59
85. Clarke ADF, Irons J, James W, Leber AB, Hunt AR (2018) Stable individual differences in strategies within, but not between, visual search tasks. https://doi.org/10.31234/osf.io/bqa5v
86. Woods AJ, Göksun T, Chatterjee A, Zelonis S, Mehta A, Smith SE (2013) The development of organized visual search. Acta Psychol (Amst) 143:191–199
87. Gauthier L, Dehaut F, Joanette Y (1989) The bells test: a quantitative and qualitative test for visual neglect. Int J Clin Neuropsychol
88. Mark VW, Kooistra CA, Heilman KM (1988) Hemispatial neglect affected by non-neglected stimuli. Neurology 38(8):1207–1211
89. Weintraub S, Mesulam MM (1988) Visual hemispatial inattention: stimulus parameters and exploratory strategies. J Neurol Neurosurg Psychiatry 51(12):1481–1488
90. Kristjánsson Á, Björnsson AS, Kristjánsson T (Submitted) Foraging with Anne Treisman: patch leaving, features versus conjunctions and memory for foraged location. Manuscript submitted for publication
91. Sokal RR, Rolff FG (1981) Biometry. W.H. Freeman and Co, New York
92. Zar JH (1974) Biostatistical analysis. Prentice-Hall, Englewood Cliffs
93. Moran MD (2003) Arguments for rejecting the sequential Bonferroni in ecological studies. Oikos 100(2):403–405
94. Hosoi E, Rittenhouse LR, Swift DM, Richards RW (1995) Foraging strategies of cattle in a Y-maze: influence of food availability. Appl Anim Behav Sci 43(3):189–196

Neuromethods (2020) 151: 23–35
DOI 10.1007/7657_2019_30
© Springer Science+Business Media, LLC 2019
Published online: 2 October 2019

Eye Tracking in Visual Search Experiments

Andrew Hollingworth and Brett Bahle

Abstract

Over the last 30 years, eye tracking has grown in popularity as a method to understand attention during visual search, principally because it provides a means to characterize the spatiotemporal properties of selective operations across a trial. In the present chapter, we review the motivations, methods, and measures for using eye tracking in visual search experiments. This includes a discussion of the advantages (and some disadvantages) of eye tracking data as a measure spatial attention, compared with more traditional reaction time paradigms. In addition, we discuss stimulus and design considerations for implementing experiments of this type. Finally, we will discuss the major measures that can be extracted from an eye tracking record and discuss the inferences that each allow. In the course of this discussion, we address both experiments using abstract arrays and experiments using real-world scene stimuli.

Keywords Eye movements, Eye tracking, Visual search, Attention, Scene perception

1 Why Track Gaze Position in Visual Search Experiments?

The vast majority of visual search experiments have used manual, end-of-trial reaction time (RT) as the dependent measure. RT provides a single data point per trial with which to draw inferences about the component operations involved in finding a particular target object. Great progress has been made using end-of-trial RT by using sophisticated experimental designs to isolate component operations and to assess them with sufficient precision (e.g., [1]). However, RT approaches provide limited insight into how search evolves over the course of the trial; that is, they provide little direct evidence about which objects in a search display were attended, for how long, and in what sequence. This is the primary benefit of eye tracking: it provides a continuous window on the allocation of attention over a display in a manner that can characterize the spatiotemporal evolution of the search process on individual trials. Thus, instead of assessing the selectivity of guidance from differences in RT as a function of set size, one can directly observe the probability that fixated objects either match or do not match a particular cued feature [2–4]. Instead of inferring attention capture from small increases in RT when a particular distractor value is present, one can directly observe the probability that distractors with that feature value are fixated [4–9]. Direct observation of the phenomena of interest leads to increased sensitivity, and it allows one to precisely characterize how a particular manipulation leads to changes in behavior. A reliable increase of, say, 50 ms in mean RT on distractor-present trials of a capture paradigm tells us only that the distractor likely interfered with search in some manner and on some proportion of trials. In contrast, eye tracking allows us to measure the probability that the distractor was fixated, how early during the course of search it was fixated, for how long, whether gaze returned to that object later in the trial, and the effect of these events on the time required to direct gaze to the target object.

As a concrete example, consider a recent study by Beck, Luck, and Hollingworth [8]. The research question concerned whether search templates in VWM could be configured to deprioritize particular feature values for selection (i.e., a *negative template* or *template for rejection*). Previous studies using end-of-trial RT as the dependent measure had produced conflicting results. On each trial of the Beck et al. study, participants saw a color cue indicating a color to be avoided: objects with this color in the search array were never targets. Gaze was monitored as participants searched through arrays of colored, circular objects for a target with an extremely small gap on the top or bottom (target feature) among distractors with gaps on the left or right. Gap discrimination required foveation. The search arrays consisted of four objects in four different colors (a total of 16) that were randomly arrayed. By monitoring gaze during search, Beck et al. were able to examine the evolution of selectivity across the trial. At the very onset of search, participants were *more likely* than expected by chance to fixate objects in the cued color (i.e., capture), but by approximately the third object fixated in the array, this pattern reversed, with cued-color objects *less likely* to be fixated than expected by chance (i.e., successful avoidance). This pattern was not observable on end-of-trial RT. Relative to a neutral-cue condition, capture early in the trial (increasing overall RT) and successful avoidance later in the trial (decreasing overall RT) largely cancelled. Thus, the ability to examine selection across the entire trial was key to understanding the mechanisms involved in the use of a negative template, and a purely RT-based design would likely have led to an erroneous conclusion (i.e., that there was no capture by matching items nor any benefit from the cue information).

The eye tracking method allowed several further analyses critical to understanding the underlying mechanisms in Beck, Luck, and Hollingworth [8]. First, fixating a cue-matching object led to a substantial increase in overall search time, reflected on both the elapsed time to the first fixation on the target object and the manual RT to report gap location (these two measures will be strongly correlated in a design such as this one). Second, there was no observable relationship between fixation of a cue-matching object early in the trial and later avoidance, indicating that fixation of the cued color was not necessary for producing later selectivity. Finally, the latency of the very first saccade on the array was longer on negative-cue trials compared with neutral and positive-cue trials, potentially indicating that it took additional time to set up the guidance operation for a negative template or that participants had to exert control to avoid oculomotor capture by matching distractors before they could initiate a saccade to a relevant array item. In sum, eye tracking produces data about the allocation of attention across time and space that can support a rich understanding of the underlying attentional mechanisms.

In addition to the increasing use of eye tracking within traditional visual search tasks, the use of eye tracking in visual search has grown more prominent as interest in real-world scene perception has increased (for a review, see [10]). Given the size and complexity of natural environments, movements of the eyes (and head and body) are typically required to obtain information from task-relevant objects. Coincident with this interest, the field of visual search has been transitioning gradually from using search paradigms as a tool for understanding basic properties of visual perception and attention (typically using abstract arrays) to investigating visual search as an important behavior to be understood in its own right: how people are able to find relevant objects within complex displays (often naturalistic scenes). The field has progressed so that current, prominent general theories of visual search seek to explain how gaze is oriented sequentially to objects within natural scenes (e.g., [11]). The same inferential advantages for eye tracking using abstract displays apply to real-world scene studies. In addition, real-world scenes have visual, spatial, conceptual, and episodic structure that are not typically found in abstract arrays and can be critical to understanding the search operation. For example, kitchens tend to contain blenders, these tend to appear on the counter rather than on the floor, and the blender will also tend to appear in the same place as it was observed previously. Thus, scenes allow additional forms of guidance that can be observed in the sequence of eye movements during search. For example, one can ask whether and how early during visual search gaze is directed to regions of the scene where a target object is likely to be found [12–19] or to locations where the target has been observed previously [20–24].

Before discussing in more detail the methods and measures involved in implementing eye tracking studies using abstract arrays and complex scenes, it is important to discuss the potential limitations in using eye tracking as means to infer the properties of *covert* attention. A relatively consistent literature (for a review, see [25]) demonstrates that saccade execution is necessarily preceded by a covert shift of spatial attention to the saccade target location [26–28]. However, attention can be shifted covertly in the absence of saccade preparation [29–34]. Thus, if the goal of the study is, specifically, to understand covert attentional processes independently of saccade preparation, then eye movements will need to be eliminated or controlled.[1] However, if a saccade is observed, one can be quite certain that, immediately preceding the saccade, there was a corresponding shift of covert attention. Thus, the sequence of eye movements provides strong evidence regarding the spatial allocation of covert attention across time, but this record may not be exhaustive; that is, there may be additional shifts of covert attention that were not associated with saccade preparation or did not ultimately lead to a saccade generation (e.g., through competition or direct inhibition). Despite this limitation, the ability of eye movement data to deliver a close approximation of the spatiotemporal properties of covert selection is a major strength of the method.

2 Methods

In this section we discuss methods for conducting experiments on eye movements in visual search, using both abstract arrays and natural scene stimuli. Then, we turn to eye tracking measures and their interpretation. Figure 1 illustrates typical designs and methods.

Abstract Arrays: Methods for conducting eye tracking studies using abstract search arrays vary only modestly from traditional RT experiments, and thus it is 'quite simple to delineate the necessary modifications, some of which are quite commonsensical (for an example of a fully implemented method, see [8]).

First, search arrays need to be appropriately constructed to minimize confusion about which object is fixated. This is mostly a simple matter of making sure that there is adequate spacing between objects. During search, if objects are well distinguished from the background, a very large majority of saccades will be directed to an object rather than to the spaces between objects. Some care should be taken in selection of the areas of interest (AOI, i.e., the spatial region around each object that is used to classify a fixation as "on" a particular object) so that they are large enough to tolerate some noise in tracking and calibration (extending beyond the physical boundaries of the objects) but not so large that they

[1] Note that instructions are rarely sufficient to ensure that participants do not make eye movements. Thus, even if the goal is to eliminate eye movements, gaze still needs to be monitored. Ideally, an eye tracker can be used, but there is another option. In covert attentions studies, we often use a simple video camera to display a large image of one of the eyes, and the experimenter monitors this image throughout the experiment (a human eye tracker). Movements of the eyes are quite easy to observe, and the experimenter both notes trials with eye movements and reminds the participant, when an eye movement is observed, to keep gaze focused on the relevant reference point. With appropriate, well-timed feedback of this sort, most participants quickly learn how to keep gaze focused centrally and rarely make eye movements after an initial practice session.

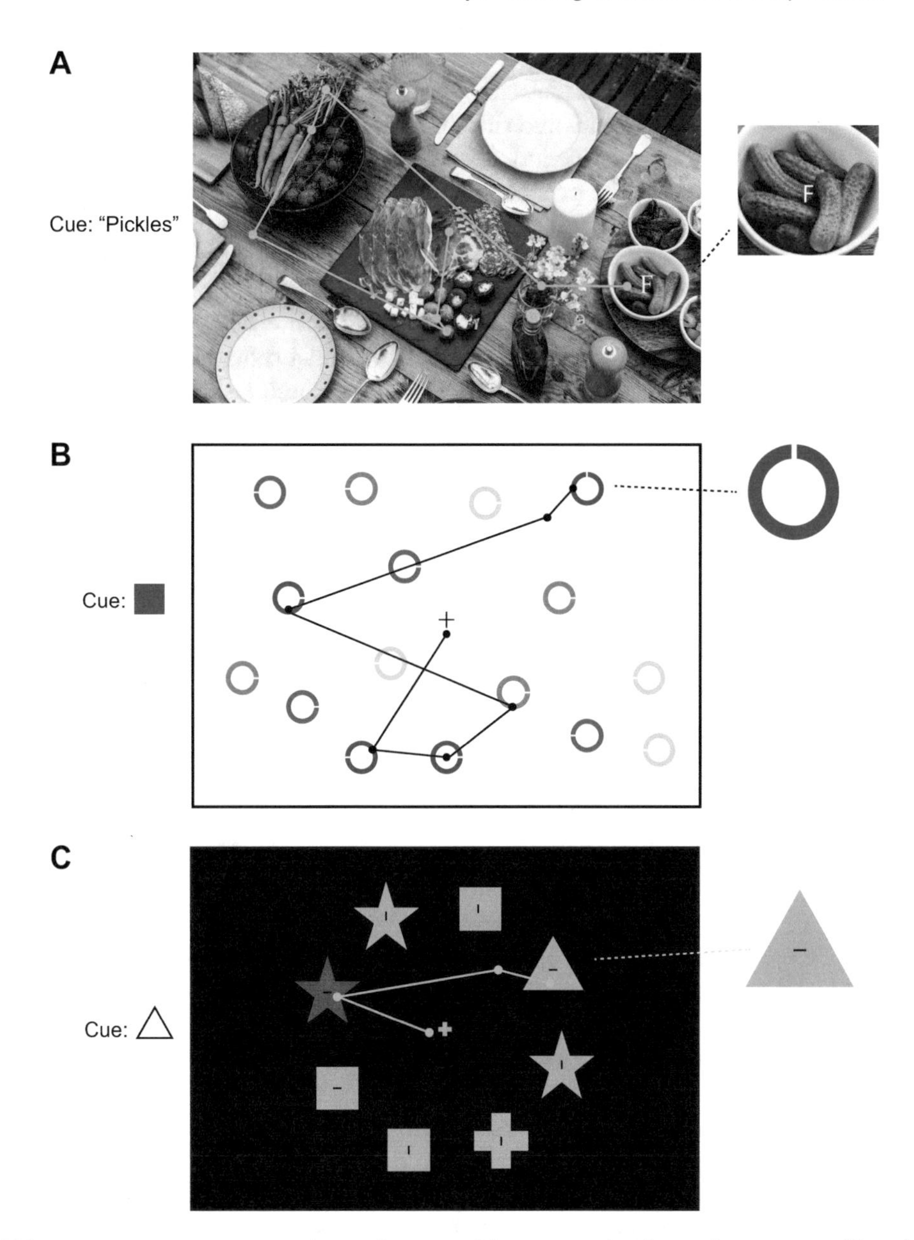

Fig. 1 Three common eye movement paradigms used to assess visual search processes. Lines indicate saccades and dots fixations. (**a**) A paradigm that probes search efficiency. The participant searches for a single cued object within a scene or array, reporting a secondary feature superimposed over it (here, the orientation of an "F"). The primary measure typically would be the elapsed time to the first fixation on the target, but, as discussed in the text, this could be subdivided into *initiation*, *scanning*, and *verification* times. (**b**) A paradigm that probes selectivity. In this example, the cued feature (blue) indicates the color of the target object, with multiple possible target objects present in the display, only one of which has the to-be-reported feature (gap on top or bottom). The primary measure typically would be the probability that a particular fixated object matches the cued value. (**c**) A paradigm that probes attention capture. Here, the participant searches for a cued shape in the presence of a physically salient distractor (uniquely colored item). The typical-dependent measure would be the probability of critical distractor fixation, either limited to the first saccade on the array or at any point before fixation of the target

miscategorize fixations that are clearly directed to the background or to other objects.

Second, and this may seem an obvious point, but search stimuli should be excluded from the immediate region around the position where the participant will be fixating when the array appears (typically the center). All stimuli should be at least a few degrees of visual angle away from this starting position so that a saccade is required for fixation of any array object.

Third, and most importantly, the search task itself should be configured so that participants must translate covert selection into overt shifts of gaze. That is, the search task should *require* fixation of individual objects.[2] This could be implemented in several ways. One could simply make the task to fixate the target. However, we typically do not want to have subjects to reflect too deeply on their gaze patterns, as this could cause them to consciously control gaze. A good alternative is to have participants report a small, secondary feature of the target object that requires foveation for discrimination [2], such as the orientation of a small bar superimposed over the object, the location of a small gap in the object, or the identity of a small letter (see Fig. 1). There is no need in this sort of method to even mention eye movements in the instructions (except that they will be monitored). One must be careful to ensure that the secondary feature is not so salient that participants can search for this feature rather than for the cued object. If concerned about this possibility, the appearance of the secondary feature can be made contingent on fixation of the target region.

Natural Scenes: Methods for constructing experiments using eye tracking on scene stimuli introduce a much greater set of challenges, most of which revolve around how to construct naturalistic images while maintaining appropriate experimental control. The suggestions outlined below apply not only to visual search experiments but to other types of experiments using natural scenes stimuli. To make the design challenges and solutions concrete, let's consider an experiment in which one asks whether objects that appear in plausible scene locations are found faster than objects that appear in implausible locations, testing the hypothesis that a key aspect of guidance in visual search through scenes is knowledge of typical object locations (e.g., that blenders tend to appear on kitchen counters).

The first consideration is how we will construct our scene stimuli to present objects in both plausible and implausible locations within images of natural scenes and how we will retain a

[2] Note that a present/absent design is not always ideal for an eye tracking study, as the mere presence or absence of the target can often be determined without foveation.

strong degree of stimulus control across this manipulation. The choice of stimulus production method will be key. First, we could use line drawings of scenes [17]. This would simplify the process of moving objects to different scene locations, as we would not need to worry about differences in illumination, and so on. But line drawings do not get us very far toward the goal of realism. Second, we could use real environments and take a photograph with a particular object in a typical location, then move the object to an atypical location, and take another photograph. However, a physically moved object will often lead to substantial differences in object appearance caused by changes in illumination, distance from the camera, perspective, and so on. Moreover, such an approach is time-consuming, and it is brittle, in that one cannot alter the manipulation post hoc: it would be practically impossible to re-create the original photographic context if, say, you later decided to add a third object location. Third, we could use a 3D modeling and rendering program that can produce near-photo-realistic images. We would then simply render the scene with the critical object in one location, move the object within the model in a manner that minimized changes in visual appearance, and then re-render the image [23, 35]. This method allows for a great deal of control over the scene stimuli and has the advantage of being robust to the addition of future manipulations. Moreover, some manipulations are possible in 3D that are virtually impossible in photographic stimuli, such as changes of object in-depth orientation, viewer perspective, illumination, and so on. Limitations to the 3D modeling approach primarily revolve around the investment of time necessary to develop expertise in 3D graphics, the availability (and perhaps cost) of 3D scene models, the computing resources necessary to render the images (especially if one uses methods such as raytracing and highly detailed models to produce near-photo-realistic images), and the time required for rendering. However, many of these limitations can be ameliorated by using programs that have been optimized for efficient rendering, such as home design programs or game engines (e.g., [20, 36]). Finally, we could simply obtain existing photographic images (from a web search or from one of the several research databases of scene images) and implement our manipulation of object position using a 2D graphics program such as Adobe Photoshop. With some expertise, it is possible to add, move, or otherwise modify objects in scene photographs in a manner that results in relatively seamless integration of the changes.[3]

[3] Note, however, that in this example experiment, if we were to manipulate photographic images, we would need to use an object image from a different source and then paste and integrate it into *both* the plausible and implausible locations within the experimental scene; this would control our two conditions for artifacts generated by the process of adding an object to a particular scene location.

In general, the choice of stimulus production method will depend on the needs of the experiment. For example, in a relatively recent study [23], we needed to have participants search for 12 different objects in each scene. This was most efficiently implemented by using a 3D modeling program to carefully place the 12 different target objects in discrete locations within each scene, so that there was sufficient object-to-object spacing. It would have been very difficult to find photographs that met these requirements (but see [22]). In another study, we had to manipulate the color of target and distractor objects in scenes [37]. This was implemented most efficiently by using Photoshop to change object color within scene photographs, which is quite easily done in a manner that looks realistic.

After ensuring control over visual properties of objects across manipulations, we also need to exert control over scene locations across manipulations. In our example experiment, we would want to ensure that, due to random variation, the locations chosen for the plausible and implausible conditions did not differ in the ease of search (independently of plausibility). To minimize such effects, we could simply make sure that, across conditions, mean eccentricity of targets was similar, but this would not control for factors such as location variability, spatial biases, or the properties of the local context at plausible an implausible positions (e.g., targets in the plausible condition might tend to be located in sparser regions of the scenes, making them easier to find). To solve this problem, we could manipulate *two* different objects within each scene [38]. For example, in an office scene, a framed picture is likely to appear on a desk and a wastebasket on the floor. Thus, we create *four* versions of the scene: picture on desk, picture on floor, wastebasket on floor, and wastebasket on desk. Now, each scene location and each target object are used in both conditions. When creating a design like this, where there are multiple versions of each scene but only one version can be shown to each subject, then scene-item-to-condition assignments will need to be counterbalanced across subjects (this will also require a relatively large number of scene items). Of course, other experimental designs will differ in the type of implementation necessary to ensure control over locations across conditions.

For the search task itself, we again need to make fixation of the target necessary for trial completion, but not the primary goal of the task. As described above, a good means to this end is to superimpose a small discrimination target over the object [20, 37]. For example, in Bahle, Matsukura, and Hollingworth [37], a small letter "F" was superimposed over the target, and this could either be normally oriented or mirror reversed. The search task was to find the cued object and then report the orientation of the superimposed "F." On the vast majority of trials, participants fixated the target immediately before reporting the orientation of the "F."

3 Measures

In this final section, we discuss four major types of eye movement measures relevant to visual search tasks: measures of search efficiency, measures of selectivity, measures of capture, and measures of memory. Before discussing these measures, we note that each depends on defining areas of interest around the array objects (and, perhaps, other relevant locations, such as the screen center). The eye tracking record is typically analyzed with respect to the fixations that occur within AOIs. In addition, it is common to eliminate trials if the starting gaze position deviates substantially from the reference point (e.g., array center)[4] or if there is no entry into the target AOI during the trial.

Measures of Search Efficiency: The most direct measure of the time taken to find a single target object in an array or natural scene is the elapsed time from the onset of the search stimulus to the beginning of the first fixation with in the target AOI (for the entry that immediately precedes the manual response).[5] We will refer to this measure at *elapsed time to target fixation* (see Fig. 1a). There are several other measures that will almost always be highly correlated with elapsed time to target fixation. *Number of saccades to target fixation* produces essentially the same data as elapsed time but with coarser grain, typically adding little to the overall analysis. *Manual RT* for discrimination of the secondary target feature (e.g., the embedded line orientation) also produces essentially the same data as elapsed time to target fixation but with increased variability due to the addition of discrimination and response processes. Finally, *Path ratio* provides a spatial, rather than temporal, measure of search efficiency. Path ratio is the sum of the amplitudes of all of the saccades before the first entry into the target region *divided by* the distance between the fixation position at search onset (e.g., the center of the scene or array) and the location of the target object. Thus, a value of 1 indicates maximally efficient search, with values increasingly greater than 1 indicating increasingly less efficient search.

A recent approach to understanding the efficiency of component processes during the search task has been to divide the elapsed

[4] An alternative is to make trial initiation contingent on central fixation.

[5] It is possible that, on a small proportion of trials, a participant fixates the target, fails to recognize it at such, leaves the target region to fixate other objects, and returns only later during search, leading to the manual response. Thus, elapsed time to target fixation should be the time until the entry that immediately precedes the response and not necessarily the elapsed time to the very first entry.

time to manual response into three different epochs [16, 38, 39]: (1) time to initiate the first saccade during search (*initiation time*), (2) time from the onset of the fist saccade to the first fixation in the target region (*scanning time*), and (3) time from the first fixation in the target region to the manual response (*verification time*). The former two constitute elapsed time to target fixation. Researchers have considered initiation time to reflect processes related to establishing the target template and to selecting the first item for scrutiny. Scanning time is considered a relatively pure measure of the search process. Finally, verification time is considered a measure of the time taken to decide that the fixated object is in, in fact, the target object (this is relevant primarily to present/absent paradigms where there are no decision processes related to a secondary target feature). Although this division can provide a mapping of eye tracking epochs to particular component processes, it is important to note that, as with any eye tracking measure, the mapping must be treated with considerable care and caution. For example, the very first saccade on an array or scene can quite often be poorly guided, driven by the sensory transient caused by the onset of the search stimulus, and thus initiation time would not necessarily reflect the time required to establish goal-directed guidance. Second, verification time assumes that target identification began only after target fixation, which does not take into account the strong possibility that verification began before the saccade that took the eyes to the target region. That is, it is likely that the target object was selected for fixation, at least in part, because it was identified as the target object in the periphery.

Measures of Selectivity: It is often of interest to attention researchers to characterize the selectivity of visual search: that is, the extent to which attention is limited to cued or otherwise goal-relevant items (Fig. 1b). For example, in a classic study, Williams [2] probed whether selective guidance was implemented more or less efficiently for different feature dimensions (color, size, and shape). This involved constructing displays in which there were multiple objects matching the cued feature value (e.g., multiple red items among items of different colors), with only one of the cue-matching objects containing a secondary target feature. The measure of selectivity was simply the probability that, for any given fixation on an array object, the fixated object matched the cued value.[6] Zelinsky [3] used a related method to show that factors influencing RT as a function of set size had corresponding effects on oculomotor selectivity. And, as noted above, it is possible to examine how selectivity changes over the course of a trial [8] by computing the probability of cued-item fixation for each ordinal fixation number (i.e., following the first saccade during search, the second, the third,

[6] An alternative would be to consider each entry and exit from an object as a single event, collapsing across multiple fixations between entry and exit.

and so on) or using some other means to temporally divide the trial into quantiles.

Measures of Capture: Similar to measuring selectivity as the probability that a cue-matching object is fixated, one can measure the probability that a *distractor* with some additional or unique feature property is fixated (relative to control items that do not have the critical attribute). That attribute could be anything from physical salience (Fig. 1c) to emotional relevance. In one of the first studies of this type, Theeuwes, Kramer, Hahn, Irwin, and Zelinsky [5] examined whether a singleton item in a search display (e.g., uniquely colored) would attract gaze preferentially in a manner consistent with RT results. Relative to non-salient distractors, singletons had a higher probability of being fixated following the first saccade on the array.[7] As in this study, researchers sometimes limit their capture analysis to the very first saccade on the array (especially if the goal is to probe processes that as hypothesized to be based on low-level stimulus properties). However, we have found that, in related tasks, differences in fixation probability can extend multiple saccades into the search process, and thus we have tended to define distractor fixation probability as an object entry at *any point* between the onset of the search array and the first fixation on the target immediately preceding the response [7, 37].

Measures of Memory: During visual search tasks, participants often have to keep track of the items that have been previously scrutinized and rejected, so that gaze can be selectively directed to possible target items [40, 41]. One direct way to measure these processes is to examine the probability of distractor refixation during search. Here, *refixation* refers to the situation in which gaze is directed to a particular region, exits that region for some period of time, and then returns. To make this type of design concrete, consider the task in Fig. 1b, but with no color cue, so that participants have to search every item in the display until the target is found. This would require keeping track of the locations of multiple, previously fixated objects, and the probability of refixation would then serve as a measure of memory-based avoidance. Using this type of method, Peterson, Kramer, Wang, Irwin, and McCarley [42] tested the hypothesis that visual search is *memoryless* in the sense that each selective operation is amnesic with respect to previous events on that trial [43]. Falsifying this hypothesis, Peterson et al. confirmed a major role for memory in search efficiency by showing that distractor refixation probability was much lower than would have been predicted by an amnesic search operation (see also [44]).

[7] It can also be useful to examine saccade latency in this context. For trials *without* oculomotor capture, several studies have observed that saccades directed to the target were delayed when a critical distractor was present versus when it was not, indicating that the programming of the saccade required additional time to resolve the competition between the salient distractor and the target.

References

1. Wolfe JM (2007) Guided search 4.0: current progress with a model of visual search. In: Gray W (ed) Integrated models of cognitive systems. Oxford University Press, Oxford, pp 99–119
2. Williams LG (1967) The effects of target specification on objects fixated during visual search. Acta Psychol 27:355–360. https://doi.org/10.1016/0001-6918(67)90080-7
3. Zelinsky GJ (1996) Using eye saccades to assess the selectivity of search movements. Vis Res 36:2177–2187. https://doi.org/10.1016/0042-6989(95)00300-2
4. Beck VM, Hollingworth A, Luck SJ (2012) Simultaneous control of attention by multiple working memory representations. Psychol Sci 23:887–898. https://doi.org/10.1177/0956797612439068
5. Theeuwes J, Kramer AF, Hahn S, Irwin DE, Zelinsky GJ (1999) Influence of attentional capture on oculomotor control. J Exp Psychol Hum Percept Perform 25:1595–1608. https://doi.org/10.1037/0096-1523.25.6.1595
6. Gaspelin N, Leonard CJ, Luck SJ (2015) Direct evidence for active suppression of salient-but-irrelevant sensory inputs. Psychol Sci 26:1740–1750. https://doi.org/10.1177/0956797615597913
7. Bahle B, Beck VM, Hollingworth A (2018) The architecture of interaction between visual working memory and visual attention. J Exp Psychol Hum Percept Perform 44:992–1011. https://doi.org/10.1037/xhp0000509
8. Beck VM, Luck SJ, Hollingworth A (2018) Whatever you do, don't look at the...: evaluating guidance by an exclusionary attentional template. J Exp Psychol Hum Percept Perform 44:645–662. https://doi.org/10.1037/xhp0000485
9. Le Pelley ME, Pearson D, Griffiths O, Beesley T (2015) When goals conflict with values: counterproductive attentional and oculomotor capture by reward-related stimuli. J Exp Psychol Gen 144:158–171. https://doi.org/10.1037/xge0000037
10. Henderson JM (2003) Human gaze control during real-world scene perception. Trends Cogn Sci 7:498–504. https://doi.org/10.1016/j.tics.2003.09.006
11. Itti L, Koch C (2000) A saliency-based search mechanism for overt and covert shifts of visual attention. Vis Res 40:1489–1506. https://doi.org/10.1016/S0042-6989(99)00163-7
12. Castelhano MS, Henderson JM (2007) Initial scene representations facilitate eye movement guidance in visual search. J Exp Psychol Hum Percept Perform 33:753–763. https://doi.org/10.1037/0096-1523.33.4.753
13. Castelhano MS, Heaven C (2011) Scene context influences without scene gist: eye movements guided by spatial associations in visual search. Psychon Bull Rev 18:890–896. https://doi.org/10.3758/s13423-011-0107-8
14. Eckstein MP, Drescher BA, Shimozaki SS (2006) Attentional cues in real scenes, saccadic targeting, and Bayesian priors. Psychol Sci 17:973–980. https://doi.org/10.1111/j.1467-9280.2006.01815.x
15. Henderson JM, Malcolm GL, Schandl C (2009) Searching in the dark: cognitive relevance drives attention in real-world scenes. Psychon Bull Rev 16:850–856. https://doi.org/10.3758/pbr.16.5.850
16. Malcolm GL, Henderson JM (2009) The effects of target template specificity on visual search in real-world scenes: evidence from eye movements. J Vis 9(8):1–13. https://doi.org/10.1167/9.11.8
17. Henderson JM, Weeks PA, Hollingworth A (1999) The effects of semantic consistency on eye movements during complex scene viewing. J Exp Psychol Hum Percept Perform 25:210–228. https://doi.org/10.1037//0096-1523.25.1.210
18. Võ MLH, Henderson JM (2010) The time course of initial scene processing for eye movement guidance in natural scene search. J Vis 10:14. https://doi.org/10.1167/10.3.14
19. Torralba A, Oliva A, Castelhano MS, Henderson JM (2006) Contextual guidance of eye movements and attention in real-world scenes: the role of global features in object search. Psychol Rev 113:766–786. https://doi.org/10.1037/0033-295X.113.4.766
20. Brockmole JR, Castelhano MS, Henderson JM (2006) Contextual cueing in naturalistic scenes: global and local contexts. J Exp Psychol Learn Mem Cogn 32:699–706. https://doi.org/10.1037/0278-7393.32.4.699
21. Brockmole JR, Henderson JM (2006) Using real-world scenes as contextual cues for search. Vis Cogn 13:99–108. https://doi.org/10.1080/13506280500165188
22. Võ MLH, Wolfe JM (2012) When does repeated search in scenes involve memory? Looking at versus looking for objects in scenes. J Exp Psychol Hum Percept Perform

38:23–41. https://doi.org/10.1037/a0024147
23. Hollingworth A (2012) Task specificity and the influence of memory on visual search: comment on Võ and Wolfe (2012). J Exp Psychol Hum Percept Perform 38:1596–1603. https://doi.org/10.1037/A0030237
24. Bahle B, Hollingworth A (2019) Contrasting episodic and template-based guidance during search through natural scenes. J Exp Psychol Hum Percept Perform 45:523–536. https://doi.org/10.1037/xhp0000624
25. Hunt AR, Reuther J, Hilchey MD, Klein R (2019) The relationship between spatial attention and eye movements. Curr Top Behav Neurosci. https://doi.org/10.1007/7854_2019_95
26. Hoffman JE, Subramaniam B (1995) The role of visual attention in saccadic eye movements. Percept Psychophys 57:787–795. https://doi.org/10.3758/BF03206794
27. Kowler E, Anderson E, Dosher B, Blaser E (1995) The role of attention in the programming of saccades. Vis Res 35:1897–1916. https://doi.org/10.1016/0042-6989(94)00279-U
28. Deubel H, Schneider WX (1996) Saccade target selection and object recognition: evidence for a common attentional mechanism. Vis Res 36:1827–1837. https://doi.org/10.1016/0042-6989(95)00294-4
29. Hunt AR, Kingstone A (2003) Covert and overt voluntary attention: linked or independent? Cogn Brain Res 18:102–105. https://doi.org/10.1016/j.cogbrainres.2003.08.006
30. Klein RM (1980) Does oculomotor readiness mediate cognitive control of visual attention? In: Nickerson RS (ed) Attention and performance VIII. Erlbaum, Hillsdale, pp 259–276
31. Klein RM, Pontefract A (1994) Does oculomotor readiness mediate cognitive control of visual attention? Revisited! In: Umilta C, Moscovitch M (eds) Attention and performance XV—conscious and nonconscious information processing. MIT Press, Cambridge, pp 333–350
32. Juan CH, Shorter-Jacobi SM, Schall JD (2004) Dissociation of spatial attention and saccade preparation. Proc Natl Acad Sci U S A 101:15541–15544. https://doi.org/10.1073/pnas.0403507101
33. Schafer RJ, Moore T (2011) Selective attention from voluntary control of neurons in prefrontal cortex. Science 332:1568–1571. https://doi.org/10.1126/science.1199892
34. Thompson KG, Biscoe KL, Sato TR (2005) Neuronal basis of covert spatial attention in the frontal eye field. J Neurosci 25:9479–9487. https://doi.org/10.1523/jneurosci.0741-05.2005
35. Hollingworth A, Henderson JM (2002) Accurate visual memory for previously attended objects in natural scenes. J Exp Psychol Hum Percept Perform 28:113–136. https://doi.org/10.1037//0096-1523.28.1.113
36. Koehler K, Eckstein MP (2017) Beyond scene gist: objects guide search more than scene background. J Exp Psychol Hum Percept Perform 43:1177–1193. https://doi.org/10.1037/xhp0000363
37. Bahle B, Matsukura M, Hollingworth A (2018) Contrasting gist-based and template-based guidance during real-world visual search. J Exp Psychol Hum Percept Perform 44:367–386. https://doi.org/10.1037/xhp0000468
38. Malcolm GL, Henderson JM (2010) Combining top-down processes to guide eye movements during real-world scene search. J Vis 10(4):1–11. https://doi.org/10.1167/10.2.4
39. Castelhano MS, Pollatsek A, Cave KR (2008) Typicality aids search for an unspecified target, but only in identification and not in attentional guidance. Psychon Bull Rev 15:795–801. https://doi.org/10.3758/pbr.15.4.795
40. Smith TJ, Henderson JM (2011) Does oculomotor inhibition of return influence fixation probability during scene search? Atten Percept Psychophys 73:2384–2398. https://doi.org/10.3758/s13414-011-0191-x
41. Klein RM, MacInnes WJ (1999) Inhibition of return is a foraging facilitator in visual search. Psychol Sci 10:346–352. https://doi.org/10.1111/1467-9280.00166
42. Peterson MS, Kramer AF, Wang RF, Irwin DE, McCarley JS (2001) Visual search has memory. Psychol Sci 12:287–292. https://doi.org/10.1111/1467-9280.00353
43. Horowitz TS, Wolfe JM (1998) Visual search has no memory. Nature 394:575–577. https://doi.org/10.1038/29068
44. Gilchrist ID, Harvey M (2000) Refixation frequency and memory mechanisms in visual search. Curr Biol 10:1209–1212. https://doi.org/10.1016/s0960-9822(00)00729-6

Neuromethods (2020) 151: 37–57
DOI 10.1007/7657_2019_20
© Springer Science+Business Media, LLC 2019
Published online: 11 May 2019

Feature Distribution Learning (FDL): A New Method for Studying Visual Ensembles Perception with Priming of Attention Shifts

Andrey Chetverikov, Sabrina Hansmann-Roth, Ömer Dağlar Tanrıkulu, and Árni Kristjánsson

Abstract

We discuss how priming of attention shifts has in recent studies proved to be a useful method for studying internal representations of visual ensembles. Attentional priming is very powerful in particular when role reversals between targets and distractors occur. Such role reversals can be used to assess how expected or unexpected a particular target is. This new method for studying representations of visual ensembles has revealed that observer's representations are far more detailed than previous studies of ensemble perception have suggested where the emphasis has been on summary statistics, i.e., mean and variance. Observers can represent surprisingly complex distribution shapes such as whether a representation is bimodal or not. We discuss the details of how this feature distribution learning (FDL) method has been used to assess internal representations of visual ensembles. We also speculate that the method can prove to be an important *implicit* way of assessing how observers represent regularities in their environments.

Keywords Perceptual representations, Visual ensembles, Visual search, Priming, Feature distribution learning (FDL)

1 Intro and Background

Priming of attention shifts has been extensively investigated over the last 25 years, mainly with various forms of visual search tasks. A key finding in this literature comes from the studies of Maljkovic and Nakayama [1]. They used a paradigm introduced by Bravo and Nakayama [2] where observers searched for an oddly colored diamond among distractors of another color and had to judge whether there was a notch on its left or right side. Bravo and Nakayama had observed that when targets maintained their color between trials (e.g., the target was always the red diamond among green distractor diamonds), search was overall faster than when the target and distractor identity reversed unpredictably (from a red target diamond among green diamond distractors to a green target diamond among red distractor diamonds). There was therefore a benefit to target and distractor consistency from one trial to the next.

Maljkovic and Nakayama [1] replicated this result, finding additionally that search became faster the more often the same target color repeated. The consistency benefit therefore reflected this "priming" effect, and importantly they found that the priming effects were only minimally affected by top-down strategies (such as whether observers knew the upcoming target color or not; but see [3, 4]). Maljkovic and Nakayama also tested the cumulative effects of priming over several trials, finding that response times decreased by 20 to 25%, with no corresponding increases in error rates (see, e.g., [5]).

This basic finding on priming during feature search has been replicated many times (see [6], for review). Here, our aim is firstly to discuss basic considerations for priming paradigms and secondly to introduce a paradigm that utilizes priming to assess how human observers represent the environment.

1.1 What Can Prime?

It is important to note that not only target characteristics can prime from one attentional allocation to the next but also the identity of the distractors [1, 7–9]. If the same distractors appear from one trial to the next, search will be speeded, irrespective of whether the target identity is unchanged [1, 7]. These distractor priming effects therefore make their independent contribution to performance [8, 10, 11] and can be just as strong as the target priming effects, although the two can interact. In a paradigm where the target and distractors do not vary independently, it is impossible to disentangle the two, and this needs to be taken into account when results from priming studies are interpreted. These two separate sources of priming effects can combine so that when an odd-one-out target contains the colors of the distractor on the preceding trials, search is slowed even more, reflecting so-called role-reversal effects. Importantly, the strength of these role-reversal effects can be used to answer other questions regarding visual perception, as we discuss below.

The priming effects have typically been thought to reflect facilitation of individual features [12]. Other findings show that such priming can occur from the repetition of more complex characteristics, such as feature combinations or objects identities [13–16]. Separate *features* of the stimuli that observers search for can cause their own priming effect, and so can whole targets, depending on the circumstances [15]. For example, color, spatial frequency, and orientation can cause independent priming effects depending on task relevance [17].

A critical feature of these priming effects is that they are so strong that they sometimes seem to be able to account for effects that have typically been attributed to explicit top-down attentional guidance in the literature [14, 18, 19]. As an example, Kristjánsson, Wang, and Nakayama showed that search times were similar when target identity was always the same and when priming effects were maximal, although target identity was not known from one trial to the next, showing that large portions of effects attributed to top-down guidance were accounted for by priming. Belopolsky et al. [18] showed that so-called contingent-capture effects [20], thought to be caused by top-down guidance, could, to a large extent, be explained by priming. Theeuwes and van der Burg [19] assessed interference effects from irrelevant distractor stimuli, finding that observers could not use top-down set (from verbal or symbolic cues) to ignore irrelevant color singletons (the interference was still present), but they further argued that when attentional priming effects were maximal, interference from irrelevant distractors was minimal.

A common interpretation of attentional priming effects is that searching for a target automatically creates a representation of that item or feature depending on context [15] which, in turn, influences subsequent attention allocation. This entails the assumption that the processing of the features that are contained in the template is facilitated. Such templates are often thought to be kept in visual working memory—and there is indeed evidence suggesting that visual working memory content can modulate priming [21, 22]. Other accounts of priming involve the dimensional weighting account [23, 24] and the relational encoding account [25] that can both surely account for priming under certain conditions. Note, however, that in their review, Kristjánsson and Campana [26] concluded that priming was so ubiquitous in attentional orienting and occurred on so many levels that no single account would probably ever explain it completely.

1.2 Key Considerations for Studies of Attentional Priming

1.2.1 The Duration of Priming Effects

Attentional priming effects are long-lasting. Maljkovic and Nakayama ([1], Exp. 5) showed that up to at least five trials in the past can cause priming, irrespective of what the target and distractor identities were on the intervening trials. Regrettably, researchers often look only at switches versus repeats in studies of priming rather than cumulative effects over several trials. This is unfortunate for two main reasons: firstly, interesting patterns of cumulative repetition are overlooked, and secondly as the effects are additive over adjacent trials with the same target and distractors, looking only at switches or repeats of target identity may not assess priming effects at their maximum strength.

Priming may have both a transient component and a longer-lasting one [27, 28]. Kruijne et al. [29] concluded however that priming involves a single facilitative memory trace that decays over approximately eight trials (see also [30]).

Maljkovic and Nakayama ([1]; see also [31]) introduced a clever way of assessing how long the priming influence from a single trial lasts independently of what occurs on intervening trials that they called *memory kernel analysis*. The method involves categorizing a given trial as the same or different as the one that appeared on the trial that preceded it, two trials preceding it, three trials preceding it, and so on (or formally, i trials preceding the given trial n). For each trial n, target color on trial $n-i$ can be the same or different as the target color on trial n. But over a large number of trials, the numbers of the same and different color trials between trial n and trial $n-i$ will even out. So, to assess the influence of the color of the target i trials in the past, performance on trials where target color (or any other property that is presumed to be primed) on trial $n-i$ was the same as on trial n and when it was different can be compared. Note that the same procedure can be carried out for *future* trials. Future trials should, of course, cause no priming effect and therefore provide a useful sanity check for the analysis and can also be used as an index of the variability in the data (see [1, 31]).

These memory kernels can then be modeled (see [27, 32]). Martini [27] found, for example, that priming effects from repeated target color and repeated position are well described by the summation of two exponential functions, one with a high gain and fast decay and another with low gain but slower decay (consistent with [28]). Kruijne et al. [29] later concluded that a single temporal function could explain priming of features.

Overall these memory kernels that describe the time course of priming effects highlight an important point, since they show that the priming effects can be subtle and long-lasting and, more importantly, how they can contaminate results in various paradigms. To take one example, unequal numbers of two different targets within blocks can cause a contaminating influence that can bias experimental results.

1.2.2 Disentangling Stimulus Priming and Response Priming

Another key consideration is that what is primed must be disentangled from what is reported to avoid the contaminating influence of response repetition effects. To take one example, Maljkovic and Nakayama [1] ensured that their response variable in their task was independent of any feature-repetition effects to avoid this. That is, while observers looked for an odd-one-out color, they responded to a location of a notch on the oddly colored item. Also, in a present versus absent visual search task tested in Kristjánsson et al. [14], the present-absent judgment was independent of the orientation of targets and distractors that either repeated or not between trials.

1.2.3 Effects of Distractor Repetition

It became clear early on that distractor priming was just as important as target priming [1] and that they have their own influence that can be disentangled in experimental design [10]. It is therefore important not to attribute priming effects to either targets or distractors, unless the effects of each can be convincingly unconfounded by design.

1.2.4 Unwanted Influences of Priming

As mentioned before, the influences of priming can be subtle. To take one example, cues are thought to summon attention, whether by cueing a location or a stimulus feature: a red stimulus may be used to alert observers to an upcoming red stimulus inducing a presumably top-down attention effect. But any benefit from the cue may be confounded with priming since the color cues can *prime* attention shifts [18]. For this reason, we recommend that priming effects should be assessed even if they are not of the main interest unless they are comprehensively ruled out by careful experimental design. Note that this actually touches upon the thorny issue of top-down effects versus priming effects [14, 33–35]. One way to avoid this in cueing studies is to use more symbolic cues (e.g., word cues), to isolate effects of top-down attention.

1.3 Using Role Reversals to Assess Probabilistic Representations of Features in the Environment

Priming effects in visual search reflect information about target and distractors that observers have accumulated over previous trials. Recently, Chetverikov, Campana, and Kristjansson [36] suggested that this allows using priming to assess how observers represent probability distributions of visual features.

The idea that the brain encodes the statistics of the environment and uses them to make inferences is well-established [37–42]. However, a given physical probability distribution of features can be represented by the visual system in different ways. While observers need to know the exact shape of distributions to make optimal inferences, approximate inferences can be made with various simplifications. For example, knowing the feature *range* of a class of objects is enough to say whether or not a new object belongs to this class. But many computational models of vision operate on the premise that a given distribution of physical features will be represented with its shape intact. For example, the core assumption of many ideal observers models is that observers accurately represent the generative model of the environment [43]. It is necessary to distinguish these possibilities to understand what kind of information the brain has access to and what it can use.

Revealing the contents of a representation of a physical probability distribution in the brain is not a trivial problem. It is possible to use traditional psychophysics to assess the representation of a single feature value (e.g., one can ask observers to adjust the orientation of the bar so that it matches a previously shown Gabor patch). It is also possible to ask observers to assess the average value of several stimuli, as has been done in "summary statistics" studies (see review in [44]). However, it is impossible to directly inquire about more complex properties, such as skewness, kurtosis, or even variance. As succinctly put by Kuriki for the case of color distributions, "there is essentially no direct approach to studying the color appearance of a multi-colored patch itself" [45, p. 249]. But by utilizing priming effects, we have found a way of circumventing this problem.

Priming effects in visual search occur because observers accumulate information about targets and distractors in order to solve the task more efficiently. In traditional visual search with known targets and distractors, an ideal observer will make inferences based on the ratio of probabilities that a given measurement originates from a target against that it originates from the distractors [46, 47]. In pop-out search, where observers do not know the targets and distractors beforehand (they simply have to find the odd-one-out), observers have to engage in a costly estimate. First, they have to compute for each stimulus, what would be the potential distractor probability distribution based on the other stimuli, and only then estimate whether or not this stimulus could be a target. Learning features of targets and distractors allow observers to avoid this and to analyze each stimulus separately. That is, an observer can look at a given stimulus and decide whether it is a target or a distractor without the need to analyze other stimuli. This shortcut comes at a cost: if observers have accumulated some knowledge of stimuli distributions, then search efficiency becomes dependent on how well these distributions describe current stimuli. In particular, when the target changes and becomes similar to distractors from previous trials, search efficiency should decrease.

Crucially, decreases in search efficiency when targets become similar to preceding distractors should depend on the degree of similarity between them. For an ideal observer, all other things being equal, search efficiency should be inversely proportional to the probability that a given stimulus belongs to a learned distractor distribution, $\mathrm{RT} \propto \frac{1}{p(x|D)}$, where x is the internal measurement of a stimulus' feature, D are the parameters of the distractor distribution, and RT is response time. Introducing targets with different degrees of similarity to the learned distractors would then enable "probing" the representation at different points. Response times in visual search should then distinguish between different representations of distractor distributions, essentially providing a continuous estimate of an internal probability density function describing the physical distribution of distractors (Fig. 1).

This idea was tested by Chetverikov et al. [36]. They found that the curve describing response times as a function of target orientation (centered on the mean of previous distractors) did indeed follow the shape of a previously presented distractor probability distribution when observers searched for an odd-one-out line target among differently oriented lines. This was observed both for distributions with the same range or standard deviation but different shapes (e.g., uniform vs. Gaussian) and for differently skewed but otherwise identical distributions. Chetverikov and colleagues later described two important limitations of this learning process. First, more complex distributions, such as bimodal ones, require more trials to be learned, while simpler ones, such as Gaussian ones, are already represented relatively accurately after one or two trials

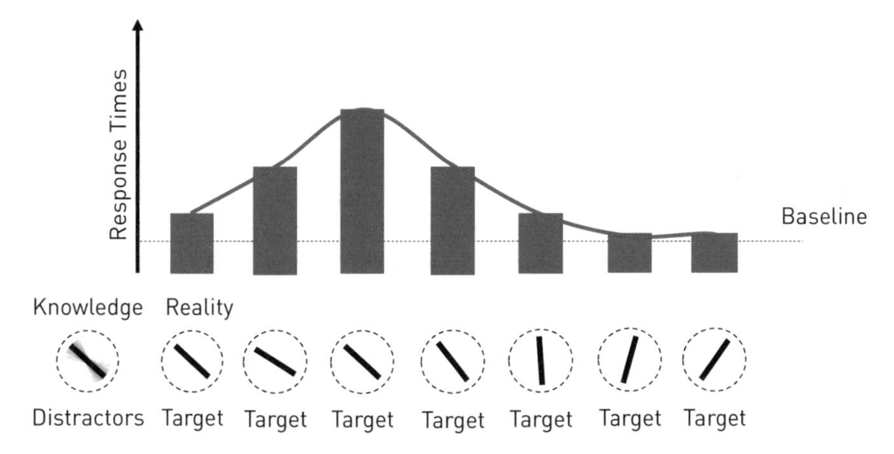

Fig. 1 Hypothetical responses to targets as a function of their similarity to a previously learned distractor distribution. After several visual search trials (here, in the orientation domain), observers obtain knowledge of distractors corresponding to the physical probability distribution of their features (bottom left). It is then possible to "probe" this knowledge by presenting different search targets (bottom central). The response times would be proportional to the degree of similarity between test target and the expected distractors (blue bars). By presenting many different targets, it is then possible to obtain a continuous estimate of a probability density function, corresponding to a representation of distractors (red line)

[48]. Second, in order to learn properties of a distribution, observers need to see a certain minimum number of individual stimuli on each trial [49]. Even mean and variance do not seem to be encoded when only eight lines are presented. Note that this is in sharp contrast to findings on explicit estimates of averages, where observers are able to judge the mean with similar precision regardless of the number of stimuli presented, as long as the overall range stays the same [50]. Finally, this implicit feature distribution learning is not limited to orientation, as similar effects were observed for colored isoluminant diamonds with different hues [51].

2 "Hands-on" Step-by-Step Walk-Through of an Application of the Feature Distribution Learning (FDL) Methodology

The methodological paradigm to assess observers' feature distribution representations has now been used in several different studies [36, 48, 49, 51–53]. Each of these studies probed different aspects of feature distribution learning. While the particular methodological details differ somewhat, they nevertheless all share the same core principle. This involves presenting subjects with a series of odd-one-out visual search trials, where the feature values of the distractors are drawn from a certain type of distribution whose shape and summary statistics stay the same throughout the learning trials. Observers are then presented with a test trial, in which the feature values of the target and the distractor distributions are swapped. This role reversal increases observer's visual search times (as seen previously in a number of studies; [7, 8, 10], and as we have discussed in the section above). When the visual search times obtained from the test trials are plotted as a function of the degree of role reversal, it reveals observer's internal representations of the distractor distribution that was used during the learning trials.

This tutorial will provide detailed information about the three main parts involved in the FDL methodology, which are the design of the visual search task, the structure of the learning trials, and the structure of the test trials. So far, the most studied features with this method are orientation and color. While this tutorial will focus on orientation, the same principles can easily be applied to color (or any other feature space). However, if there are any feature-specific requirements for using color as the main feature, then these are also noted. The next section will focus on the details of the visual search display, while the following ones will focus on how the feature values of the target and distractors are determined.

2.1 Visual Search Task

The main task is an odd-one-out visual search where observers try to find the item whose feature value differs from the rest of the items in the display. When this feature is orientation, the search array includes 36 white lines displayed in a six by six grid on a gray background (Fig. 2b). This method has been tested with smaller set sizes (e.g., 8, 14, 24 lines), but distribution shape learning has only been observed with a larger set size (36 lines, [74]; see discussion above).

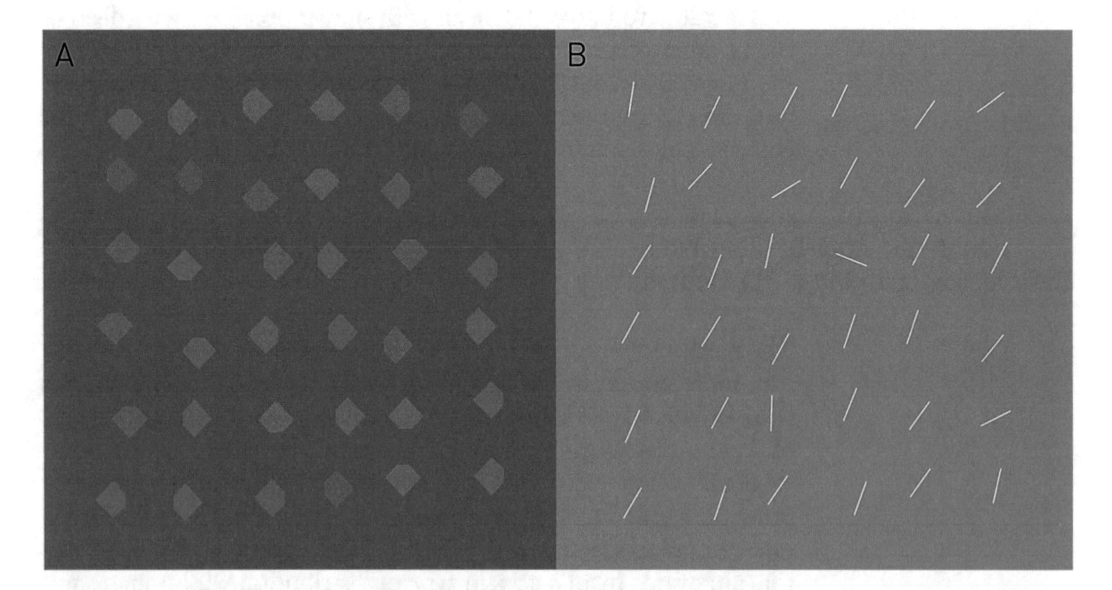

Fig. 2 (**a**) An example visual search display for assessment of the learning of color distributions. Participants search for the oddly colored diamond and report the location of the cutoff on that diamond. (**b**) An example visual search display for testing learning of orientation distributions. Participants search for the oddly oriented line and report whether that line is in the upper or lower half of the display

The size of each line in the search array is set to approximately 1°. The search array subtends approximately 15° × 15° and is positioned at the center of the screen. A random jitter (in the range of ±0.5°) is added to both the vertical and horizontal coordinates of each line. This is done to decrease the precision of the orientation estimate from each individual line, which presumably decreases the viability of serial processing. The position of the target (i.e., the oddly oriented bar) in the search array is randomized.

Participants indicate the location of the target by pressing the "up" button if the target is found in the upper three rows (upper half) and the "down" button if the target is found in the bottom three rows (bottom half) of the search array. The rationale for this choice of response is that the observers respond to the location but not the feature of the target. In other studies, observers responded to a quadrant in which a target was presented [52] or to a shape of a target in a study on colors (see an example below, and further details in [51]). If the participant responds incorrectly, then a feedback display is presented for 1 s, which includes the word "ERROR" in red at the center of the screen. If the response is correct, the search array for the next trial appears immediately. The rationale for not providing feedback after correct responses is to avoid interrupting the between-trial continuity with a feedback display. Feedback screens for incorrect responses slow down the experiment, which, in turn, functions as a motivation for the participant to respond correctly.

However, the main motivating factor for the participant is the score calculated based on accuracy and reaction time. Participants are encouraged to respond as fast as possible to increase their scores. For each trial the score is calculated as follows (where RT is the response time in seconds):

$$\text{For correct answers}: \text{Score} = 10 + (1 - \text{RT}) \times 10$$

$$\text{For incorrect answers}: \text{Score} = -|10 + (1 - \text{RT}) \times 10| - 10$$

On each trial, the score from the previous trial can be shown in one of the corners of the screen. Positive scores are displayed in green and negative in red. When a break is reached during the experiment, the participants' current total score is shown, along with information about what percentage of the experiment has been completed. The only function of keeping score in this experiment is to motivate participants. The particular choice of the score formula is arbitrary, but as is evident from the equation, it is positive for accurate responses faster than 1 s.

When participants perform this odd-one-out visual search task for the first time, their reaction times generally turn out to be too long (>2 s). Therefore, training sessions are needed for naïve participants to get used to the task. The duration of the training might vary, but typically a hundred training trials or more are necessary. The goal of the training is to reduce the average response times and increase accuracy (on a version of the paradigm, such as the one used by [36], for the well-trained observers, the average RT are below 1 s, and accuracy is above 85% correct). The rationale for excluding observers with average response times above a certain threshold is to exclude those who engage in serial processing of the display. As discussed above, that might diminish the learning.

The same principles indicated above can also be applied when color is used as the main feature in visual search. An example display for color search can be seen in Fig. 2a. Instead of lines, participants see 36 diamonds each with a different hue. Each diamond contains a cutoff in any of their four corners. As explained previously, it is important to prevent response repetition from interfering with perceptual priming [1]. Therefore, participants are asked to find the diamond with a hue unlike all the other and report the location of the cutoff (i.e., up, down, left, right).

2.2 Learning Trials

Feature distribution learning experiments consist of blocks. Each block includes streaks of learning trials that are each followed by one or two test trials. Since we are interested in observers' ability to learn the shape of a distribution of feature values, the key aspect of the learning trials is that the shape of the distribution, from where the distractor orientation is drawn, is constant throughout the learning streak.

The length of the learning streaks can be from 1–2 trials up to 10–11 trials depending on the complexity of the distribution used for distractor orientations. For simpler distributions (e.g., Gaussian, uniform), it has been shown that even one to two trials can suffice to uncover learning of distribution shapes. However, when more complex distributions are tested (e.g., bimodal), longer learning streak (7–10 trials) seems to be needed [48]. Even though with simpler distributions one or two trials might be enough to see the learning effects, we recommend keeping the learning streak length to at least three to four trials in order to reduce carry-over effects from preceding streaks. Generally, the length of a learning streak randomly varies during an experiment within a very brief range (e.g., from five to seven trials) in order to break the regularity of the learning and test trials so that participants do not build any expectations about when a block starts and ends. From extensive querying of observers performing this task, they never report having any knowledge of the nature of the sequential trial structure.

The mean of the distractor distribution is randomly determined between −90° and +90° for each learning streak and kept constant within that streak (since this is the distribution that observers are supposed to learn). The target orientation is randomly determined for each trial within a learning streak but is always at least 60° away from the mean of the distractor distribution. This is to ensure that the target is sufficiently dissimilar from distractors, keeping the task relatively easy for observers [54]. In cases where the distractor distribution has high variance and/or has long tails (e.g., Gaussian), some of the distractor lines could turn out to have a similar orientation to the target line. This, in turn, would make the visual search task impossible to carry out. To avoid this, the range of the distractor distribution can be restricted. For example, a Gaussian distractor distribution can be truncated such that any outlier orientation outside of the two-standard deviation range can be removed and then resampled accordingly. The same principles mentioned above for orientation can be applied to color space as well.

2.3 Test Trials

On test trials, the feature values that have been assigned to the target and the distractors during the learning trials are switched in order to reveal observer's internal expectation of distractor orientations (or colors). This, in turn, exposes observer's internal representation of the distractor distribution used in the learning trials. While this is achieved with one test trial, the number of test trials can randomly vary between one and two during an experiment, in order to prevent observers from building expectations about when a block begins and ends. However, only the first test trial is usually included in the data analysis as the effects of the learning streak are expected to dissipate quickly.

The most important consideration for test trials is the selection of the target orientation. The main variable that determines the extent of role-reversal effects is the distance between the target orientation on the test trial and the mean orientation of the previous distractors that are used on the learning trials. This variable, which we refer to as CT-PD ("current target-previous distractor" distance), is essentially the main factor that is manipulated in FDL studies. Plotting reaction times from the test trials as a function of CT-PD reveals observers' internal representation of the previous distractor distribution (Fig. 3). In order to reveal this representation in the range of all possible orientation values, CT-PD values have to uniformly cover this whole range. In order to do that, the orientation space is divided into bins (e.g., 12 bins from $-90°$ to $+90°$ such that each bin covers a range of $15°$). Then, a CT-PD value is randomly chosen for each test trial in such way that at the end of the experiment, the number of CT-PD values chosen from each bin would be equal. Once the CT-PD value is determined for a test trial, the target orientation is selected in a way so that the distance between the current target and the previous distractor mean is equal to this CT-PD value (Fig. 3).

Once the target orientation is determined, the mean of the distractor distribution is chosen randomly given that the distance between the target orientation and the mean orientation of the distractor distribution is at least $60°$. The distractor distribution on the test trial has to be chosen such that the difficulty of the search on the test trial should be intermediate. If the search is too

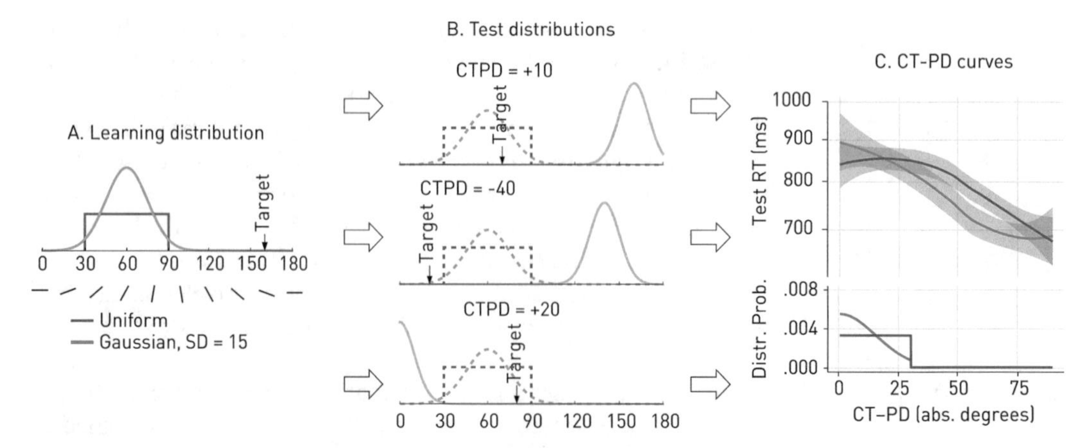

Fig. 3 An example of how CT-PD curves are generated. CT-PD refers to the distance between current target (CT) and previous distractor (PD) distribution mean. After a few trials with distractors drawn randomly from a given learning distribution (**a**), observers are presented with a test trial (**b**). CT-PD distances on test trials are manipulated, and then reaction times obtained from test trials are plotted as a function of these CT-PD distances (**c**, upper). CT-PD curves reveal the observers' internal model of the distractor distribution used on the learning trials and are compared to the physical distribution of stimuli during learning trials (**c**, lower). The CT-PD curves in this figure are based on the results of Experiment 3C from Chetverikov et al. [36] and are replotted from the data available at https:/osf.io/3wcth/

easy or too difficult, floor and ceiling effects might override the role-reversal effects. Previous studies have shown that using a Gaussian distribution with a standard deviation of 10° for the distractor distribution on the test trial provides good testing conditions, but note that this can differ strongly by the aim and characteristics of individual experiments. This Gaussian distribution is also truncated so that no orientation appears as a distractor outside of the two standard deviation ranges.

The number of learning and test streaks used in the experiment is another important factor in this methodology. When the goal is to test whether observers learn the shape of a feature distribution, the experiment should include enough test trials (hence, enough CT-PD values) to fully reveal observers' inner representation of that distribution. We suggest having at least ~350 CT-PD values per observer distributed evenly over the possible feature values. The same principles described here for orientation also apply to color feature space by using just-noticeable differences (JND) as the basic unit (see details on using JND to create a color space for search in [51]).

3 Data Analysis

The goal of the FDL as a method is to infer the characteristics of the ensemble representations based on the search efficiency on test trials. One of the important questions is to analyze whether the visual system represents features as probability distributions rather than only representing the mean and variance (as normally described in summary statistics account of ensemble perception).

In order to reveal observer's internal representation, search times of the first test trial are analyzed. As previously mentioned, the main manipulation factor is the distance between the target on the test trial and the previous distractor distribution mean. Search times of the first test trial are plotted against the distance between the current target and the previous distractor mean (CT-PD) as shown in Fig. 3c. For symmetrical distractor distributions, plotting the absolute distance is sufficient. Figure 3c plots the RT–CT-PD functions using a local regression fit (upper graph) and the probability density function of the distractor distribution (lower graph).

Since search times are analyzed, statistical analyses should be done on log-transformed data due to the skewed search time distributions of the raw data [55–57] and maybe remove outliers due to very slow responses. In addition, only the correct trials are analyzed. One should also make sure that there are enough correct trials for each observer so that the analyses satisfy the usual criterion of more than 85% correct responses.

If role reversals affect search times, then search times of targets within the previous distractor distribution should be slower than search times for targets outside the previous distractor distribution. And, if the actual probabilities of distractors and therefore the distractor distribution shape are represented, the RT function should resemble the shape of the previous distractor distribution.

How can we judge how well the observed RT curve corresponds to the distribution shape? There are several ways for quantitative analysis of RT patterns, including segmented regression and model-fitting. For simple distributions, such as Gaussian or uniform, a useful tool for evaluating these RT functions is to use segmented regression [58, 59]. Following a Gaussian distractor distribution, a monotonically decreasing RT curve is expected, but following, for example, a uniform distractor distribution an RT curve that consists of a flat segment within the distribution range and a steep decrease and faster RT's outside the distribution range is expected, since the probabilities of all feature values within the distribution range are equal (see Fig. 3c). A segmented regression analysis involves searching for significant changes in RT at some particular CT-PD distance. Previous data has shown [36, 48, 49, 51, 53] that the representation of a uniform distribution results in significant breakpoints around the "edge" of the uniform distribution. Search times suddenly change and become faster as the edge of the distribution is reached. The slopes of the individual regression segments are used to support this pattern. Following a uniform distractor distribution, a slope around zero before the breakpoint and a negative slope after the breakpoint resemble the two parts of a uniform distribution. Following a Gaussian distribution, a negative slope of the single segment resembles the monotonic decrease in search time as the distance between the target and the previous distractor mean increases. Statistical tests like the Davies' test [60] compare a two-line model with a single line model that has no breakpoint and provides information about whether the two slopes of the different segments are significantly different. Further analyses could be done on individual subjects' data by comparing the average regression slopes before and after the breakpoint determined on a group basis or based on a priori assumptions (such as the range of the distribution).

A second method of analyzing CT-PD curves is to compare the observed data with pre-defined models that correspond to different distribution shapes. Data can be tested against these pre-defined models of different distribution shapes, and the quality of the different fits can be assessed with the Bayesian Information Criterion. Model fits can be done across subjects or for each subject individually. The model fits seem to be in agreement with segmented regression data in previous studies [51], and they provide a more principled way of testing the hypotheses about encoding of distribution shape. On the other hand, they might lack sensitivity given that the perceptual space might be different from the physical feature space.

We have previously used the following set of models to distinguish between Gaussian and uniform distractor distributions:

1. Half-Gaussian model with a SD $= \sigma$:

$$\mathrm{RT} = c_0 + 2a \times e^{-\frac{\mathrm{CTPD}^2}{2\times\sigma^2}},$$

where a defines the height of the peak, c_0 corresponds to the RT outside the distribution range, and σ corresponds to the standard deviation of the feature distribution that has been used in the experiment (note that the model is half-Gaussian because the Gaussian distribution is symmetric and can be analyzed using the absolute orientation values).

2. Uniform model with a range of $2 \times \sigma$:

$$\mathrm{RT} = \begin{cases} c_0, & \mathrm{CTPD} \leq 2 \times \sigma \\ c_1, & \mathrm{CTPD} > 2 \times \sigma \end{cases}$$

where c_0, c_1 determine the RT inside and outside the distribution range, respectively.

3. "Uniform with decrease" model:

$$\mathrm{RT} = \begin{cases} c_0, & \mathrm{CTPD} \leq 2 \times \sigma \\ c_0 + b \times \mathrm{CTPD}, & \mathrm{CTPD} > 2 \times \sigma \end{cases}$$

where c_0 determines the RT within the distribution range and b is the decrease outside the distribution range.

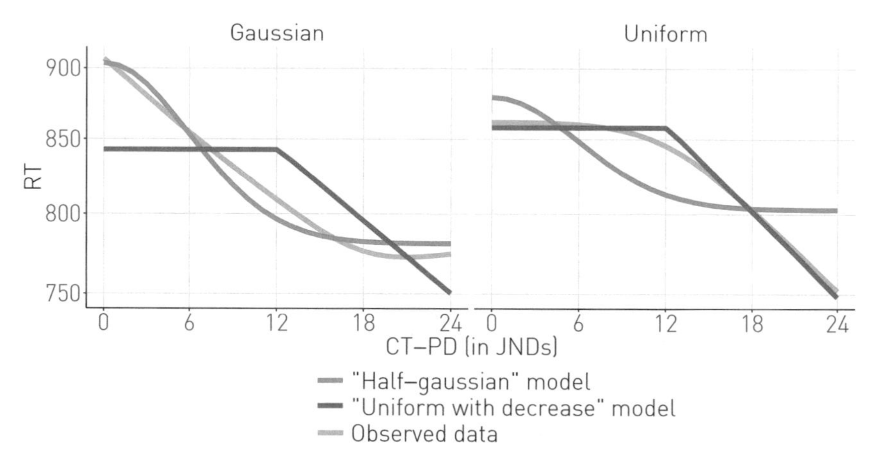

Fig. 4 Observed data and modeling fits for maximum likelihood estimation. Observed data (green) and the best fits of half-Gaussian (blue) and "uniform with decrease" (red) models are based on the results from Chetverikov et al. [51] and are replotted from the data available at https:/osf.io/t2856/

4. Linear model:

$$\mathrm{RT} = c_0 + b \times \mathrm{CTPD}.$$

Each model includes a Gaussian-distributed error term. Different models can be fitted to the data and the best-fitting parameters are obtained using Maximum Likelihood Estimation.

Previous results have shown that search times following a uniform distractor distribution are best fit by a "uniform with decrease" model and following a Gaussian distractor distribution were best fit with either a half-Gaussian or a linear model [51]. Figure 4 shows two of the best-fitting models and the observer data for two different distribution shapes. Together, the segmented regression and the model fitting can provide clear evidence of feature distribution learning.

Note that the goals of data analysis depend on the aims of the study. While here we concentrated on testing distribution shape learning, that is by no means the only possibility. For example, Chetverikov et al. [52] tested whether observers can encode two of the modes of a mixture distribution simultaneously following a single learning block. It is also possible to test, for example, how observers encode the mean or variance of a distribution by analyzing locations of the peaks on a CT-PD curve and its width. Different analytic approaches might be suitable for different research questions.

4 Potential Benefits of FDL

Chetverikov et al. [49] measured whether evidence of FDL leads to higher search efficiency. In that sense it might have a similar effect as contextual cueing. They found some evidence of this, indicating that participants could use the learned distributions, guiding attention across the visual scene. We should also note some preliminary evidence that shows that the distribution learning can be specific to certain portions of the visual scene [61]. But in some sense, the proof is in the pudding, since this paradigm shows how search is affected by the role reversals and that this effect follows the shape of the curves. In other words, the distribution learning improves the efficiency of the search both in terms of speed and accuracy. The paradigm therefore reveals ways in which our visual system makes use of regularities in the visual environment to aid navigation and object recognition.

A notable aspect of feature distribution learning is that it does not require observers to report the feature of interest in any way. This stands in direct contrast to summary statistics studies that require observers to explicitly report or compare some property of the distribution [62–67] which makes FDL more similar to methods relying on other versions of priming or adaptation [68–70]. Importantly, the absence of explicit judgment removes a potential bottleneck in processing of the feature distributions. When making explicit judgments, observers first have to transform the information about the feature likelihood into a single value, i.e., make a "readout" from the information about feature probabilities (akin to how an explicit confidence judgment might involve a readout from probabilistic information on uncertainty [71]). It is also possible that observers do not have explicit access to information about feature probabilities at all and rely on some heuristics when asked to estimate summary statistics [72, 73]. Furthermore, in contrast to other similar methods, FDL allows for the mapping of the feature distribution representations at different points of feature space. In essence, FDL can be considered a behavioral alternative to currently available neurophysiological decoding methods, enabling understanding of the way information is represented in the brain.

5 Summary

Priming effects from repeated target and distractor features are very strong. Their influences last for a long time and they are very hard to willingly overcome [1]. When target repetition effects and distractor repetition effects are combined, role-reversal effects occur that are very large in the context of visual psychophysics and research on visual attention. This makes these effects highly useful since their statistical power enables various applications, including the assessment of more subtle effects, such as how distractor distributions are encoded, just as we described above. But we also emphasize that the priming effects can contaminate experimental results if their influence is not taken into account.

Here we have also shown how priming effects (in particular role reversals) have been used to assess our representations of features in the environment through the implicit assessment of feature distribution learning (FDL). The key insights that our methods have provided involve that our representations are more sophisticated and include far more detail than had previously been assumed in the literature. We believe that this methodology will be of use in assessing many aspects of the function of visual attention, visual working memory, and statistical representations of the environment.

The exact mechanisms of feature distribution learning remain to be studied. However, some characteristics of this process are already known (*see also* [74]). First, as already noted above, there are lower limits on both the number of trials and the number of stimuli necessary for the effect to appear. This suggests that the information about distractors is not accumulated purely locally (otherwise a lower number of stimuli could be compensated for by higher number of trials, which was not the case) and also that this effect involves the accumulation of information over time rather than being a passive aftereffect of stimulus presentation (in the latter case, the bimodal distribution should be represented as unimodal when stimuli from different modes are presented at the same location over several trials). Second, observers can gather information about several different subsets of stimuli on a single trial [52]. This again highlighted the point that information about features is gathered in parallel across the visual field. Third, the generalization from orientation to color suggests that the mechanisms of FDL are dimension-independent. However, whether observers can gather information about the distributions of more complex features (e.g., motion direction) or noncircular ones (e.g., lightness) remains to be studied. There are also a lot of unknowns. The question whether information about feature distribution shape is represented in the cortex or simply readout from a population of differently tuned neurons cannot be answered without neurophysiological studies. A related question is how do observers update the information about the distribution from trial to trial? Do they update the weights of different feature values, or do they update the parameters of the distributions? Also, it is not yet known whether and how the information about different features can be combined. It seems natural to assume that it should be possible, and this combined information can potentially allow for probabilistic object representations. For example, an apple could be a combination of representations of probability distributions for colors and shapes. There is certainly a lot of space for exploring the mechanisms of this unique phenomenon. The method has the additional asset of being an implicit measure of representations of the stimuli in our environment. Statistical representations (such as in the summary statistics literature) are typically assessed with explicit methods, and recently we have provided preliminary evidence that implicit methods may uncover representations of moments of statistical distributions of stimulus ensembles that are masked by explicit methods [75].

Finally, we would like to emphasize that this method demonstrates that observers can learn feature probability distributions in the environment as opposed to a more traditional emphasis on discrete feature values originating in the feature-integration theory of attention [76]. Furthermore, the method shows that such probability distributions are used to guide observer's attention toward (or away from) targets, hindering or facilitating visual search. Recent studies also demonstrate the probabilistic nature of attentional and working memory templates for discrete features [77–81]. Thus, it is likely, in our opinion, that such probabilistic language is more naturally suited to describe representations in the human brain. FDL, in turn, is naturally suited to study representations as it provides an assumption-free continuous description of them. We hope that the addition of this method to the repertoire of commonly used techniques for studying visual attention, memory, or learning processes will bring exciting new discoveries.

Acknowledgments

SHR, ODT, and AK were supported by grant IRF #173947-052 from the Icelandic Research Fund and by a grant from the Research Fund of the University of Iceland. AC was supported by a Radboud Excellence Fellowship.

References

1. Maljkovic V, Nakayama K (1994) Priming of pop-out: I. Role of features. Mem Cognit 22:657–672
2. Bravo MJ, Nakayama K (1992) The role of attention in different visual-search tasks. Percept Psychophys 51:465–472
3. Pascucci D, Mastropasqua T, Turatto M (2012) Permeability of priming of pop out to expectations. J Vis 12:21
4. Shurygina O, Kristjansson Á, Tudge L et al (2019) Expectations and perceptual priming in a visual search task: evidence from eye movements and behavior. J Exp Psychol Hum Percept Perform 45(4):489–499. https://doi.org/10.1037/xhp0000618
5. Sigurdardottir HM, Kristjánsson Á, Driver J (2008) Repetition streaks increase perceptual sensitivity in visual search of brief displays. Vis Cogn 16(5):643–658. https://doi.org/10.1080/13506280701218364
6. Kristjánsson Á, Ásgeirsson ÁG (2019) Attentional priming: recent insights and current controversies. Curr Opin Psychol 29:71–75
7. Wang D, Kristjánsson Á, Nakayama K (2005) Efficient visual search without top-down or bottom-up guidance. Percept Psychophys 67:239–253
8. Lamy DF, Antebi C, Aviani N et al (2008) Priming of Pop-out provides reliable measures of target activation and distractor inhibition in selective attention. Vision Res 48:30–41
9. Saevarsson S, Jóelsdóttir S, Hjaltason H et al (2008) Repetition of distractor sets improves visual search performance in hemispatial neglect. Neuropsychologia 46:1161–1169
10. Kristjánsson Á, Driver J (2008) Priming in visual search: separating the effects of target repetition, distractor repetition and role-reversal. Vision Res 48:1217–1232
11. Chetverikov A, Kristjánsson Á (2015) History effects in visual search for monsters: search times, choice biases, and liking. Atten Percept Psychophys 77:402–412
12. Ásgeirsson ÁG, Kristjánsson Á, Bundesen C (2014) Independent priming of location and

color in identification of briefly presented letters. Atten Percept Psychophys 76:40–48
13. Hillstrom AP (2000) Repetition effects in visual search. Percept Psychophys 62:800–817
14. Kristjánsson Á, Wang D, Nakayama K (2002) The role of priming in conjunctive visual search. Cognition 85:37–52
15. Kristjánsson Á, Ingvarsdóttir Á, Teitsdóttir UD (2008) Object- and feature-based priming in visual search. Psychon Bull Rev 15:378–384
16. Huang L, Holcombe AO, Pashler H (2004) Repetition priming in visual search: episodic retrieval, not feature priming. Mem Cognit 32:12–20
17. Kristjánsson Á (2006) Simultaneous priming along multiple feature dimensions in a visual search task. Vision Res 46:2554–2570
18. Belopolsky AV, Schreij D, Theeuwes J (2010) What is top-down about contingent capture? Atten Percept Psychophys 72:326–341
19. Theeuwes J, van der BE (2011) On the limits of top-down control of visual selection. Atten Percept Psychophys 73:2092–2103
20. Folk CL, Remington RW, Johnston JC (1992) Involuntary covert orienting is contingent on attentional control settings. J Exp Psychol Hum Percept Perform 18:1030–1044
21. Carlisle NB, Kristjánsson Á (2017) How visual working memory contents influence priming of visual attention. Psychol Res 82:833–839
22. Kristjánsson Á, Saevarsson S, Driver J (2013) The boundary conditions of priming of visual search: from passive viewing through task-relevant working memory load. Psychon Bull Rev 20:514–521
23. Muller HJ, Reimann B, Krummenacher J (2003) Visual search for singleton feature targets across dimensions: stimulus- and expectancy-driven effects in dimensional weighting. J Exp Psychol Hum Percept Perform 29:1021–1035
24. Found A, Müller HJ (1996) Searching for unknown feature targets on more than one dimension: investigating a "dimension-weighting" account. Percept Psychophys 58:88–101
25. Becker SI (2010) The role of target-distractor relationships in guiding attention and the eyes in visual search. J Exp Psychol Gen 139:247–265
26. Kristjánsson Á, Campana G (2010) Where perception meets memory: a review of repetition priming in visual search tasks. Atten Percept Psychophys 72:5–18
27. Martini P (2010) System identification in Priming of Pop-Out. Vision Res 50:2110–2115
28. Brascamp JW, Pels E, Kristjánsson Á (2011) Priming of pop-out on multiple time scales during visual search. Vision Res 51:1972–1978
29. Kruijne W, Brascamp JW, Kristjánsson Á et al (2015) Can a single short-term mechanism account for priming of pop-out? Vision Res 115:17–22
30. Kruijne W, Meeter M (2015) The long and the short of priming in visual search. Atten Percept Psychophys 77:1558–1573
31. McPeek RM, Maljkovic V, Nakayama K (1999) Saccades require focal attention and are facilitated by a short-term memory system. Vision Res 39:1555–1566
32. Maljkovic V, Martini P (2005) Implicit short-term memory and event frequency effects in visual search. Vis Res 45(21):2831–2846. https://doi.org/10.1016/j.visres.2005.05.019
33. Theeuwes J, Reimann B, Mortier K (2006) Visual search for featural singletons: no top-down modulation, only bottom-up priming. Vis Cogn 14:466–489
34. Folk CL, Remington RW (2008) Bottom-up priming of top-down attentional control settings. Vis Cogn 16:215–231
35. Wolfe JM, Butcher SJ, Lee C et al (2003) Changing your mind: on the contributions of top-down and bottom-up guidance in visual search for feature singletons. J Exp Psychol Hum Percept Perform 29:483–502
36. Chetverikov A, Campana G, Kristjánsson Á (2016) Building ensemble representations: How the shape of preceding distractor distributions affects visual search. Cognition 153:196–210
37. Girshick AR, Landy MS, Simoncelli EP (2011) Cardinal rules: visual orientation perception reflects knowledge of environmental statistics. Nat Neurosci 14:926–932
38. Rao RP, Olshausen BA, Lewicki MS (2002) Probabilistic models of the brain: perception and neural function. MIT Press, Cambridge, MA
39. Pouget A, Beck JM, Ma WJ et al (2013) Probabilistic brains: knowns and unknowns. Nat Neurosci 16:1170–1178
40. Ma WJ (2012) Organizing probabilistic models of perception. Trends Cogn Sci 16:511–518
41. Fiser J, Berkes P, Orbán G et al (2010) Statistically optimal perception and learning: from behavior to neural representations. Trends Cogn Sci 14(3):119–130
42. Feldman J (2014) Probabilistic models of perceptual features. In: Wagemans J (ed) Oxford handbook of perceptual organization. Oxford University Press, Oxford, pp 933–947

43. Vincent BT (2015) A tutorial on Bayesian models of perception. J Math Psychol 66:103–114
44. Whitney D, Yamanashi-Leib A (2018) Ensemble perception. Annu Rev Psychol 69:105–129
45. Kuriki I (2004) Testing the possibility of average-color perception from multi-colored patterns. Opt Rev 11(4):249–257. https://doi.org/10.1007/s10043-004-0249-2
46. Ma WJ, Navalpakkam V, Beck JM et al (2011) Behavior and neural basis of near-optimal visual search. Nat Neurosci 14:783–790
47. Ma WJ, Shen S, Dziugaite G et al (2015) Requiem for the max rule. Vision Res 116:179–193
48. Chetverikov A, Campana G, Kristjánsson Á (2017) Rapid learning of visual ensembles. J Vis 17:1–15
49. Chetverikov A, Campana G, Kristjánsson Á (2017) Set size manipulations reveal the boundary conditions of distractor distribution learning. Vision Res 140:144–156
50. Utochkin IS, Tiurina NA (2014) Parallel averaging of size is possible but range-limited: a reply to Marchant, Simons, and De Fockert. Acta Psychol (Amst) 146:7–18
51. Chetverikov A, Campana G, Kristjánsson Á (2017) Representing Color Ensembles. Psychol Sci 28:1–8
52. Chetverikov A, Campana G, Kristjánsson Á (2018) Probabilistic rejection templates in visual working memory. Submitted for Review. doi: https://doi.org/10.31234/osf.io/vrbgh. Preprint available at https://psyarxiv.com/vrbgh/
53. Hansmann-Roth S, Chetverikov A, Kristjánsson Á (2019) Representing color and orientation ensembles: can observers learn multiple feature distributions? Submitted for Review
54. Duncan J, Humphreys GW (1989) Visual search and stimulus similarity. Psychol Rev 96:433–458
55. Palmer EM, Horowitz TS, Torralba A et al (2011) What are the shapes of response time distributions in visual search? J Exp Psychol Hum Percept Perform 37:58–71
56. Kristjánsson Á, Jóhannesson ÓI (2014) How priming in visual search affects response time distributions: analyses with ex-Gaussian fits. Atten Percept Psychophys 76:2199–2211
57. Luce RD (1986) Response times: their role in inferring elementary mental organization. Oxford University Press, New York, NY
58. Muggeo VMR (2003) Estimating regression models with unknown break-points. Stat Med 22:3055–3071
59. Muggeo VMR (2008) Segmented: an R package to fit regression models with broken-line relationships. R News 8:20–25
60. Davies RB (1987) Hypothesis testing when a nuisance parameter is present only under the alternative. Biometrika 74:33–43
61. Chetverikov A, Campana G, Kristjánsson Á (2018) Probabilistic perceptual landscapes. J Vis 18:529
62. Atchley P, Andersen GJ (1995) Discrimination of speed distributions: sensitivity to statistical properties. Vision Res 35:3131–3144
63. Morgan MJ, Chubb C, Solomon JA (2008) A "dipper" function for texture discrimination based on orientation variance. J Vis 8:9–9
64. Im HY, Chong SC (2014) Mean size as a unit of visual working memory. Perception 43:663–676
65. Chong SC, Treisman A (2003) Representation of statistical properties. Vision Res 43:393–404
66. Webster J, Kay P, Webster MA (2014) Perceiving the average hue of color arrays. J Opt Soc Am A Opt Image Sci Vis 31:A283–A292
67. Attarha M, Moore CM (2015) The capacity limitations of orientation summary statistics. Atten Percept Psychophys 77:1116–1131
68. Norman LJ, Heywood CA, Kentridge RW (2015) Direct encoding of orientation variance in the visual system. J Vis 15:1–14
69. Michael E, de Gardelle V, Summerfield C (2014) Priming by the variability of visual information. Proc Natl Acad Sci 111:7873–7878
70. Corbett JE, Melcher D (2014) Stable statistical representations facilitate visual search. J Exp Psychol Hum Percept Perform 40:1915–1925
71. Meyniel F, Sigman M, Mainen ZF (2015) Confidence as Bayesian probability: from neural origins to behavior. Neuron 88:78–92
72. Solomon JA (2010) Visual discrimination of orientation statistics in crowded and uncrowded arrays. J Vis 10:19
73. Lau JS, Brady TF (2018) Ensemble statistics accessed through proxies: range heuristic and dependence on low-level properties in variability discrimination. J Vis 18:3
74. Chetverikov A, Campana G, Kristjánsson Á (2017) Learning features in a complex and changing environment: a distribution-based framework for visual attention and vision in general. Prog Brain Res 236:97–120. https://doi.org/10.1016/bs.pbr.2017.07.001
75. Hansmann-Roth S, Kristjansson Á, Whitney D et al (2018) Explicit and implicit judgments of distribution characteristics: Do they lead to different results? Oral presentation at European Conference on Visual Perception 2018, Trieste, Italy. Abstract available at https://guidebook.com/guide/123359/poi/10443998/

76. Treisman AM, Gelade G (1980) A feature-integration theory of attention. Cogn Psychol 12:97–136
77. Won BY, Geng JJ (2018) Learned suppression for multiple distractors in visual search. J Exp Psychol Hum Percept Perform 44:1128–1141
78. Geng JJ, Witkowski P (2019) Template-to-distractor distinctiveness regulates visual search efficiency. Curr Opin Psychol 29:119–125
79. Hout MC, Goldinger SD (2014) Target templates: the precision of mental representations affects attentional guidance and decision-making in visual search. Atten Percept Psychophys 77:128–149
80. Ma WJ, Husain M, Bays PM (2014) Changing concepts of working memory. Nat Neurosci 17:347–356
81. Bays PM (2015) Spikes not slots: noise in neural populations limits working memory. Trends Cogn Sci 19:431–438

Neuromethods (2020) 151: 59–72
DOI 10.1007/7657_2019_19
© Springer Science+Business Media, LLC 2019
Published online: 13 April 2019

Contextual Cueing

Yuhong V. Jiang and Caitlin A. Sisk

Abstract

Contextual cueing refers to the facilitation of visual search by the occasional repetition of a visual context. In standard spatial contextual cueing, a visual search target appears in a consistent location within a repeated array of objects. Search time is faster on repeated displays relative to novel displays, even in participants who do not explicitly recognize the repeated displays. Contextual cueing exemplifies the importance of statistical learning and the resulting memory in guiding attention. Because it involves implicit, relational learning, it has been instrumental in understanding brain functions, cognitive changes across the life span, as well as effects of various neurological, neurodevelopmental, and psychiatric conditions. To stimulate further behavioral and brain research on memory-guided attention and to facilitate comparisons across studies, here we provide a methodological guide on the experimental paradigm of contextual cueing and review key findings. We identify factors that influence the strength of the effect and underscore potential pitfalls in experimental design, data analysis, and interpretation.

Keywords Contextual cueing, Visual attention, Statistical learning, Visual search, Implicit learning, Spatial cognition, Medial temporal lobe, Striatum

1 Introduction

Visual objects typically appear within the context of other objects and background scenes. The stability of this context across multiple encounters allows people to learn the consistent association between an object and its context. Contextual cueing refers to the use of visuospatial contexts to facilitate the localization of important objects—the "targets" [1]. An example in daily experience is the rapid orienting toward the locations of traffic signals when driving along a familiar route. Contextual cueing underscores the sensitivity of visual attention to previous experience and search history [2].

Contextual cueing has been instrumental in understanding the intersection between memory and attention. Although certain types of repeated visual contexts, such as meaningful background scenes, are usually learned explicitly [3], other contexts can speed up visual search in participants who have little explicit awareness of the context repetition [4, 5]. As a form of implicit, relational learning, contextual cueing provides a unique opportunity for understanding brain function, cognitive development, and the effects of neurological, neurodevelopmental, or psychiatric disorders. Though many studies have obtained consistent findings, variability in methodology and contradiction in findings also exist. Here we review key findings and provide a methodological guide on contextual cueing.

The most commonly used variant of contextual cueing involves visual search for a target letter "T" among distractor letter "Ls" (Fig. 1). The experiment is divided into 30, 1- to 2-min blocks. Trials in each block use unique search arrays in terms of item location and identity. Half of the displays presented in a block repeat once in each subsequent block, whereas the other half are newly generated. The contextual cueing effect refers to an improvement in search reaction time (RT) in the repeated condition, relative to the novel condition (Fig. 1). In healthy adults, this effect emerges quickly, after just five to eight blocks. It produces an RT gain of about 50–100 ms. Once acquired, contextual learning is retained for at least 1 week. Because all search arrays are generated from the same randomization scheme and lack distinctiveness, few participants spontaneously notice the repetition. Explicit recognition rate is at or slightly above chance [4, 6], and the size of contextual cueing is unrelated to explicit awareness [5].

Because contextual cueing involves implicit, relational learning, it provides an informative methodological tool for neuroscience, as well as psychological, research. Although contextual cueing involves predominantly implicit learning, it is impaired in both transient, midazolam-induced amnesia [7] and amnesic patients with medial temporal lobe (MTL) damage [8, 9]. This highlights the role of the MTL in relational memory. Parkinson's and

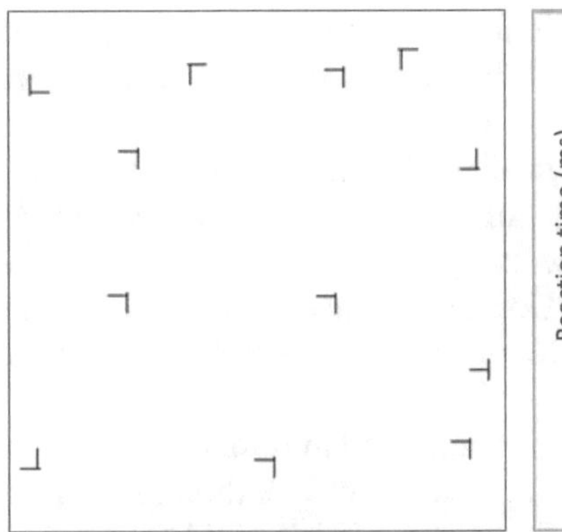

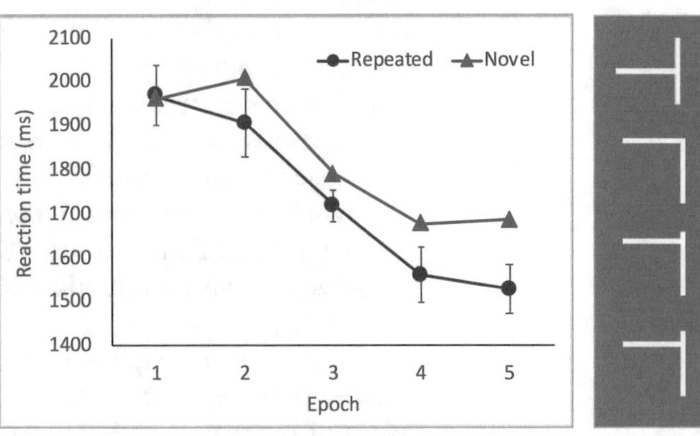

Fig. 1 Contextual cueing. Left: a sample visual search display. Middle: representative finding in reaction time (RT). The error bars for the repeated condition show +/−1 S.E.M of the RT difference between the repeated and novel conditions (data source: [7]). Right: examples of a target T (top) and distractor Ls (bottom three images). Distractor Ls may vary in their similarity to the target depending on the degree of offset between the end of the bisected line and the point of bisection

Huntington's patients are also impaired, implicating the involvement of the striatum [10, 11]. Loss of central vision, as in age-related macular degeneration, negatively affects contextual cueing [12]. fMRI studies have revealed differential activation between repeated and novel displays in the hippocampus [13, 14] and the posterior parietal cortex [15], among other regions.

Two competing theories have been proposed to account for contextual cueing. The attentional guidance account suggests that the repeated context expedites visual search. This pinpoints the onset of the effect at a relatively early point in the search process—before the target is found [1]. On this account, through repeated encounters with a target in a consistent location, participants develop a "context map" that specifies the probability of finding a target in various locations within a given context. Similar to the "saliency map," the context map guides visual attention to locations likely to contain the target, given the context. Alternatively, the response selection account proposes little to no enhancement of the search process, placing the locus of contextual cueing at a late stage—after the target is found [16]. On this account, repeated and novel conditions do not differ during visual search. Upon the detection of the target, participants are faster in responding to the target in repeated displays owing to reduced double checking, verification, and other response-related processes. These two accounts are not mutually exclusive, but they disagree on whether repeated context primarily affects search or post-search processes.

In EEG and MEG, the repeated and novel conditions diverge as early as 50–100 ms after presentation of the search display [17], with multiple studies showing a greater N2pc (or N210) component for repeated than novel displays [18–21]. Eye tracking shows a reduction in the number of fixations on repeated displays prior to target localization [22, 23] but little to no difference in the time to respond after target localization [22]. These findings provide strong evidence that repeated contexts expedite search. At the same time, the response selection account receives some support from EEG and eye tracking [21, 24, 25]. The strongest evidence for this account is the lack of consistent improvements in search slope [16], though contradicting findings [24, 26] and alternative explanations [27] exist. Thus, contextual cueing reflects primarily memory-based attentional guidance and may additionally enhance response selection.

The implicit nature of contextual cueing and its robustness in healthy young adults have led many investigators to ask whether it also occurs in other populations. In addition to deficits in amnesic patients and striatal patients, diminished contextual cueing is found in some studies on older adults, school-aged children, patients with depression, and children with ADHD. Other studies, however, reveal intact contextual cueing in older adults, infants and school-aged children, children with ADHD, and children with ASD [28]. Not all studies use the same experimental paradigm, and some contain data anomalies, such as an unexplained RT disadvantage in the repeated condition, relative to the novel condition, early in the experiment [10, 29]. Additional research using a consistent method of experimentation and analysis is needed to resolve these discrepancies.

The following sections provide a methodological guide on contextual cueing, focusing primarily on the standard T-among-L search task. We will also discuss other variants of contextual cueing that may be used in future behavioral and brain research.

2 Materials

The visual search task is administered on a computer in a quiet room with normal interior lighting. Viewing distance is approximately 57 cm, at which distance 1 cm on the screen is ~1° visual angle. Responses are recorded using the keyboard.

3 Methods

This section focuses on the standard array-based contextual cueing paradigm using the T-among-L search task [1]. An example script written in MATLAB and Psychtoolbox can be found at http://jianglab.psych.umn.edu/ContextualCueing/ContextualCueing.html.

3.1 Stimuli

Each trial of the T-among-L search task displays one target letter T among a variable number of distractor letter Ls (Fig. 1). Most typical is one T and 11 Ls. In experiments that measure the slope of the search RT as a function of the number of items, the displays may contain 7, 11, or 15 Ls, for set sizes of 8, 12, and 16, equally distributed across quadrants. The tail of the T points either to the left or to the right. The Ls are equally likely to be rotated 0°, 90°, 180°, or 270°, determined randomly.

The background of the display is gray (RGB values {127 127 127}). The search items can appear in various colors, but typically they appear in a single color, such as white. The use of a single color reduces color-based contextual cueing [30], allowing the paradigm to more clearly index spatial context learning.

The target T is composed of two perpendicular segments of the same length, with one bisecting the other in the middle. In displays where all items are uniform in size, the T subtends 1.5° × 1.5°. The Ls are also composed of two perpendicular segments of the same length (1.5°), except that the bisecting segment is near one end of the other segment. The point of bisection affects the speed and the nature of search (Fig. 1). When the offset is 0°, meaning that the two segments meet at the corner to form a perfect L, visual search is relatively efficient. In fact, the task is closer to feature search than conjunction search (or "configuration search," [31]). Because the role of attentional guidance is reduced in feature search, experiments using perfect Ls typically produce fast RTs and small contextual cueing effects (e.g., [20]). We do not recommend the use of such stimuli because they are insensitive to contextual guidance. As the offset increases, the Ls become increasingly similar to the T, increasing search difficulty. We recommend an offset of 0.1° for Ls. Increasing the offset further raises search RT without necessarily enhancing the size of contextual cueing. Though search lasts longer, the additional search time is mainly spent on item inspection, rather than formation of contextual associations [32].

3.2 Placement of Items

The screen is divided into "cells" that fall within an invisible grid; the sample MATLAB script uses a 10×10 grid. The search space subtends approximately $27° \times 27°$. Items are placed inside randomly selected cells, with constraints specified below. To avoid accidental formation of a virtual square or other familiar configurations [33], each item is placed at a location slightly off the center of a cell. The amount of this "jitter" is maintained across blocks for repeated displays, ensuring that repeated displays are identical in terms of target and distractor locations.

The experimental session is divided into 30 blocks of 24 trials each. Item placement is unique for each trial within a block, but half of the displays repeat once per block, and the other half are novel. To achieve this design, at the beginning of the experiment, 24 locations are chosen from the grid as target locations—half for the repeated and half for the unrepeated displays. Although most studies select these 24 locations randomly, we recommend two constraints that limit the extraneous impact on search RT. First, for each condition, the 12 target locations should be evenly distributed across quadrants (i.e., three per quadrant). Second, given the large effect of target eccentricity on search RT, we recommend equating the eccentricity of the target location between the repeated and novel conditions. Suppose the target's eccentricity for one of the 12 repeated trials is $v°$; then there should be a novel display with a target location of the same eccentricity.

Once the target location is chosen for a given display, distractor items are placed in random locations within the invisible grid, provided that an equal number of items appear in each quadrant. The orientation of each distractor is random. For repeated displays, the selection of distractor locations occurs at the beginning of block 1. For new displays, the distractor locations are generated anew at the beginning of each block. The repeated and novel displays are intermixed randomly in presentation order.

3.3 Experimental Design

The experiment is composed of 30 blocks of trials. Each block includes 12 repeated and 12 novel displays randomly intermixed in presentation order. To increase data stability, every five blocks is combined into an epoch in data analysis. The experimental design therefore involves a two (context condition: repeated vs. novel) by six (epoch) within-subject design. Each of the 12 repeated displays is presented once per block; the configuration of the array is repeated, including the locations and orientations of the Ls and the location (but not orientation) of the target. For novel displays, the distractor locations are newly generated at the start of each block. However, the target's location (but not orientation) is repeated across blocks in the novel condition. Figure 2 illustrates the study design.

To minimize motor and response learning, participants are asked to press a button to indicate the orientation of the target. Its value—left or right—is randomly determined, with the constraint that an equal number of left and right orientations is used in each condition. The randomization of the target's orientation ensures that display repetition is not predictive of the target's identity or the response. We discourage the use of a touchscreen response because it associates a repeated context with limb movement and introduces a confounding element of motor learning.

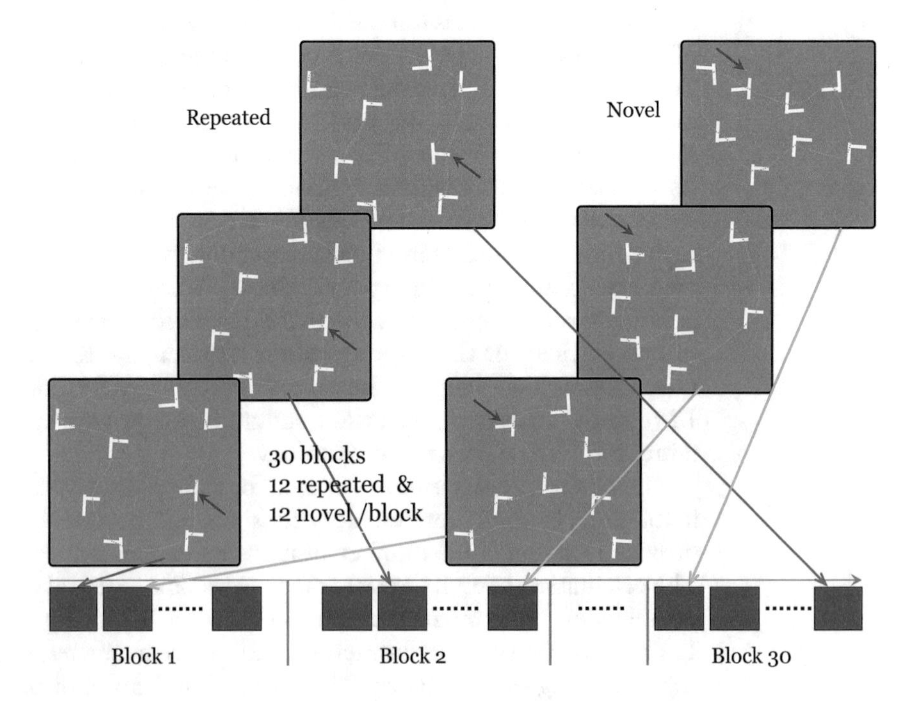

Fig. 2 An illustration of the contextual cueing design. The green outlines on the display and the red arrows indicating the target's location are for illustrative purposes only

3.4 Procedure

The study starts with a short practice phase of 12 trials, all of which use novel displays. Though longer practice may be needed in some circumstances, we discourage the use of long practice of over 100 trials due to the finding that prolonged initial exposure to novel trials reduces subsequent learning [34]. Following practice, participants engage in the actual experiment, with chances to take a short break after each block. On each trial, a central fixation point is presented for 500 ms, followed by the visual search display. Upon participants' response, the display is erased, and accuracy feedback in the form of a high-pitched chirp (for correct trials) and a low buzz (for incorrect trials) is presented for 200 ms. To reduce the negative impact of an incorrect response on subsequent trials, a 2 s pause is inserted after an incorrect response. The next trial starts after an inter-trial interval of 0.5 s.

Sample instructions used in incidental learning tasks are as follows:

Welcome to the visual search task. You will be searching for a rotated letter T presented among L-shaped letters. The experiment is divided into several hundred trials, each taking seconds to complete. On each trial, you will first see a small dot at the center of the display. Please keep your eyes on this dot. Then, a display of letters will appear and remain on the display until you make a response. You can move your eyes if you want to, to help you find the T. There is one T turned sideways with its tail pointing either to the left or to the right. As soon as you spot the T, press an arrow key to indicate its direction: left arrow if the tail points to the left or right arrow if the tail points to the right. Please respond as accurately and as quickly as you can. Minimize errors to no more than one per block.

We will start with a few trials as practice. In the experiment, you will be given a short break once every 1–2 min. The entire task takes about 50 min to complete.

Please ask the experimenter now if you have questions. Otherwise, press the spacebar to start practice.

Studies that examine explicit learning may wish to consult the method section of [4] for specific instructions.

3.5 Assessment of Explicit Awareness

At the completion of the visual search task, participants' explicit awareness of the display repetition can be assessed using either the configuration recognition [1] or the target generation task [4]. Following an informal query about whether participants have noticed that some displays repeated, they will be informed of the display repetition. In the configuration recognition test, participants will be shown the 12 repeated displays, along with 12 novel displays, presented one at a time in a randomly intermixed order. Their task is to press a button to indicate whether a given display is "old" or "new." Alternatively, in the target generation task, the 24 displays will be presented one at a time, but the letter T will be replaced by a letter L. The participants' task is to select the quadrant that should have contained the target, based on their previous search experience.

Of note, previous studies have found a low level of explicit awareness [6]. The size of the effect is small, with an estimated effect size *d* of 0.31. Unless the sample size is high (e.g., 84 or more), the recognition test will not be sensitive enough to detect this effect. Nonetheless, collecting recognition data facilitates an analysis on the association between awareness and contextual cueing.

3.6 Sample Size Considerations

The basic contextual cueing effect—RT difference between the repeated and novel conditions in the second half of the experiment—is associated with a large effect size, with Cohen's *d* ranging from 0.84 to 0.99 in [5]. Based on G*Power analysis, a within-subject design testing this effect can reach a power of .80 with a sample size of 14 and a power of .95 with a sample size of 21. In other words, a typical sample size of 14 is associated with a false-negative rate of .20, yielding negative findings in the literature, such as [35]. A larger sample size is recommended (e.g., $N = 20$) to reduce false negatives.

In addition, most studies are likely to explore aspects of contextual cueing with a smaller effect size, such as the effect of partial display repetition, differences between a neurological group and a control group, or effects of an additional manipulation (e.g., working memory load). The effect size associated with these comparisons is likely smaller, rendering a typical sample size of 14 inadequate. We recommend conducting a pilot study to estimate statistical power or consulting existing studies on similar effects to determine sample size.

3.7 Data Analysis

Because the display is presented until a response is made, accuracy in contextual cueing studies is high—around 98%. The main index of contextual cueing is RT, excluding incorrect trials. As is typical of RT measures, the distribution of responses will be positively skewed, with a small number of extremely long RTs. The "outlier" RTs are commonly excluded, but existing studies differ widely in the definition of outliers. Some use a fixed upper cutoff such as 4000 ms, whereas others eliminate trials longer than three standard deviations of each participant's mean RT. The exact cutoff is likely to vary depending on the difficulty of the search task. Although a fixed number cannot be given, we recommend using a conservative approach, removing no more than 1% of the data.

After incorrect trials and outliers are excluded, an average RT is obtained for each participant in each condition (repeated or novel) of each epoch. A repeated measures ANOVA using condition (repeated vs. novel) and epoch as factors will be conducted. A robust contextual cueing effect will manifest both as a main effect of condition—faster RT in the repeated than the novel condition—

and as an interaction between condition and epoch—a larger RT difference between repeated and novel conditions in later epochs than earlier ones. However, because of the gradual increase in contextual cueing with training, the interaction effect in the omnibus *F* often misses statistical significance. Therefore, contextual cueing is most reliably indexed by a significant RT difference between the repeated and the novel conditions in the second half of the experiment, as well as an interaction between condition and epoch when contrasting data from the first and the last epoch. When participants have different baseline RT, it is possible to compute normalized RT [36] or to express contextual cueing as a proportion [37].

4 Notes

4.1 Eye Movements

Participants are free to move their eyes in the standard version of contextual cueing. However, in studies that prevent eye movements during search, items in peripheral locations are difficult to identify. To compensate for the effect of eccentricity, in these situations, we recommend enlarging items according to the cortical magnification factor [38]. Care should be taken in choosing the proper target-distractor similarity to ensure that search remains inefficient.

Contextual cueing is found even when eye movements are prevented [4, 39], suggesting that it reflects high-level changes in attention rather than low-level procedural learning of the eye movements. Although eye movements are not a significant concern for behavioral studies measuring RT, a nuanced interpretation of neuroimaging findings must take them into consideration. Because participants make fewer fixations on repeated than novel displays, brain regions sensitive to eye movements are likely to show differential activity, even if they play no role in contextual cueing. Studies using fMRI should carefully consider whether to restrict eye movements.

4.2 Analysis Approaches to Avoid

We caution against a few analytical approaches, even though they have been used previously. First, some studies restrict analysis to just participants who show a contextual cueing effect. For example, a study on proactive interference from Session 1's learning on Session 2's learning includes only those who showed contextual cueing in Session 1. The rationale is that if learning did not occur in Session 1, it would not induce proactive interference [40]. However, the filtering of participants introduces biases, owing to the phenomenon of "regression to the mean"—by limiting the analysis to those who showed a larger effect in Session 1, the same participants are likely to show a smaller effect in Session 2. In addition, this practice assumes that people who did not show contextual cueing had not acquired contextual learning. This need not be the case—it is possible that fluctuations in the randomization of stimuli make search in the repeated displays harder than novel ones in some participants, thus masking the learning effects.

Second, we caution against the analysis of individual configurations as a means of indexing the number of repeated displays that participants have learned. For example, several studies have analyzed the mean RT for each of the 12 repeated displays and compared it with the 99% confidence interval of the mean RT in the novel condition. The rationale is that the number of configurations a participant learns can be uncovered by identifying the individual repeated configurations that show significantly faster RT than the novel condition [41]. This analytical approach is flawed because it assumes that all of the repeated and all of the novel displays are equally demanding. This is not the case: visual search RT is strongly influenced by target eccentricity and other accidental stimulus factors (e.g., crowding near the target). The repeated and novel displays are "equivalent" only when averaged across a relatively large number of displays. Each individual display, however, will vary in target eccentricity and other factors. The approach of characterizing an individual configuration as "learned" if it is faster than a set of novel configurations confounds contextual cueing with stimulus differences. Given that the novel displays vary widely in target eccentricity, the 99% confidence interval is likely very wide, allowing perhaps only the easiest repeated displays (e.g., those with a target near fixation) to pass this criterion. It is not surprising, then, that this analysis has produced a very low estimate of the number of learned configurations.

Finally, studies that rely on individual differences in contextual cueing should exert caution. The reliability of contextual cueing within an individual has not been systematically examined. One study measured contextual cueing in five sessions, using different displays in each session. It found a lack of consistency in the size of contextual cueing for a given individual across sessions, even though the baseline RT showed strong consistencies [37]. One implication of this inconsistency is that the capacity for implicit learning is preserved in all individuals, though its effects may not be observed in every instance [42]. Individual differences measured in a single session may be attributed primarily to stimulus characteristics (e.g., the use of different random displays for each participant) rather than to participant characteristics. If this is the case, then studies that rely on individual difference approaches, such as analysis of the correlation between contextual cueing and individual preferences for multiple media use or individual differences in hippocampal volume, should be interpreted with caution. At the very least, such studies should aim to establish the reliability of contextual cueing within an individual, based on data across sessions or based on a split-half analysis.

4.3 Factors that Influence the Size of Contextual Cueing

Contextual cueing has been reported in more than 100 studies using various stimuli and tasks. In addition to search difficulty (see Sect. 3.1), several factors influence the size of contextual cueing and should be considered in future work.

First, contextual cueing is subject to a local spatial constraint [27]. The repetition of distractors adjacent to the target affects RT more than the repetition of distractors far from the target [43]. In one demonstration, repetition of just the item locations within the target quadrant induced as much contextual cueing as repetition of the entire display [27].

Second, as an associative learning mechanism, contextual cueing is reduced if a single context is associated with more than one target location. Some studies find a complete abolition of the effect when a repeated context is associated with three different target locations. Along a similar vein, if the target is never present on a given display, repetition of that display yields little (or inconsistent) benefit [25, 44, 45].

Third, because contextual cueing relies on the detection of repeated contexts presented among novel ones, the effect is somewhat sensitive to the proportion or number of repeated and novel displays [46]. A lower ratio of repeated to novel displays generates less contextual cueing, owing either to decreased signal-to-noise ratio or to other factors correlated with a change in proportion (e.g., immediacy of repetition).

Fourth, the expression of contextual learning is sensitive to working memory load and selective attention [47, 48]. Learning may not be adequately expressed if the repeated context is not currently attended or if the concurrent working memory load is high. This factor should be considered when testing children, older adults, or neurological patients, as an objectively identical task may induce different working memory load in different populations.

Fifth, contextual cueing reflects not just spatial learning but also nonspatial associative learning. When the identities of the distractor objects are predictive of the target's location or shape, RT is also facilitated. Because spatial and nonspatial properties may, in some situations, be integrated to form a rich context [49], spatial contextual cueing may be disrupted if the object identity is variable or if it ceases to be predictive [50].

4.4 Scene-Based Contextual Cueing

Along with spatial and identity-based contextual cueing, scene-based contextual cueing is a common variant of this paradigm (for a more comprehensive list of variants, see [28]). In Brockmole and Henderson [3], participants search for a target letter superimposed on a natural scene. A set of scenes are repeated across blocks, with the target appearing in the same location within the scene each time. This manipulation induces a large RT benefit for repeated displays. In Summerfield et al. [51], participants first learn the placement of a target object in a large number of scenes.

Subsequently, the presentation of the scene provides an attentional orienting cue, allowing participants to more quickly find the target than in unfamiliar scenes. Scene-based contextual cueing has also been implemented in tests of nonhuman animals and with the use of EEG and fMRI.

The underlying mechanism likely differs between the standard, array-based contextual cueing and scene-based contextual cueing. The latter involves explicit learning, induces an effect five to ten times greater than array-based contextual cueing, and produces strong global learning rather than, or in addition to, local learning. Studies should carefully consider which type of contextual cueing (and which mechanism) they are evaluating when meaningful scenes or objects are employed.

In summary, by reviewing key findings and providing a methodological guide on contextual cueing, we hope to stimulate future neuroscience and psychological research on the role of memory in guiding spatial attention.

References

1. Chun MM, Jiang Y (1998) Contextual cueing: implicit learning and memory of visual context guides spatial attention. Cogn Psychol 36:28–71. https://doi.org/10.1006/cogp.1998.0681
2. Awh E, Belopolsky AV, Theeuwes J (2012) Top-down versus bottom-up attentional control: a failed theoretical dichotomy. Trends Cogn Sci 16:437–443. https://doi.org/10.1016/j.tics.2012.06.010
3. Brockmole JR, Henderson JM (2006) Recognition and attention guidance during contextual cueing in real-world scenes: evidence from eye movements. Q J Exp Psychol (Hove) 59:1177–1187. https://doi.org/10.1080/17470210600665996
4. Chun MM, Jiang Y (2003) Implicit, long-term spatial contextual memory. J Exp Psychol Learn Mem Cogn 29:224–234
5. Colagiuri B, Livesey EJ (2016) Contextual cuing as a form of nonconscious learning: theoretical and empirical analysis in large and very large samples. Psychon Bull Rev 23:1996–2009. https://doi.org/10.3758/s13423-016-1063-0
6. Vadillo MA, Konstantinidis E, Shanks DR (2016) Underpowered samples, false negatives, and unconscious learning. Psychon Bull Rev 23:87–102. https://doi.org/10.3758/s13423-015-0892-6
7. Park H, Quinlan J, Thornton E, Reder LM (2004) The effect of midazolam on visual search: implications for understanding amnesia. Proc Natl Acad Sci U S A 101:17879–17883. https://doi.org/10.1073/pnas.0408075101
8. Chun MM, Phelps EA (1999) Memory deficits for implicit contextual information in amnesic subjects with hippocampal damage. Nat Neurosci 2:844–847. https://doi.org/10.1038/12222
9. Manns JR, Squire LR (2001) Perceptual learning, awareness, and the hippocampus. Hippocampus 11:776–782. https://doi.org/10.1002/hipo.1093
10. van Asselen M, Almeida I, Andre R et al (2009) The role of the basal ganglia in implicit contextual learning: a study of Parkinson's disease. Neuropsychologia 47:1269–1273. https://doi.org/10.1016/j.neuropsychologia.2009.01.008
11. van Asselen M, Almeida I, Júlio F et al (2012) Implicit contextual learning in prodromal and early stage Huntington's disease patients. J Int Neuropsychol Soc 18:689–696. https://doi.org/10.1017/S1355617712000288
12. Geringswald F, Herbik A, Hoffmann MB, Pollmann S (2013) Contextual cueing impairment in patients with age-related macular degeneration. J Vis 13. https://doi.org/10.1167/13.3.28
13. Greene AJ, Gross WL, Elsinger CL, Rao SM (2007) Hippocampal differentiation without recognition: an fMRI analysis of the contextual cueing task. Learn Mem 14:548–553. https://doi.org/10.1101/lm.609807
14. Preston AR, Gabrieli JDE (2008) Dissociation between explicit memory and configural

memory in the human medial temporal lobe. Cereb Cortex 18:2192–2207. https://doi.org/10.1093/cercor/bhm245

15. Manginelli AA, Baumgartner F, Pollmann S (2013) Dorsal and ventral working memory-related brain areas support distinct processes in contextual cueing. Neuroimage 67:363–374. https://doi.org/10.1016/j.neuroimage.2012.11.025
16. Kunar MA, Flusberg S, Horowitz TS, Wolfe JM (2007) Does contextual cuing guide the deployment of attention? J Exp Psychol Hum Percept Perform 33:816–828. https://doi.org/10.1037/0096-1523.33.4.816
17. Chaumon M, Drouet V, Tallon-Baudry C (2008) Unconscious associative memory affects visual processing before 100 ms. J Vis 8:10.1–10. https://doi.org/10.1167/8.3.10
18. Johnson JS, Woodman GF, Braun E, Luck SJ (2007) Implicit memory influences the allocation of attention in visual cortex. Psychon Bull Rev 14:834–839
19. Olson IR, Chun MM, Allison T (2001) Contextual guidance of attention: human intracranial event-related potential evidence for feedback modulation in anatomically early temporally late stages of visual processing. Brain 124:1417–1425
20. Schankin A, Schubö A (2009) Cognitive processes facilitated by contextual cueing: evidence from event-related brain potentials. Psychophysiology 46:668–679
21. Schankin A, Schubö A (2010) Contextual cueing effects despite spatially cued target locations. Psychophysiology 47:717–727. https://doi.org/10.1111/j.1469-8986.2010.00979.x
22. Harris AM, Remington RW (2017) Contextual cueing improves attentional guidance, even when guidance is supposedly optimal. J Exp Psychol Hum Percept Perform 43:926–940. https://doi.org/10.1037/xhp0000394
23. Peterson MS, Kramer AF (2001) Attentional guidance of the eyes by contextual information and abrupt onsets. Percept Psychophys 63:1239–1249
24. Zhao G, Liu Q, Jiao J et al (2012) Dual-state modulation of the contextual cueing effect: evidence from eye movement recordings. J Vis 12:11. https://doi.org/10.1167/12.6.11
25. Schankin A, Hagemann D, Schubö A (2011) Is contextual cueing more than the guidance of visual-spatial attention? Biol Psychol 87:58–65. https://doi.org/10.1016/j.biopsycho.2011.02.003
26. Couperus JW, Hunt RH, Nelson CA, Thomas KM (2011) Visual search and contextual cueing: differential effects in 10-year-old children and adults. Atten Percept Psychophys 73:334–348. https://doi.org/10.3758/s13414-010-0021-6
27. Brady TF, Chun MM (2007) Spatial constraints on learning in visual search: modeling contextual cuing. J Exp Psychol Hum Percept Perform 33:798–815. https://doi.org/10.1037/0096-1523.33.4.798
28. Goujon A, Didierjean A, Thorpe S (2015) Investigating implicit statistical learning mechanisms through contextual cueing. Trends Cogn Sci (Regul Ed) 19:524–533. https://doi.org/10.1016/j.tics.2015.07.009
29. Weigard A, Huang-Pollock C (2014) A diffusion modeling approach to understanding contextual cueing effects in children with ADHD. J Child Psychol Psychiatry 55:1336–1344. https://doi.org/10.1111/jcpp.12250
30. Kunar MA, Flusberg SJ, Wolfe JM (2006) Contextual cuing by global features. Percept Psychophys 68:1204–1216
31. Wolfe JM (1998) What can 1 million trials tell us about visual search? Psychol Sci 9:33–39
32. Rausei V, Makovski T, Jiang YV (2007) Attention dependency in implicit learning of repeated search context. Q J Exp Psychol (Hove) 60:1321–1328. https://doi.org/10.1080/17470210701515744
33. Conci M, von Mühlenen A (2009) Region segmentation and contextual cuing in visual search. Atten Percept Psychophys 71:1514–1524. https://doi.org/10.3758/APP.71.7.1514
34. Jungé JA, Scholl BJ, Chun MM (2007) How is spatial context learning integrated over signal versus noise? A primacy effect in contextual cueing. Vis Cogn 15:1–11. https://doi.org/10.1080/13506280600859706
35. Lleras A, Von Mühlenen A (2004) Spatial context and top-down strategies in visual search. Spat Vis 17:465–482
36. Faust ME, Balota DA, Spieler DH, Ferraro FR (1999) Individual differences in information-processing rate and amount: implications for group differences in response latency. Psychol Bull 125:777–799
37. Jiang Y, Song J-H, Rigas A (2005) High-capacity spatial contextual memory. Psychon Bull Rev 12:524–529
38. Carrasco M, Evert DL, Chang I, Katz SM (1995) The eccentricity effect: target eccentricity affects performance on conjunction searches. Percept Psychophys 57:1241–1261
39. van Asselen M, Castelo-Branco M (2009) The role of peripheral vision in implicit contextual cuing. Atten Percept Psychophys 71:76–81. https://doi.org/10.3758/APP.71.1.76

40. Conci M, Sun L, Müller HJ (2011) Contextual remapping in visual search after predictable target-location changes. Psychol Res 75:279–289. https://doi.org/10.1007/s00426-010-0306-3
41. Smyth AC, Shanks DR (2008) Awareness in contextual cuing with extended and concurrent explicit tests. Mem Cognit 36:403–415
42. Reber AS, Walkenfeld FF, Hernstadt R (1991) Implicit and explicit learning: individual differences and IQ. J Exp Psychol Learn Mem Cogn 17:888–896
43. Olson IR, Chun MM (2002) Perceptual constraints on implicit learning of spatial context. Vis Cognit 9:273–302. https://doi.org/10.1080/13506280042000162
44. Kunar MA, Wolfe JM (2011) Target absent trials in configural contextual cuing. Atten Percept Psychophys 73:2077–2091. https://doi.org/10.3758/s13414-011-0164-0
45. Beesley T, Vadillo MA, Pearson D, Shanks DR (2015) Pre-exposure of repeated search configurations facilitates subsequent contextual cuing of visual search. J Exp Psychol Learn Mem Cogn 41:348–362. https://doi.org/10.1037/xlm0000033
46. Yang Y, Merrill EC (2015) The impact of signal-to-noise ratio on contextual cueing in children and adults. J Exp Child Psychol 132:65–83. https://doi.org/10.1016/j.jecp.2014.12.005
47. Annac E, Manginelli AA, Pollmann S et al (2013) Memory under pressure: secondary-task effects on contextual cueing of visual search. J Vis 13:6. https://doi.org/10.1167/13.13.6
48. Jiang Y, Leung AW (2005) Implicit learning of ignored visual context. Psychon Bull Rev 12:100–106. https://doi.org/10.3758/BF03196353
49. Jiang Y, Song J-H (2005) Hyperspecificity in visual implicit learning: learning of spatial layout is contingent on item identity. J Exp Psychol Hum Percept Perform 31:1439–1448. https://doi.org/10.1037/0096-1523.31.6.1439
50. Makovski T (2016) What is the context of contextual cueing? Psychon Bull Rev 23:1982–1988. https://doi.org/10.3758/s13423-016-1058-x
51. Summerfield JJ, Lepsien J, Gitelman DR et al (2006) Orienting attention based on long-term memory experience. Neuron 49:905–916. https://doi.org/10.1016/j.neuron.2006.01.021

Neuromethods (2020) 151: 73–103
DOI 10.1007/7657_2019_32
© Springer Science+Business Media, LLC 2019
Published online: 30 October 2019

Contextual Cueing in Virtual (Reality) Environments

Nico Marek and Stefan Pollmann

Abstract

Search in repeated contexts can lead to increased search efficiency due to contextual cueing. Contextual cueing has mainly been investigated in two-dimensional displays. However, contextual cueing is affected by depth information as well as by the "realism" of the search environment. To investigate these aspects further, we present a guide to design contextual cueing experiments in virtual three-dimensional environments. Specifically, we provide a general introduction to the Unity gaming engine and scripting in C#. We will focus on experimental workflows, but also cover topics like timing precision, how to process and handle participants' input, or how to create visual assets and manipulate aspects like color. Ultimately, we will turn the entire project into a virtual reality experiment.

Keywords C#, Contextual cueing, Object-oriented programming, Unity, Virtual reality, Visual search

1 Introduction

The efficiency of visual search can benefit from an individual's search history. Search times have been found to be shorter in displays that have been searched previously, compared with new displays. Faster search in repeated displays has not only been observed in semantically meaningful environments (like rooms; [1, 2]) but also in symbolic search displays – often search for a rotated T-shaped target among L-shaped distractors – without semantic connotation [3]. Particularly in the latter, the search time reduction is often unrelated to the explicit recognition of repeated search displays, implying implicit learning [4].

Repetitions occur typically over blocks of trials, with many intervening displays, distinguishing this contextual cueing effect [3] from priming. In fact, search time reduction in repeated displays has been observed even after delays as long as a week [5]. A detailed description of the contextual cueing paradigm can be found in the chapter by Jiang and Sisk [6] in this volume. In this introduction we concentrate on spatial aspects of contextual cueing that motivate the investigation of contextual cueing in virtual reality, as a background for a step-by-step guide for developing contextual cueing experiments in three-dimensional space with virtual reality equipment.

So far, contextual cueing has mainly been investigated in two-dimensional search displays. However, a study that used pseudo-3D scenes showed that rotations in the depth plane reduced the contextual cueing effect [7] potentially indicating that depth cues contribute to contextual cueing. In another study, attending to a depth plane defined by binocular disparity was vital for contextual cueing [8]. The target was presented consistently in a specific depth plane. Contextual cueing was only observed when the consistent distractor configuration was presented in the attended plane that contained the target and not when the consistent configuration was presented in an unattended plane. Moreover, reaction times for consistent configurations were increased when the disparity of the distractor configurations (but not the target) was reversed, so that the repeated configurations were moved from the back to the front plane or vice versa. The conclusion drawn from these studies that depth cues contribute to contextual cueing was challenged by a recent study that again used binocular disparity cues and found that contextual cueing was intact after changes from the back to the front plane or vice versa (in contrast to in-plane changes of the left versus right display halves) when the distractor configuration and the target remained in the same depth plane [9]. These studies show that depth information affects contextual cueing at least under some circumstances.

While binocular disparity cues allow the manipulation of depth cues, they do not allow movement in space. Interestingly, however, observer movement appears to affect contextual cueing. Tsuchiai, Matsumiya, Kuriki, and Shioiri [10] observed contextual cueing for 3D configurations rotated in depth only when this rotation was caused by the motion of the observer to a new viewpoint. Similar restrictions were observed for target probability cueing [11]. To our knowledge, contextual cueing has not yet been investigated in virtual environments that enable movements in artificial spaces that may allow more comprehensive control of environmental variables than movements in front of a 2D monitor (as in [10, 11]).

Most contextual cueing experiments have been carried out with standard monitors, leading to search displays covering, e.g., ca. 40° × 30° visual angle [3] or less in diameter. Contextual cueing was also reported for larger search displays, e.g., roughly 50° × 40°, using a back projection screen [12, 13]. However, to our knowledge, it has not yet been investigated if contextual cueing will also be observed if, e.g., the target is presented outside the initial field of view of the observer, requiring head and/or body movements to find it.

These examples are not meant to be a comprehensive overview of 3D effects on contextual cueing. They show, however, that virtual environments offer interesting new ways of investigating the contextual cueing effect that have the potential to increase our knowledge about the mechanisms of contextual cueing in 3D space. In addition, studying contextual cueing in virtual environments may also be worthwhile because of their immersive nature. Search in realistic two-dimensional displays [1, 2] leads to greater search benefits and more explicit memory for repeated displays compared with search in meaningless symbolic displays. It might be that search in virtual environments leads to even faster and more explicit memory of repeated presentations. On the other hand, the higher complexity of a three-dimensional environment that needs to be explored with head movements increases working memory load during exploration which may affect contextual cueing [14].

In summary, there are plenty of reasons to explore contextual cueing in virtual environments. In the following, we present a step-by-step guide to the creation of a virtual reality contextual cueing experiment. As an example, we use a classical T-among-L search paradigm transformed to 3D space, for better comparability with the classical 2D contextual cueing literature. However, the methods that we describe allow the creation of more realistic virtual environments as well.

2 Methods

At its core, a behavioral experiment follows a predetermined order which is very similar to video games. First, we place participants in an artificial environment. Just like gamers, they receive visual, auditory, or other sensory input. Depending on the research question, some experiments require fast responses to complex stimuli or the ability to plan ahead to achieve future goals. Video games and experiments alike detect and collect responses. Participants might receive feedback about their actions via changes in the user interface (UI), they compete or cooperate with other humans, and changes to the environment are applied according to the experimental manipulation.

Everything players and participants alike see and every interaction or changes to the experimental conditions can be set up with software development environments (*engines*), like Xenko Game Engine,[1] Unreal[2], or Unity.[3] These frameworks simplify the developmental process by unifying *rendering*[4] engines and support for different underlying programming languages. But engines are not limited to graphics and coding. They play sound, handle physic simulations, and control non-player agents; they provide built-in

[1] https://github.com/xenko3d/xenko

[2] https://www.unrealengine.com/

[3] https://unity.com/

[4] *Rendering* is a term established by computer graphics that refers to an automated process, which combines different aspects like lighting, color, and shape to generate an image that can be presented to a viewer.

solutions for environment lighting or web interfaces to collect data. A game engine gives software engineers and scientists alike the opportunity to focus on their game design/experimental workflows by providing out of the box solutions.

Software packages like PsychoPy [15] or Psychtoolbox [16] are designed with the special needs of the scientific community in mind and are great for 2D experiments with simple visual or auditory stimulation and precise timing. With little effort and a few tweaks, everything you need for a behavioral experiment can be done in all of the previously mentioned engines as well and more. We have chosen Unity for this implementation, because it is well documented and supports two high-level programming languages that are easy to learn (C# or JavaScript).

Since three-dimensional virtual environments are a common part of modern video game development, Unity was designed to implement 3D environments and provides features for most of the common topics (e.g., object placement, object rotation, input detection) out of the box. You can also build 3D objects within Unity, but the engine allows to import models from other software tools too. For instance, you can build naturalistic visual objects with open-source tool sets like Blender[5] and treat them as game objects. Moreover, you can develop your experiment on one platform (e.g., Windows) and deploy it to another (e.g., Linux, macOS) or even other devices like smartphones. You can also benefit from middleware developments in the gaming industry, which might simplify your development pipeline by providing easy-to-use solutions. Middleware is a general term for software that serves a special purpose and can be imported into an engine, like 3D human animations, sound processing or pathfinding, and collision detection.

Last but not the least, you can learn from tutorials provided by a large community. Of course, the target audience for most tutorial videos are not neuroscientists, but most of the topics can help you with your own implementation.

2.1 Unity and Game Objects

At first glance Unity may seem unfamiliar to researchers. Some concepts derive from the gaming industry; others are special to the engine itself. We will address topics like prefabs, materials, and children throughout the upcoming chapter whenever necessary. But one fundamental principle needs to be explained in advance: in Unity different layers of development come together and are summarized as *game objects*. Your code will be written in an integrated development environment (IDE) and complex visual or auditory stimuli are created using specialized software. In an

[5] www.blender.org

upcoming section, we will show you how to create simple visual assets within the Unity engine itself.

The first and most important concept of Unity is called a game object. Game objects are used as containers for different components of your experiments. Let's illustrate this concept with a simple example. Say you want to present a car to your participants. The 3D model of the car itself is one component; the sound of the engine and the brakes are another. If you want to move your car, you have to change its location in virtual space by accessing the *Transform* component. If you want to change its surface color, you have to attach a material component to it. If you want your participants to control the car, you will need to write a script that detects input and changes the Transform component accordingly. The simulation of the driving physics is another (complex) component and interacts with other game objects of your scene. From the participant's representation in virtual space to lighting, keyboards, and gamepads, basically everything is a game object.

In terms of the contextual cueing paradigm, a target and all distractors can be treated as children of a game objects themselves. Instead of one car, we will have several distractors and one target. Every search display, old or novel, is called a *parent* game object, and every distractor and target are its children. Those children refer back to the parent and are generated with distinct properties (e.g., Transform components that ensure that they are not overlapping). In the upcoming sections, we will show you all key implementations of a contextual cueing task. We will code C# scripts to generate novel and old displays; you will learn how to process user input and how to save response time data in a three-dimensional virtual environment.

Due to the size of the project, we cannot walk you through the entire code, but we will cover all relevant techniques and workflows. Once you are finished with this chapter, we invite you to download the complete project at: https://github.com/nimarek. The GitHub page is also the best place to ask questions and to make suggestions on how to improve our code.

2.2 Getting Started with Unity

You can download a free version of the Unity engine at https://unity3d.com/get-unity/download. We will use the latest personal version 2019.1.0f2 along with Unity hub 2.0.0 for the following tutorial. Unity hub is an additional application that helps you to manage different projects on the same machine. You can run older projects with older versions of the engine or test your experiment with new versions if you need to migrate. After the installation is finished, you need to create a new project.

Unity imports all basic scripts, and when it is done, you will see a first minimalistic sample scene (Fig. 1) with two game objects in the Hierarchy tab on the right. The Hierarchy tab lists all game objects in the scene and their relationship to each other. This concept is called parenting, and every game object attached to

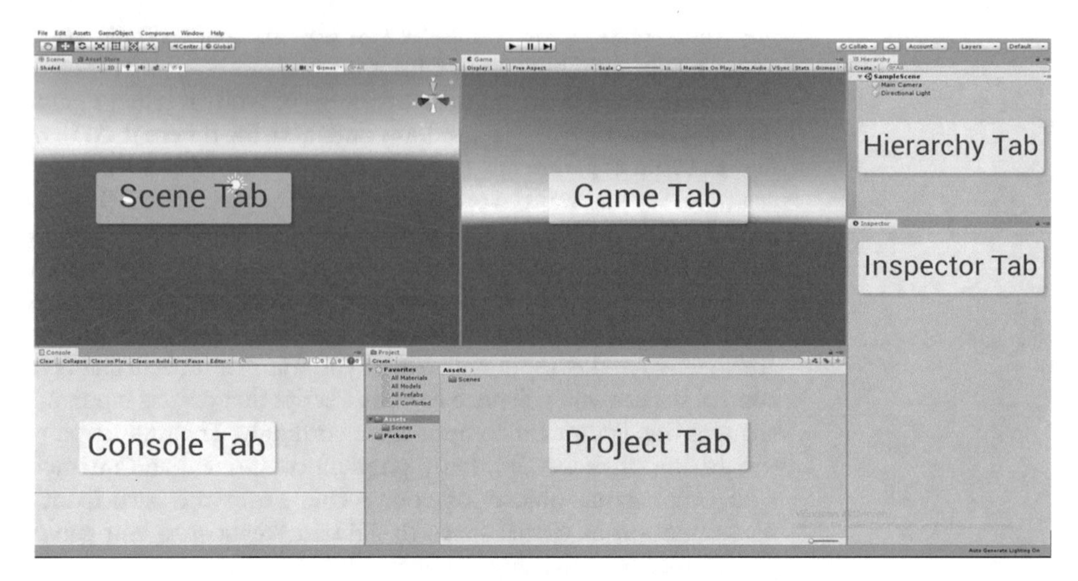

Fig. 1 The default interface configuration of Unity. You can resize and move every window with simple drag and drop. The Scene view can be seen in the upper left part of the figure; the window in the middle displays the Game view. You can access game object components via the two right tabs. The upper one is called Hierarchy; the Inspector view underneath displays all components attached to a game object. Console logs are printed to the window down left, while all folders and their respective files can be accessed through the Project tab in the middle

another is called child. You can declare a game object a child of another by simple drag and drop. In an upcoming section, we will use this concept to generate and delete our contextual cueing displays. In the first scene, there are only two game objects with no children. One is the Main Camera object, and the other is the Directional Light object that illuminates the scene.

On the far left of the interface, you can see the Scene view. This window provides an overview of the current scene, and you can navigate it by using the W, A, S, and D keys. Both the Main Camera and the Directional Light object have their default place in our virtual space. If you click on one of the two objects in the Hierarchy tab, Unity will highlight their location in the Scene view and vice versa. The Main Camera object is equal to your participant's point of view in the experiment. The Main Camera input is again displayed in the center of the interface, the so-called Game view.

Click on the Main Camera object in the Hierarchy next, and look down to the Inspector view underneath. You will see an overview of all components attached to the object with the Transform component on top. Every game object (visible or invisible) holds a position, rotation, and a scaling parameter in virtual space. We will learn how to manipulate these values in the upcoming sections via C# scripts, but for now we will focus on the Camera component underneath (Fig. 2).

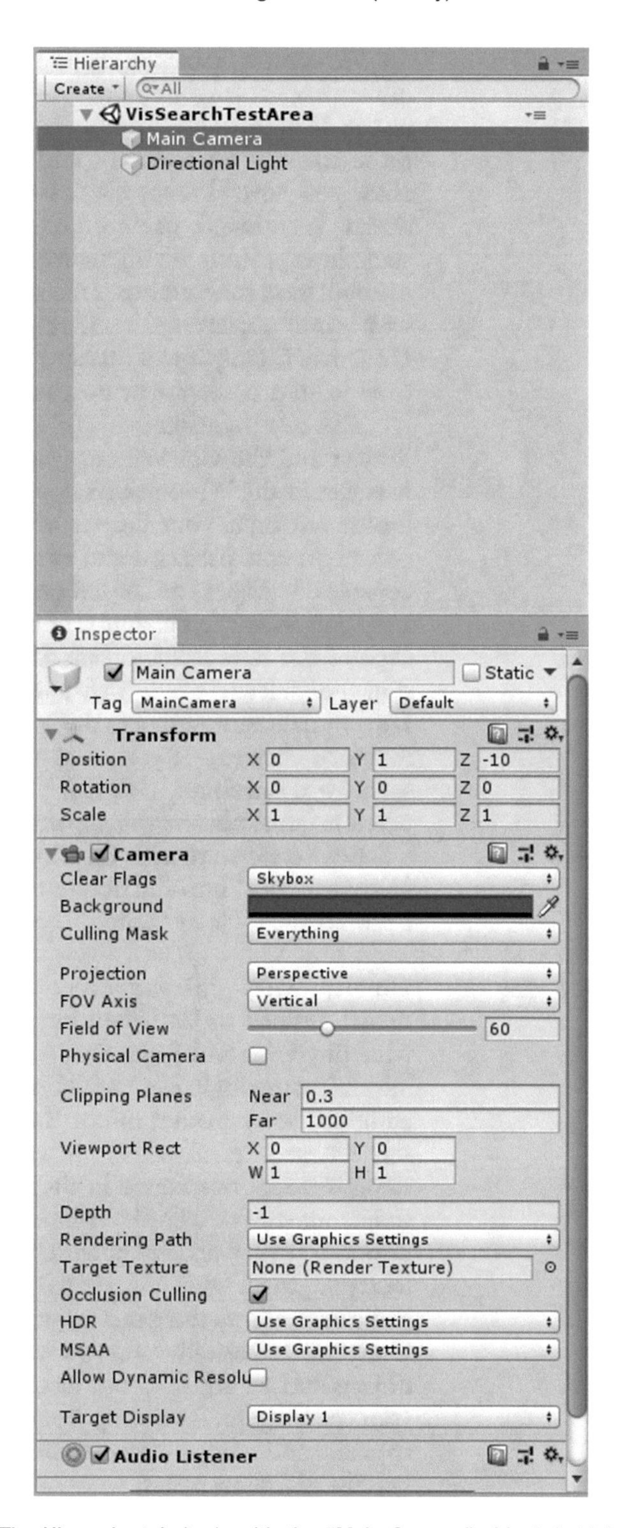

Fig. 2 The Hierarchy tab (top), with the “Main Camera” object, is highlighted. Adjustments of all components can be made via the Inspector tab (bottom)

We apply a first small change to our scene. Select "solid color" in the drop-down menu next to "Clear Flags," and choose a color in the "Background" option. The Main Camera object captures these changes immediately, and the Game view shows only the color you have chosen, instead of the default skybox texture. As already mentioned, game objects are a central concept to understand how projects are implemented in Unity. If you have a specific manipulation for your experiment in mind, it is best practice to start with game objects and their respective components provided by Unity itself. Only if you do not find what you are looking for, you need to start coding your own solutions.

The two remaining major windows in the interface are the Project and the Console tab (Fig. 1). You can create folders and files like in the Windows Explorer, just by clicking with the right mouse button in your Project tab. We will store the 3D models of our target and distractor and all the scripts in the Asset folder. The console tab (Fig. 1) on the left provides error messages oder Debug Log(); messages from your scripts. You should take your time and explore the interface and especially the Inspector and scene window. For starters you can change the transform parameters of the Directional Light or delete the Main Camera object to see how the interface changes. If you want to bring an object back into the scene, you can simply right click in the Hierarchy tab, and a list of possible game objects will appear.

Before we start with our first self-coded component, we will give you a short introduction to the most fundamental principles of a script's life cycle in Unity.

2.3 Basic Coding Overview

Unity provides different event functions that are executed in a predetermined order. When hitting the Play button, Unity starts with physics simulations first (e.g., gravity, collision detection), internal game logic is executed second, and the UI comes last. In addition, several other minor functions can be called in between those steps (e.g., Awake();, Start();, OnDestroy();). Once everything is neatly positioned in the virtual environment, the internal scripts need to detect player input first and trigger an appropriate response (e.g., walking animation) second. At the end of a script's life cycle, the engine collects feedback about internal states of the game and adjusts the game objects accordingly. Keep in mind that Unity treats basically everything as a game object, not only 3D objects can be adjusted but also UI elements or variables within your scripts (e.g., Booleans). Some of the most important methods to access and change information during an experiment are Update ();, FixedUpdate();, and LateUpdate();.

The Update(); function is used to keep track of the internal logic and is called once per frame by every object with this particular script component attached to it. This way parts of the environment can be changed (e.g., moving non-physics objects) or user input can be detected. In game development this function is utilized to implement simple timers, but the intervals between two frames might vary based on the time it takes to process the previous frame. As a result Update(); is not called on a regular timeline and should therefore not be used to present stimuli which require high-timing precision.

FixedUpdate(); does not depend on the individual frame rate and is called on a regular basis. Thus the function can be executed once, twice, or even more often per frame. In the Unity interface, these time steps can be changed: Edit → Project Settings → Time → Fixed Timestep. Due to its nature, FixedUpdate(); might not be responsive enough to capture every possible user input. If a button is pressed between two time steps, input for a given frame is not recorded, and data is lost. Therefore the Update(); function is commonly used to detect user input. The LateUpdate(); function is very similar to Update(); but is called once per frame after Update (); is finished. It can be used to create highly interactive virtual environments that immediately respond to user input (e.g., changes in the UI, trigger animations).

2.4 How to Create Simple Prefabs

Every contextual cueing experiment needs (at least) one target and several distractors. Instead of creating the target and all distractors at the beginning of every trial, we will simply construct prototypical three-dimensional objects once and clone, copy, and alter them via C# script commands. In Unity those object containers, their corresponding scripts, materials, and property values are summarized as *prefabs*. Prefabs are very similar to game objects; they contain all previously attached components plus the Hierarchy information (e.g., which object is a parent, which objects are children to the parent), but they can be stored as a reusable asset on the hard drive. Basic visual assets can be created using only Unity, but more immersive and visually complex environments require additional software. Since we create artificial-looking search displays composed of basic shapes, no further software installation will be necessary.

We will use an empty game object and two cubes as children to create a target prefab. In the editor view, click on GameObject → Create Empty, and create two cubes using GameObject → 3D Object → Cube. Change the transform coordinates of the first cube to X = 0, Y = 0, Z = 0 and the scale to X = 2, Y = 0,5, Z = 0,5, and change the coordinates of the second cube to X = −1, Y = 0, Z = 0 and the scale to X = 0,5, Y = 2, Z = 0,5. Now move both cubes into the empty game object, and rename it to "Target." After everything is finished, you can store the newly created prefab simply by dragging it from the Hierarchy into a folder of your choice (e.g., prefabs). Now the color displayed in the Hierarchy tab should have changed from white to blue, and a > symbol has appeared on the right. This confirms that you have created your first prefab!

Our distractor is created in a similar way, or you can take the existing game object and lower the y-coordinate of the first cube until it resembles an L. Simply rename the object afterward to "Distractor," and store it in your Prefab folder as well. The prefab acts like a prototype, it can be altered, and all changes applied to the original will affect all clones in the same way. Finally we will create an inter-trial fixation game object: GameObject → 3D Object → Sphere. The sphere will be placed in the middle of every search display to ensure equal head and eye positions throughout the trial starts. Call it "IT-Fixation." To indicate that a new trial is about to begin, we will change the color of the sphere gradually from red to green.

2.5 A First C# Script

Instead of the traditional print "Hello World" to screen introduction, we will change the color of a game object with some very simple lines of code. This is also a good opportunity to make yourself familiar with C#, Unity's fundamental class, and the connection between code and game objects. First create a Scripts folder, and right click → Create → C# script in the Project tab. Name it ColorChanger, and open the script afterward in the code editor of your choice. Unity will load a template with a class name according to the name of your script. Therefore our new class is called ColorChanger.

```
using System.Collections;
using System.Collections.Generic;
using UnityEngine;

public class ColorChanger : MonoBehaviour
{
  // Start is called before the first frame update
  void Start()
  {

  }

  // Update is called once per frame
  void Update()
  {

  }
}
```

This class inherits from MonoBehaviour, the base class of Unity, which grants you access to useful public methods like Update();, Start();, static methods like Instantiate();, and many more. Most of the scripts used in this project will inherit from MonoBehaviour; therefore it is worth looking at the documentation (https://bit.ly/2UVpL5f) to make yourself familiar with the entire concept. Next, write the following variables into your MonoBehaviour class.

```
public float speed = 1.0f;
public float startTime;
```

Every time a variable is declared public, it becomes accessible to other scripts and the Unity Inspector. A float of one (1.0f) is preassigned to our variable speed, but startTime remains empty for now. Unity automatically assigns a value of zero when the script is executed. If you want to introduce a delay, you can change the value in the script itself or via the Unity editor.[6] With the next two lines of code, we create two colors: One for the beginning of the inter-trial interval (ITI) and one for the end. You can define a Color with r, g, b,and a components or simply use the Inspector again.

```
public Color startColor;
public Color endColor;
```

As one of the first steps in the script's life cycle, it will begin measuring the time when it is executed. Time.time is set equal to the float startTime whenever the Start(); method is called.

```
private void Start()
{
   startTime = Time.time;
}
```

Remember that Unity calls the Update(); method once per frame. Now we can take advantage of this feature to ensure that the transitions between both colors are smooth. We will calculate the float t once per frame and change its value times the speed variable. This way a bigger speed value leads to faster transitions and vice versa. A bigger startTime value on the other hand leads to a later onset. At last we tell Unity to change the color of our fixation object. The color can be accessed through:

```
GetComponent<Renderer>().material.color.
```

As mentioned before colors are only one aspect of materials; therefore we must call the Renderer component, and access the material and finally the color. To interpolate between the colors, we call the Color.Lerp method and set it equal to our renderer component. This requires three arguments as input: a color a, a second color b, and a float t. Since we created all of them before,

[6] Note that changes made in the Inspector have higher priority and values in the script will be overwritten.

we can pass all arguments to the method Color.Lerp(startColor, endColor, t), and the engine will manage the rest!

```
private void Update()
  {
    float t = (Time.time - startTime) * speed;
    GetComponent<Renderer>().material.color = Color.Lerp(startColor, endColor, t);
  }
```

Let us return to Unity now. Click on the Sphere ("IT-Fixation") game object in the Hierarchy tab, then click on Add Component, and attach our ColorChanger script to it. Thus our public variables become accessible in the Inspector. You can change the values of startTime and speed or assign a color of your choice as starting and end color. Similar to the target and distractors, we will frequently use the fixation object for the ITI, so make a prefab out of it! Keep in mind that the values you have assigned via the Inspector before creating the prefab are now its default parameters. If you want to change them, drag your fixation sphere back in the Hierarchy tab, click on the > symbol, and save the prefab once you have made all your adjustments.

2.6 Passing Variables from One Script to Another

Before we write larger scripts in the upcoming sections, we need to make ourselves familiar with another recurring topic. All our scripts implement one aspect of the project alone, but they need to work together. This is only possible if their respective variables are accessible from other components whenever needed.

By now you know how to create C# scripts and empty game objects yourself, so create two of them each. Name one script and the associated game object ProvideVariable and the other script and game object AccessVariable (Fig. 3). Use the Hierarchy tab to add the script component to the game object. Next you need to tell Unity which class you want to access and which variable to pass. Open the ProvideVariable script, delete both the Start(); and Update(); method, and replace it with:

```
public class ProvideVariable : MonoBehaviour
{
  public AccessVariable accessVariable;

  public int myInt2Pass = 42;
}
```

After saving the script, an empty slot will appear in the Inspector. This is an easy way to link scripts in Unity. Simply drag the AccessVariable game object from the Hierarchy to the empty slot in the Inspector. Now open up the AccessVariable script, but delete

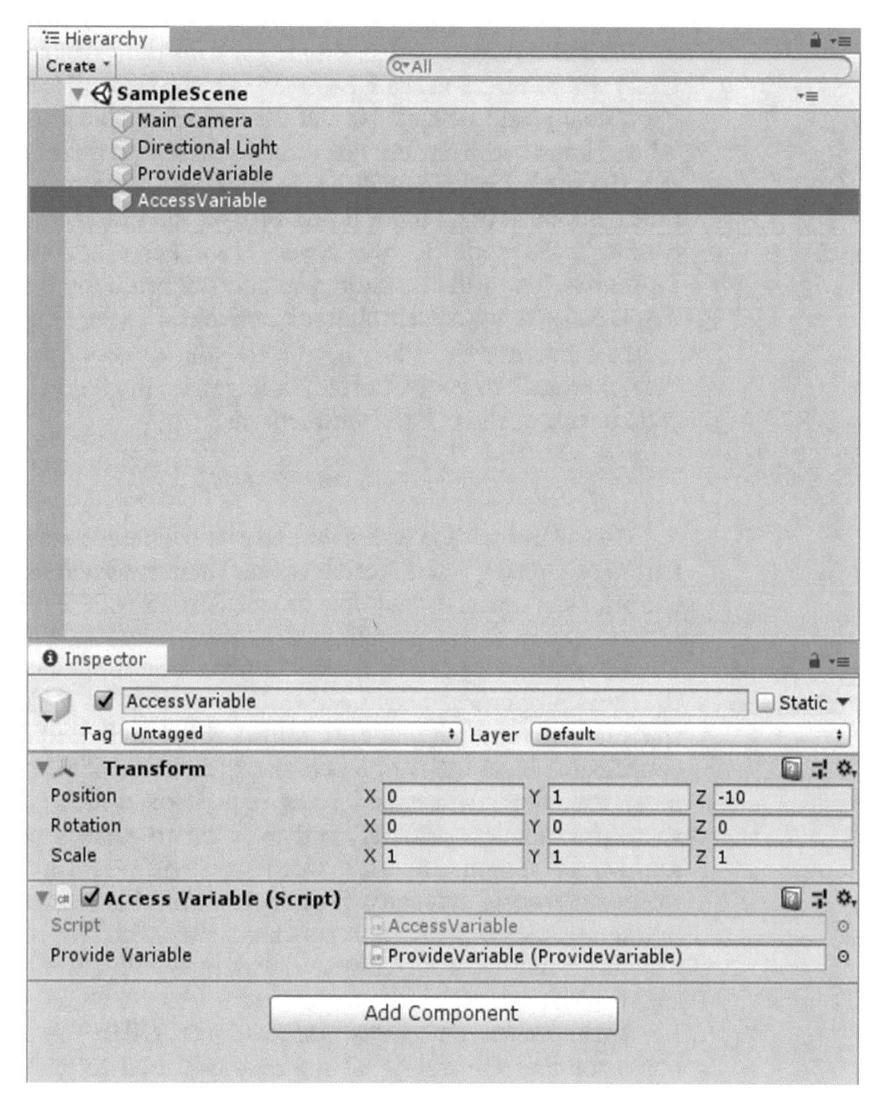

Fig. 3 The Inspector view is one of the most important features. Every public variable in your script can be accessed here. Unity lists every other linked component, in this example our Transform component. You can attach your self-coded scripts to the game objects via Add Component (bottom)

only the Update(); method from the template, and write the following lines of code:

```
public class AccessVariable : MonoBehaviour
{
   public ProvideVariable provideVariable;

   private void Start()
   {
      Debug.Log("You shall pass: " + provideVariable.myInt2Pass);
   }
}
```

Start the project with a simple click on the Play button, and Unity will print: "You shall pass: 42" in the console. Note that the "You shall pass:" string is part of the AccessVariable class, but the 42 was imported from ProvideVariable. In later parts of this tutorial, the same workflow will be used to pass floats (e.g., reaction time), Booleans (e.g., input detection), or other integers (e.g., trial counter). As soon as we cover "Processing and Collecting Responses," we will also show you another brute force technique that is easier to implement, but might slow down your experiment.

Last but not the least, if you are still interested in printing "Hello World" to your Console, simply write the following line of code to one of the two default methods.

```
Debug.Log("Hello World");
```

We will not tell you which, but take your time and reflect on the nature of Update(); and Start(); again. There is a good reason why the print statement should only be called in one of the two.[7]

2.7 Scripting Search Displays from Scratch

Even though every search display looks a little different, all share the same properties. They need to have a target to look for, distractors to distract (or guide attention), the stimuli need transform coordinates in virtual space and the generation of these vectors must follow predetermined rules (e.g., two distractors cannot share the same coordinates) and we want to place a predefined number of stimuli. In other words, we will perform the same calculations again and again for each trial. In object-oriented programming, classes are used to perform these tasks. We create game objects (in our case search displays) with the same components, but slightly different values.

Again create an empty game object called Spawner and corresponding C# script. All the methods used to generate and delete the visual search displays will be written inside of this script. In this tutorial we will only cover the basics necessary to create novel and old search displays. Most of the details of our actual implementation are not covered here, like counterbalancing the number of distractors per quadrant or the location of the target. If you have questions regarding our code, feel free to contact us via GitHub. For now, create another new script, call it Trial, and enter

[7] Spoiler alert: If you place the line of code in the Update(); method, Unity will write "Hello World" once per frame to your console. This may lead to many (many) "Hello Worlds" in seconds.

the code of the public class Trial. This time we do not inherit from any other class.

```
public class Trial
{
        public bool isTargetFlipped;
        public int targetChooseCircle;
        public int targetChoosePosition;
        public List<List<int>> randomPositionVoidLists;
           public List<List<int>> randomRotationVoidLists;
}
```

Every time we utilize Trial, several variables will be called. The isTargetFlipped Boolean allows us to check whether the target is oriented to the left or to the right. In an upcoming section, we will use this Boolean for input detection. Every search display consists of one target and a predefined number of distractors. Hence both integers targetChooseCircle and targetChoosePosition are used for target placement, while both the randomPositionVoidLists and randomRotationVoidLists are reserved for distractor placement and rotation. Since the novel and old search displays are generated the same way, we can use the class later to generate both conditions. The only real difference is we will roll the dice once for each old display configuration, but again and again at the beginning of every novel trial.

Close the script afterwards and open the Spawner script again. To increase the search difficulty, we will randomly rotate our distractors in four different ways. If you switch to the Inspector, you can see that the Transform component of every game object determines the position, rotation, and scale of the object. Switch back to the Spawner script, and create the getRotationDistractor( ); method below. Again this method is not a void and returns either an int of 0, 90, 180, or 270, which is then interpreted as a Euler angle.

```
private int getRotationDistractor()
  {
    int rndRotateDistractor = Random.Range(0, 99);
    if (rndRotateDistractor >= 0 && rndRotateDistractor <= 25)
    {
      return 0;
    }
    else if (rndRotateDistractor > 26 && rndRotateDistractor <= 50)
    {
      return 90;
    }
    else if (rndRotateDistractor > 51 && rndRotateDistractor <= 75)
    {
      return 180;
    }
    else
    {
      return 270;
    }
  }
```

In our example implementation, we spawn every object on a circle. Using this technique we can scale the distance between virtual objects and participant using the radii as distance. You need to add two public variables to our Spawner script, the public floats scaleCircles and radius. This method again is not a void, it returns a Vector3 containing three different values and requires two arguments as input.

```
private Vector3 circle(Vector3 center, float ang)
  {
    Vector3 pos;
    pos.x = center.x + radius * Mathf.Sin(ang * Mathf.Deg2Rad / scaleCircles);
    pos.z = center.z + radius * Mathf.Cos(ang * Mathf.Deg2Rad / scaleCircles);
    pos.y = center.y;
    return pos;
  }
```

The process of cloning a previously created prefab in Unity is called instantiating, and we will use this method to generate and place our target-distractor configurations in a virtual environment. Let's start with the target and add the following variables to your script:

```
public GameObject Target;
public Vector3 targetPos;
public float scaleTarget;
public bool flipTarget;
```

We will create one instance of the target, using the Instantiate method. The first parameter tells Unity which existing object will be copied, in this case of course our target prefab. The second is a Vector3 containing the position. The third variable rot is a Quaternion and is used to represent rotations. The last transform parameter can be used to scale the target according to your needs. We will assign a tag to each of our distractors and to our target. Tags are an easy way to identify game objects. Just switch back to the Unity interface, click on one of your game objects in the Inspector tab, and click on Add tag (Fig. 4).

After all the necessary input variables are processed, we will introduce the flipTarget Boolean. Independently from the novel or old condition, the direction our target will point to is always decided at the beginning of every trial. This can be achieved by accessing the instance of our target's Transform component, and we either multiply the localScale ∗ -1 (target is flipped to the right) or we leave it as it is (target is flipped to the left).

Some of the variables introduced in the upcoming lines of code will only be used by subsequent steps. If your IDE points out some errors, ignore them for now! Later we will call the createTarget(); method from another method; we will toss a coin for the flipTarget

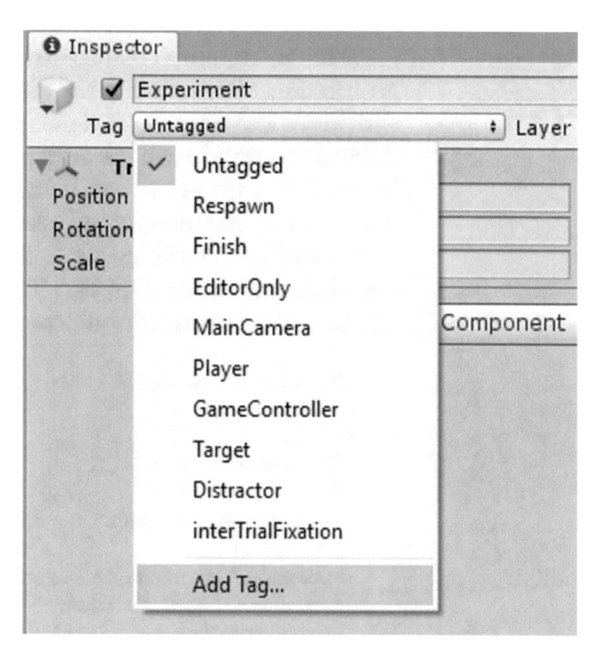

Fig. 4 You can use the Inspector again to create and assign tags. Tags can be used to find objects via script commands

Boolean and use the Vector3 returned by the circle method to spawn our objects.

```
public void createTarget(Vector3 pos, Quaternion rot)
  {
    targetPos = pos;
    GameObject instanceOfTarget = Instantiate(Target, pos, rot, transform);
    instanceOfTarget.gameObject.tag = "Target";

    if (flipTarget)
    {
      instanceOfTarget.transform.localScale = new Vector3(-1 * scaleTarget,
      scaleTarget, scaleTarget);

    }
    else
    {
      instanceOfTarget.transform.localScale = new Vector3(scaleTarget,
      scaleTarget, scaleTarget);
    }
  }
```

Add the following variables to your script next. Again we will use a simple public float to change the size of the distractors via the Inspector. Since we want all distractors to have the same size, we will simply use the same float again and again whenever a game object is instantiated.

```
public GameObject Distractor;
public float scaleDistractor;
```

The upcoming createDistractor method handles distractor positioning. We will place the distractors depending on a list filled with zeros and ones. If an element in the randomPositionVoidLists is equal to one, we will use Instantiate to spawn a distractor prefab. If not, the position will remain empty. The same logic is applied to the rotation component of every single distractor. Whenever a distractor is instantiated, we will use one of the values stored in the randomRotationVoid list. These values will be added one by one to our transform.Rotate component for each distractor.

```
private void createDistractor(int randomPositionVoid, int randomRotationVoid, Vector3
pos, Quaternion rot)
  {
    if (randomPositionVoid == 1)
    {
      GameObject instanceOfDistractor = Instantiate(Distractor, pos, rot,
      transform);
      instanceOfDistractor.transform.localScale = new Vector3(scaleDistractor,
      scaleDistractor, scaleDistractor);
      instanceOfDistractor.gameObject.tag = "Distractor";
      instanceOfDistractor.transform.Rotate(new Vector3(0, 0,
      randomRotationVoid));
    }
  }
```

The upcoming variables serve several purposes. The numObjects integer refers to the maximum number of visual objects per circle; the numCircle refers to the maximum number of circles. If you want some of the distractors to appear outside of the initial field of view, you can change the size of all circles with the scaleCircles float. Last, the spacingCircles float determines the distance between each circle.

```
public int numObjects;
public int numCircles;
public float scaleCircles;
public float spacingCircles;
```

Every iteration of the upcoming loop instantiates a prefab or creates empty space in our Trial class, depending on your desired number of objects and circles. This is one of the most essential parts for setting up search displays. Since we want our participants to only look for one target, we will call createTarget(); only once. Every remaining object will be a distractor spawned by the createDistractor(); method. Our last task will be a short calculation of our search display center. We will need this value to create an InterTrialFixation object.

```
private void drawCircle(Vector3 center, bool targetIsOnCircle, Trial trial, int circle)
{
  for (int i = 0; i < numObjects; i++)
  {
    Vector3 pos = Circle(center, i * angle);
    Quaternion rot = Quaternion.LookRotation(pos - center);

    if (targetIsOnCircle && i == trial.targetChoosePosition)
    {
      createTarget(pos, rot);
    }
    else
    {
      List<int> randomPositionVoidList =
            trial.randomPositionVoidLists[circle];
      int randomPositionVoid = randomPositionVoidList[i];

      List<int> randomRotationVoidList =
            trial.randomRotationVoidLists[circle];
      int randomRotationVoid = randomRotationVoidList[i];

      createDistractor(randomPositionVoid, randomRotationVoid, pos, rot);
    }

    if (i == numObjects / 2)
    {
      centerPos = pos;
    }
  }
}
```

From inside of startTrialWithRandomValue();, we will call our previously written methods. All methods fill our Trial class with numbers, which are subsequently interpreted by Unity. In our example implementation, we excluded all possible target locations on the outside borders of our search displays. All adjustments of the display configurations are applied by restrictions of the Random. Range method. You can use Random.Range(1, numObjects); to exclude the first possible target position. To exclude the last position, we simply subtract the maximum number of distractors by one

(Random.Range(1, numObjects - 1);). The same logic can be applied to the targetChooseCircle integer.

```
public void startTrialWithRandomValue()
  {
    Trial randomTrial = new Trial();

    randomTrial.isTargetFlipped = (Random.value > 0.5f);
    randomTrial.targetChooseCircle = Random.Range(1, numCircles - 1);
    randomTrial.targetChoosePosition = Random.Range(1, numObjects - 1);

    randomTrial.randomPositionVoidLists = new List<List<int>>();
    randomTrial.randomRotationVoidLists = new List<List<int>>();

    for (int z = 0; z < numCircles; z++)
    {
      List<int> randomRotationVoidList = new List<int>();

      for (int i = 0; i < numObjects; i++)
      {
        randomRotationVoidList.Add(getRotationDistractor());
      }

      randomTrial.randomRotationVoidLists.Add(randomRotationVoidList);
    }

    Debug.Log("Trial: " + chooseTrial.countTrialGlobal + " is a random trial.");
    startTrial(randomTrial);
  }
```

The last method passes the flipTarget Boolean to our Trial class, this allows us to vary the space between circles and starts our Timer (which is covered in detail in the upcoming section). One important remark before you push the play button: switch back to the Unity interface and add our previously created prefabs to our script via the Inspector tab. Call the startTrial(); method from Start(); for now and enjoy your first search display (Fig. 5)!

```
private void startTrial(Trial trial)
  {
    flipTarget = trial.isTargetFlipped;
    Vector3 center = transform.position;

    for (int i = 0; i < numCircles; i++)
    {
      bool targetIsOnCircle = trial.targetChooseCircle == i;
      drawCircle(center, targetIsOnCircle, trial, i);
      center.y += spacingCircles;
    }

    timer.StartRecord();
  }
```

Since every contextual cueing task highly depends on the repetition of search displays, we want to save specific target-distractor

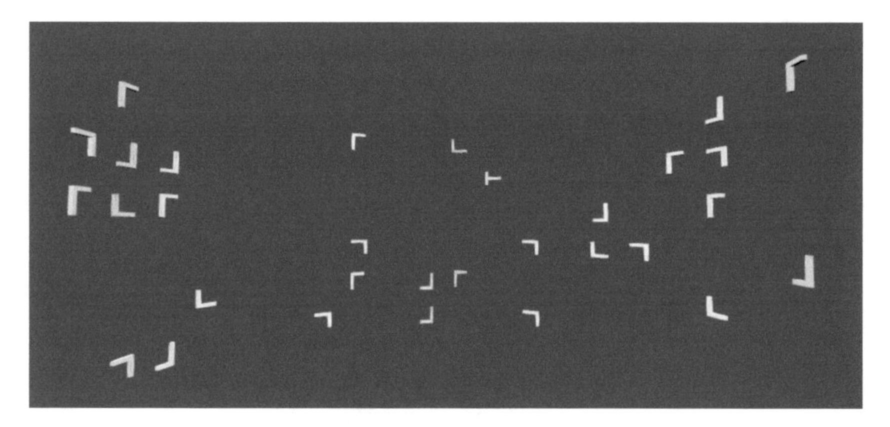

Fig. 5 An example of a randomly generated search display as seen in the Game view tab

configurations for our actual experiment. If you are familiar with the different event functions of MonoBehaviour, creating repeated search displays is straightforward. Add the following lines of code to your script:

```
// Initialize with random trial parameters
public Trial rndTrial = new Trial();

// Initialize with fixed trial parameters
public Trial FixedTrial = new Trial();
```

Instead of rolling the dice again and again whenever a novel trial starts, we will simply call the following method from Start();.

```
    public void generateOldTrial()
    {
        FixedTrial.targetChooseCircle = Random.Range(0,
numCircles);
        FixedTrial.targetChoosePosition = Random.Range(0,
numObjects);
        FixedTrial.randomPositionVoidLists = new List<List<int>>();
        FixedTrial.randomRotationVoidLists = new List<List<int>>();
    }
```

This will generate one repeated trial and also ensures that all our old search displays are generated right at the beginning of the session. Since we won't call the method again, the transform coordinates in the Trial class will not be changed over the course of the experiment. If you have specific types of display configurations for your experiment in mind, you can always assign constants to the variables within the Trial class instead of calling the Random.Range method. Say you have six circles and you want your target to appear always on the fourth, just replace the FixedTrial.targetChooseCircle line with:

```
FixedTrial.targetChooseCircle = 3;
```

Using this method as a template, you can generate as many old configurations as necessary for own experiment. If you want to see how you can implement a method to shuffle and execute the trials in a specific (or randomized) order, you can find an entire script on our GitHub page (Fig. 5).

2.8 Timing and Cleanup

Unity provides several different implementations when it comes to timing. You either use these to move visual assets in virtual space, to trigger events based on time intervals, or to measure the responses of your participants. As usual, create a script called Timer first.

One of the methods provided by the engine is called deltaTime. It will calculate the time based on the difference of two values. If you call this method from Update();, the calculation is based on the time it took to generate the last frame. As mentioned earlier, this may lead to inconsistent time intervals.[8] Depending on your hardware and the complexity of your experiment, you may see huge differences.[9] To achieve adequate timing, we will again inherit from MonoBehaviour, but this time we will use the InvokeRepeating(); method. The method requires three arguments: (1) a string referring to the name of the method (here a timer), (2) a float for time, and (3) an increment indicating the time steps. If you want to add a global time stamp for your experiment, just call the method from Start();, and create a public float called globalTime! To keep it simple, we create only two floats and a Boolean as follows:

```
public float increment = 1;
public float currentTime;
private bool record;
```

Next we will write two methods, the first is called whenever a trial is started, and the second will stop the timer when user input is detected. You will need to call both methods from the Spawner script. When the startRecord(); method is called, we will set the currentTime variable back to zero and the record Boolean to true.

```
public void StartRecord()
  {
    currentTime = 0.0f;
    record = true;
  }
```

[8] If you can ensure that your hardware produces, for example, constant 100 frames per second, you can (in theory) measure with a precision of 10 ms.

[9] You can find a script that collects time data from three different methods here: https://bit.ly/2W4g0XN.

The StopRecord(); method is straightforward. We will set the Boolean to false and print the currentTime variable to our console (in red) to keep the experimenter entertained during the session.

```
public void StopRecord()
  {
    record = false;
    Debug.Log("<color=red>GlobalTime</color>: " + globalTime);
    Debug.Log("<color=red>TrialTime</color>: " + currentTime);
  }
```

You can put the following line of code in a void of your choice. InvokeRepeating(); needs to be called whenever a trials starts, and it ends when user input is detected (this topic is covered in the upcoming section). Simply add an if-statement with the record Boolean and call the method with the following parameters:

```
InvokeRepeating("InvokeTimer", 0, increment);
```

The upcoming method is called from InvokeRepeating();.

```
void InvokeTimer()
{
   currentTime = currentTime + increment;
}
```

One of our last methods will be placed inside the Spawner script again and acts as our private janitor. It cleans up the environment whenever we switch from one part of the experiment to another. If you plan to have multiple scripts that spawn objects, it is advisable to create a separate script and call the method from there.

In the internal logic of our experiment, our janitor method is called when (1) the start text disappears, (2) when we switch from a search display to the ITI, (3) when the ITI has ended and the fixation objects disappear, (4) when the block pause ends and we start a new trial, and (5) when our experiment is finished. Let's call the method deleteAllChildren();. Our janitor needs only limited information to know exactly what to clean. To detect all necessary visual assets, we will use the foreach loop as an easy way to iterate over elements in a list or an array. Since all our game objects are children of the Spawner script with an individual Transform component attached to them, we simply iterate over all Transform components. For each iteration we will call the Destroy(); method to delete the visual assets.

```
private void deleteAllChildren()
  {
    foreach (Transform child in transform)
    {
      GameObject.Destroy(child.gameObject);
    }
  }
```

Now we know how to generate old and novel search displays and also how to delete them from our screen. Since we need to ensure that initial head placement becomes comparable for every trial, we will finally introduce our IT-Fixation game object to the experiment. By now you know how to use the Instantiate(); method. You can either adjust the createTarget(); or the createDistractor(); method to spawn the game object. Since the position of your IT-Fixation sphere highly depends on the number of circles and objects you want to use, you either have to calculate its position based on the maximum number of circles and objects, or you can simply assign a Vector3 when instantiating the object.

2.9 Processing and Collecting Responses

Correct input detection and data collection is of major interest for every research project. We will show you our implementation in the upcoming sections, using both traditional keyboard inputs and in a subsequent section using gaming controllers. Again we need to inherit from MonoBehaviour, and we will finally use Update();. This method provides the shortest time interval possible for input detection and therefore the best timing accuracy. As described in the first section, Update(); will minimize potential data loss and maximize the timing precision. The following scripts will also be used to keep track of the number of trials and blocks in our experiment.

As before, we create a new C# script which handles all the controller inputs and one that writes the data to a file. We name the first HandleSubInput and the second SaveData. To manage communication between the engine and our scripts, we need to tell Unity which input tag corresponds to which button on our keyboard, joystick, or gamepad. Again, using different pieces of hardware is easy, as long as you stick to conventional gaming devices (e.g., PlayStation, Xbox, Valve Controllers). For the sake of simplicity, we will only use keyboard input for now.

Go to Edit → Project Settings ... → Input, and Unity will list standard input configurations like mouse axis, jump, etc. These configurations are saved and can be called from our scripts whenever needed. Let us choose Fire1 to indicate that our target is rotated to the left and Fire2 for right. Go to Positive Button and replace the default setting with y button. Do the same for Fire2, but use x button instead. Close the tab afterward, and we can do some C# scripting!

Open the HandleSubInput script and create your handleSubInput(); method. We will use Input.GetButtonDown(); in an if-statement to detect user input. Now a lot of our previous work comes together. After our participants have made their decision, we will stop the timer by calling the stopRecord(); method, and we will check whether their input was correct or not.

Since we are using this script not only to detect input but to trigger the ITI as well, we have to keep one thing in mind: Input detection is only required when a search display is presented. Input at the very beginning, between blocks, or during the ITI may affect the internal logic of the experiment, leading to unexpected consequences (e.g., a trial is started within the ITI due to an accidental button press). To prevent controller input for a limited time, we will simply introduce two new Booleans.

```
public bool isBlockPause = false;
public bool isInterTrial = false;
```

Both Booleans check via a nested if-statement whether our participant is (1) currently between blocks (if false: continue) and (2) if the inter-trial fixation check is running (if false: continue). Every following if-statement records user input. If Fire1 (y button) is pressed and our flipTarget Boolean is true, we will stop the timer, and the correctResponse Booleans is not changed. If the target is flipped, but our participant says otherwise, we will change the Boolean from true to false and stop the timer as well. The same logic within the block and inter-trial checks applies to Fire2 (x button), but we introduce a simple negation operator (!flipTarget) to check the state of the Boolean!

```
if (isBlockPause == false)
  {
    if (isInterTrial == false)
      {
        if (Input.GetButtonDown("Fire1"))
          {
           if (flipTarget)
             {
               timer.StopRecord();
             }
             else
             {
               correctResponse = false;
               timer.StopRecord();
             }
          }
      // Insert the remaining methods here!
      }
  }
```

Now we need to switch the Booleans from false to true whenever necessary. Of course we could set the Booleans back to their default state by hand, but instead we will write another two short methods. They are called right at the beginning of every trial. This will make sure that input becomes detectable as fast as possible. We will also need to distinguish between novel (rnd) and old (fixed) trials in our subsequent data analysis. Therefore the isOldTrial Boolean will always be set to false when this method is called by the startTrialWithRandomValue(); method. The only difference between setBooleansRND(); and setBooleansFixed(); is isOldTrial equals to true.

```
// Call at the beginning of every novel display configuration
public void setBooleansNovel()
  {
    isOldTrial = false;
    isInterTrial = false;
    correctResponse = true;
  }

// Call at the beginning of every old display configuration
public void setBooleansOld()
  {
    isOldTrial = true;
    isInterTrial = false;
    correctResponse = true;
  }
```

Now that the keyboard input will be properly detected, we can write a script to finally collect some response data! But we are lazy programmers, so let's start with a small script to change the participant info via the Inspector. As long as you use Unity's interface to conduct your experiment, you can change the values using the Inspector. If you want to use a Build version of your experiment, you will need to implement an UI menu to change the values! We will keep it simple for this tutorial. By now you know what to do: create a script called SubInfo, attach it to an empty game object with the same name, and simply add the following lines of code:

```
public class SubInfo : MonoBehaviour
  {
    public string subID;
    public string age;
    public string gender;
    public string corrected;
  }
```

For now we will only collect age, gender, and if a person wears glasses or not. We will also assign an individual ID per participant. Since every string is declared public, we can access it from every other script or the Inspector. Now create a new folder called 1_Data in your Asset folder, and open your SaveData script next.

Our first method will build a string that contains not only the current path from which it is executed but also the information of our newly created 1_Data folder. We will also look for the subID string in our SubInfo script and finally store it as a comma-separated value (CSV) file. Note that this method is again not a void, but returns a string whenever it is called.

```
private string getPath()
  {
    return Application.dataPath + "/1_Data/" + "sub-" +
        FindObjectOfType<SubInfo>().subID + "_data.csv";
  }
```

To actually write the data, we will use using System.Collections. Generic; using System.Text;, and using System.IO;. Make sure that you access the namespace at the beginning of your script; otherwise you cannot use all methods needed for the upcoming code. This method will be to write each file's header as well as to collect the actual user input. To avoid confusion, we will use tabs (indicated by "\t") as delimiters, even though the data is saved as a CSV file. The writeData()[10]; method below needs to be called whenever a participant creates an input.

```
private void writeData()
  {
    string[][] output = new string[rowData.Count][];

    for (int i = 0; i < output.Length; i++)
    {
      output[i] = rowData[i];
    }

    int length = output.GetLength(0);
    string delimiter = "\t";

    StringBuilder sb2 = new StringBuilder();

    for (int index = 0; index < length; index++)
      sb2.AppendLine(string.Join(delimiter, output[index]));

    string filePath = getPath();

    StreamWriter outStream = File.CreateText(filePath);
    outStream.WriteLine(sb2);
    outStream.Close();
  }
```

For the sake of this tutorial, we will only store limited amounts of data,[11] precisely everything from our SubInfo script, response times from the timer, a Boolean containing the target orientation (correctResponse), a Boolean for the actual user input (subResponse), and one to distinguish between novel and old conditions (isOldTrial). You can add as many rows as you need, simply by

[10] This method was supported by the friendly people over at Stack Overflow: https://bit.ly/2YyF46j.

[11] The complete project on the GitHub page contains all the variables needed for inference.

changing the value in new string[ ] according to your needs (e.g., a global time stamp, target position, etc.). But we have to make sure that the header is written first! By now you definitely know how to do that! Spoiler alert: we will call the writeHeaderRT(); method in Start();. You may have noticed that writeData(); is called at the end of the method. This workflow ensures that the header is only written once to our file. Now go to the subInfo game object in Unity and change the subID to one. If you click the Play button, Unity will create a file within your 1_Data folder, which is called sub-1_data.csv. If you open the file, you should see only the strings of our headers.

```
public void writeHeaderRT()
  {
    string[] rowDataTemp = new string[7];
    rowDataTemp[0] = "sub-ID";
    rowDataTemp[1] = "sex";
    rowDataTemp[2] = "corrected";
    rowDataTemp[3] = "age";
    rowDataTemp[4] = "rt";
    rowDataTemp[5] = "isOldTrial"; // true for old, false for novel
    rowDataTemp[6] = "correctResponse"; // true for correct, false for incorrect
    .
    rowData.Add(rowDataTemp);
    writeData();
  }
```

The final method will simply access values/Booleans from other scripts and convert them into strings. Just make sure that header and every single cell in a row is filled in the right order. You will call this method every time our handleSubInput(); detects input.

Remember how we imported an integer from the ProvideVariable script? You can either use this example to access everything you need (which we recommend) or you can brute force your way with the FindObjectOfType<>(); method. In the upcoming code snippet, we will use this method for demonstration only, even though it is costly and may affect the performance of your experiment.[12]

```
public void saveTrial()
  {
    string[] rowDataTemp = new string[7];
    rowDataTemp[0] = FindObjectOfType<SubInfo>().subID;
    rowDataTemp[1] = FindObjectOfType<SubInfo>().sex;
    rowDataTemp[2] = FindObjectOfType<SubInfo>().corrected;
    rowDataTemp[3] = FindObjectOfType<SubInfo>().age;
    rowDataTemp[4] = FindObjectOfType<Timer>().currentTime.ToString();
    rowDataTemp[5] = FindObjectOfType<HandleSubInput>().isOldTrial.ToString();
    rowDataTemp[6] = FindObjectOfType<HandleSubInput>().correctResponse.ToString();
    rowData.Add(rowDataTemp);

    writeBehavData();
  }
```

By now, all relevant parts of the experiment are implemented. We have shown you how to create prefabs within Unity, how to pass variables from one script to another, how to generate old and novel display configurations on the fly, and how input can be properly detected and recorded. Please keep in mind that not all small pieces of the project could be covered in this tutorial. Feel free to download, adjust the experiment to your needs, or contact us if you have questions via email or GitHub.

2.10 Adding Virtual Reality Components

Virtual reality (VR) is a promising way to expand upon existing research topics as well as to address completely new questions. While the presence of an additional in-depth dimension and new hardware for data collection is an exciting opportunity for researchers, experiments may require new solutions for long solved problems (e.g., object placement and rotation, soft- and hardware integration, etc.). Unity was designed to simplify this process and is frequently used in the gaming industry for fast prototyping. With some additional tweaks and two very useful packages called SteamVR (or OpenVR) and Virtual Reality Toolkit (VRTK), we can turn the complete project into a virtual reality experiment for an HTC Vive or Oculus Rift. Both systems are supported by VRTK, but we will focus on the HTC Vive for now. If you want to turn the project into an experiment for your Oculus Rift, I highly recommend VRTKs great documentation: https://vrtoolkit.readme.io/.

First we will go to Unity's asset store and download VRTK (in Unity → Asset Store → search: VRTK → Download → Import). To this day, VRTK 4.0 is still in beta, but you can use the latest stable release of the toolkit (Version 3.3.0) and an older

[12] Best practices for performance optimization of your experiment is provided here: https://bit.ly/2VGVw2k.

release of SteamVR (Version 1.2.3, https://bit.ly/2DC3gw0) for now.[13] After you downloaded SteamVR, you have to manually import the package (Assets → Import Package → Custom Package ...). Once Unity is done, all hardware-specific assets, like threedimensional models of your controllers, the application programming interface (API) and a camera object to communicate between Unity and the head-mounted display (HMD) become accessible.

Make sure that VR support is enabled for your project (Edit → Project Setting → Player → XR Setting → Virtual Reality Supported). Since Unity does not support multiple cameras in one scene, the existing Main Camera has to be replaced by the [CameraRig] prefab (Project → SteamVR → Prefabs) in the Hierarchy tab. The [CameraRig] comes with two children: controller (left) and controller (right) and their respective 3D models.

Add the VRTK; namespace right at the top and create the following variables:

```
public GameObject subInputLeft;
public GameObject subInputRight;
```

This gives you access to all the variables needed for this tutorial. Again we will use a Boolean for input detection, but the ones provided by VRTK_ControllerEvents are automatically switched back to their default states. Go back to your HandleSubInput script, delete the if-statements for your Fire1 and Fire2 code and replace the first with:

```
(subInputLeft.GetComponent<VRTK_ControllerEvents>().triggerClicked == true)
```

Replace Fire2 with:

```
(subInputRight.GetComponent<VRTK_ControllerEvents>().triggerClicked == true)
```

If you start the experiment now, Unity will detect the hardware, and you can explore your search displays with an HMD!

Acknowledgments

This work was supported by a grant of the Deutsche Forschungsgemeinschaft (DFG PO-548/14-2 to S.P.). We thank Rebecca Burnside for her help editing a draft of this chapter.

[13] Keep in mind: The code provided here may not work properly with future versions, but we will migrate everything to the newest stable release and update the project on our GitHub page: https://github.com/nimarek.

References

1. Brockmole JR, Hambrick DZ, Windisch DJ, Henderson JM (2008) The role of meaning in contextual cueing: evidence from chess expertise. Q J Exp Psychol 61(12):1886–1896
2. Brockmole JR, Henderson JM (2006) Recognition and attention guidance during contextual cueing in real-world scenes: evidence from eye movements. Q J Exp Psychol 59 (7):1177–1187
3. Chun MM, Jiang Y (1998) Contextual cueing: implicit learning and memory of visual context guides spatial attention. Cogn Psychol 36 (1):28–71
4. Colagiuri B, Livesey EJ (2016) Contextual cuing as a form of nonconscious learning: theoretical and empirical analysis in large and very large samples. Psychon Bull Rev 23 (6):1996–2009
5. Chun MM, Jiang Y (2003) Implicit, long-term spatial contextual memory. J Exp Psychol Learn Mem Cogn 29(2):224
6. Jiang YV, Sisk CA (2019) Contextual cueing. In: Pollmann S (ed) Spatial learning and attention guidance, Neuromethods. Humana Press, Totowa
7. Chua KP, Chun MM (2003) Implicit scene learning is viewpoint dependent. Percept Psychophys 65(1):72–80
8. Kawahara JI (2003) Contextual cueing in 3D layouts defined by binocular disparity. Vis Cogn 10(7):837–852
9. Zang X, Shi Z, Müller HJ, Conci M (2017) Contextual cueing in 3D visual search depends on representations in planar-, not depth-defined space. J Vis 17(5):17–17
10. Tsuchiai T, Matsumiya K, Kuriki I, Shioiri S (2012) Implicit learning of viewpoint-independent spatial layouts. Front Psychol 3:207
11. Jiang YV, Swallow KM (2013) Spatial reference frame of incidentally learned attention. Cognition 126(3):378–390
12. Schmidt A, Geringswald F, Sharifian F, Pollmann S (2018) Not scene learning, but attentional processing is superior in team sport athletes and action video game players. Psychol Res:1–11
13. Schmidt A, Geringswald F, Pollmann S (2018) Spatial contextual cueing, assessed in a computerized task, is not a limiting factor for expert performance in the domain of team sports or action video game playing. J Cognit Enhancement:1–12
14. Geringswald F, Pollmann S (2015) Central and peripheral vision loss differentially affects contextual cueing in visual search. J Exp Psychol Learn Mem Cogn 41(5):1485–1496
15. Kleiner M, Brainard D, Pelli D, Ingling A, Murray R, Broussard C (2007) What's new in psychtoolbox-3. Perception 36(14):1
16. Peirce JW (2007) PsychoPy—psychophysics software in python. J Neurosci Methods 162 (1–2):8–13

Part II

Psychophysiological Methods

Neuromethods (2020) 151: 107–128
DOI 10.1007/7657_2019_22
© Springer Science+Business Media, LLC 2019
Published online: 11 May 2019

Using the Contralateral Delay Activity to Study Online Processing of Items Still Within View

Halely Balaban and Roy Luria

Abstract

In recent years, there has been growing research regarding the online nature of visual working memory (VWM). These online aspects are arguably the defining attributes of working memory, but they are challenging to study using traditional behavioral paradigms. One powerful tool to examine online processing in VWM is the contralateral delay activity (CDA), the ERP marker of VWM. We review studies that convincingly demonstrated that the CDA is a unique marker of VWM activity. This specificity joins the excellent temporal resolution of the CDA and the fact that it can be measured not only during memory retention but also when items are visible on the screen, to make the CDA an ideal tool for studying the online processing of items still within view. We present several lines of research that successfully utilized the CDA to uncover the role of VWM in online processing. Finally, we present basic guidelines for using the CDA to study online processes, along with examples from our recent research. We hope that this will enable more researchers to capitalize on the CDA's advantages, allowing new discoveries to be made regarding VWM as an online workspace.

Keywords Contralateral delay activity, Visual working memory, ERP, Online processing, Updating, Resetting

1 Introduction

Visual working memory (VWM) is our online workspace, responsible for the storage and manipulation of visual information [1]. VWM can hold a limited amount of information in an active state, ready to be manipulated by higher cognitive functions. Its storage capacity is extremely limited, typically estimated at only about 3–4 simple items' worth of information [2]. Two lines of work regarding this capacity limit corroborated the importance of VWM in guiding everyday behavior. First, capacity is specifically damaged in a range of conditions, including Alzheimer's disease, normal aging, attention deficit hyperactivity disorder (ADHD), and schizophrenia (e.g., [3–6]). Second, stable individual differences in capacity are tightly correlated with measures such as attentional control and fluid intelligence (e.g., [7–9]). The obvious importance of storage capacity motivated many researchers to investigate the nature of capacity limits, producing a wealth of interesting finding, sometimes even leading to fierce debates (e.g., as to whether capacity is limited by the number of items or by the overall information load; [10–12]).

Most research in the field focused on classic memory paradigms, such as change detection (e.g., [13–15]) or delayed continuous response (e.g., [16–18]) tasks, whose hallmark is the retention interval. However, VWM is involved not only when information is maintained over a retention period, but whenever we must hold visual representations in an accessible state. One example is indeed when we try to hold in mind visual information that is then removed from view, but a similar need arises in a range of situations in which the items remain visible. A prominent example of this is a recent study [19] which showed that performance on change detection tasks with and without a retention interval (i.e., a "pure" memory component) is highly correlated ($r \geq 0.8$) and they reach extremely similar capacity limits.

These results corroborate the argument that VWM's active maintenance is critical whenever we handle task-relevant perceptual input that is not stable from one moment to the next. For example, as we go about in the world, the incoming visual input changes constantly, and VWM is necessary for connecting the representations from one moment to the next, even though the relevant information still surrounds us. Moreover, every time we move our eyes, each part of the visual input changes its position across the retina, again necessitating VWM to bridge the gap [20]. One might even argue that the active nature of the representations is the defining characteristic of VWM, and in recent years, the "online" aspects of VWM were studied more closely, highlighting the "processing" aspects of VWM and not only its "storage" aspects.

Studying these online abilities creates a challenge that is hard to overcome with traditional paradigms. This is because they usually reflect only the end result of the process a representation went through. Additionally, they reflect only the outcome of a series of processing stages, most of which are actually external to working memory, e.g., perception or response selection [15]. To overcome this challenge, some researchers chose to use behavioral measures that were specifically developed to uncover the ongoing dynamics of VWM processing (e.g., [21, 22]). Here, we focus on another approach: utilizing an electrophysiological marker that allows uncovering the online nature of VWM processing, namely, the contralateral delay activity (CDA; [23–25]).

1.1 The CDA as a Marker of VWM

The CDA is an ERP marker, first introduced by Vogel and Machizawa in 2004 ([23]; although an earlier attempt to study VWM using ERPs was made by [26]). They employed a lateralized change detection task, in which the visual input was equated on both sides of the screen, but participants were asked to attend to only one side, as indicated by arrow cues presented before the memory array onset. ERPs were time-locked to the presentation of the memory array, and difference waves were computed by subtracting the activity in electrodes ipsilateral to the attended side from those contralateral to the attended side. About 300 ms after the onset of the memory array, a negative slow wave emerged in occipital-parietal electrodes (originally, at the OL/OR sites, corresponding to PO7/8 in the extended 10–20 system), persisting throughout the retention interval (hence, "contralateral delay activity").

Importantly, the CDA allows isolating VWM activity from processes that precede or follow it. Because the CDA is a difference wave, any activity that is common to the ipsilateral and contralateral sides (e.g., low-level visual processing of the similar items that are presented in both sides) is cancelled out by the subtraction, meaning that processes that precede attentional focusing, such as perception, will not be reflected in the CDA. Because the CDA is measured before a response can be prepared (e.g., during the retention interval), it cannot reflect response-related processes.

The key characteristic of the CDA is that its amplitude increases (i.e., becomes more negative) as more items are held in VWM [23]. Critically, this is exclusive to items that are actually held in VWM and does not simply reflect task difficulty: the set-size-dependent increase stops when average capacity is reached. Furthermore, the set-size effect (i.e., the difference in amplitude between set-sizes) is tightly related to individual capacity estimates, producing a correlation of ~0.6 in a meta-analysis [25]. Thus, the more information an individual can hold in VWM, the more the CDA will continue to rise. Similarly, the CDA has a reduced amplitude in incorrect trials compared to correct trials [24], in which participants presumably hold more information in VWM. All of these findings suggest that the CDA amplitude can be used as an ERP marker of VWM, creating a fruitful new way to study this cognitive process.

Over the years, several studies supported the notion that the CDA reflects VWM activity, by ruling out several confounding factors. First, the CDA does not reflect the number of active locations but rather the number of items held in VWM (although these two factors are usually confounded). When four items were presented across two consecutive memory arrays (two in each), the CDA amplitude was the same regardless of whether each item had a unique location or the items in the second memory array employed the same locations as those in the first array [27]. Thus, the CDA reflected the number of items (four in both cases) and not the number of locations (either two or four). Furthermore, when two distinct items are placed one on top of the other (e.g., an oriented bar on top of a colored square), the CDA is as high as when each of these items has a unique location [28–31]. Second, despite the spatial cue employed for extracting the difference waves, the CDA does not reflect the focus of spatial attention, because when participants had to encode items in one hemifield and then in the other, the CDA corresponded to the memory load across both hemifields [32]. Similarly, the CDA is unaffected by the distance between items [24], even though when items occupy a larger space, more spatial attention is needed. Third, even though the spatial cues might induce microsaccades in the relevant direction, these small eye movements are not the source of the CDA [33]. Fourth, the CDA is unaffected by the contrast of the items in the task [27, 34, 35]. Because brighter colors produce better change detection performance without affecting the CDA, this further shows that the amplitude does not reflect task difficulty. All of these results confirm that the CDA is specific to VWM processes, a unique feature which allowed it to serve as an excellent marker of VWM.

Aside from its general applicability in VWM research due to its specificity, the CDA has two critical attributes that make it ideal for studying online processing and manipulation. As mentioned above, the CDA is unaffected by whether or not the items are visible on the screen: similar amplitudes are found in a change detection that lacks a retention interval [19], and even in classic change detection memory tasks, the CDA does not abruptly change when the memory array disappears (although its amplitude might slightly and gradually decrease throughout the retention interval). The second feature is the precise temporal resolution of the ERP technique, meaning that the CDA tracks the development of representations over time instead of only giving a static view (e.g., [9, 28, 31]). Taken together, this enables the CDA to be used to uncover the moment-by-moment dynamics of the processing of information that continues to be visible.

1.2 Studies Utilizing the CDA in Online Processing Paradigms

Over the past few years, the CDA has been successfully used to study the processing of information still within view. Research on the storage and manipulation of visible information in a range of paradigms offered new insights on the role of VWM in these processes. Some of the studies revealed hidden involvement of VWM that was too delicate to study using behavioral tools alone, and others could identify several subprocesses that could not be identified with a single response measure.

1.2.1 Visual Search

While VWM had a central role in theories of visual search, behavioral evidence surprisingly suggested that search can be efficient even without relying on VWM to store the search items [36]. This puzzle was solved by research showing that the CDA is present when participants perform a lateralized visual search task, indicating the involvement of VWM in the search process [37]. This study further demonstrated that higher VWM capacity leads to more efficient search, which is portrayed not only behaviorally but also in faster-rising CDA amplitudes and, as shown later [38], also by lower CDA amplitudes. Accordingly, VWM involvement in visual search increases as the task becomes more difficult [38, 39], either "objectively" (when the task itself changes, e.g., from localization to identification) or "subjectively" (on slower reaction times trials of the same task). Subsequent research relied on the CDA to examine the processing of targets versus distractors in visual search, finding that although distractors draw attention, only targets elicit a CDA component, suggesting that only targets are stored in VWM during search [40]. The CDA was later used to study the influence of VWM training on visual search performance, again finding that improved search efficiency following VWM training decreases the reliance on VWM during search [41].

1.2.2 Multiple Object Tracking (MOT)

When participants track a subset of identical moving items, VWM is logically involved in their ongoing maintenance even though the items remain visible. Indeed, the CDA can be measured during MOT, with an increase in amplitude as more targets are tracked, until the individual capacity limit is reached [42]. The maintenance of items could be dissociated from the active updating of their locations, based on a surplus activity when the items are moving that was spatially distinct from the CDA [43].

The CDA reflects the current number of items being tracked, such that when targets are added or removed during the tracking period, the CDA amplitude increases or decreases accordingly [44]. The fact that the CDA can change online with the actual number of items being tracked allowed it to serve as an index for the potential sources of task difficulty in MOT, which behavioral performance could not distinguish between (because lower performance can be the result of any number of reasons). Specifically, when more distractors were added, the CDA indicated that the difficulty was due to swapping targets and distractors, while when movement speed increased, the CDA indicated the difficulty was due to a higher chance of targets being "dropped" [45]. While dropping was reflected in a lower CDA amplitude because less information was held in VWM, swapping did not affect the CDA, because participants were tracking the same number of items, simply the wrong ones.

1.2.3 Online Grouping

The CDA amplitude rises when more information is stored in VWM, but it seems to reflect the number of objects or "chunks," instead of the number of features. For example, the amplitude of a colored shape when participants maintain both color and shape is the same as the amplitude of a black shape in a shape-only task [30]. This makes the CDA an excellent tool for studying grouping in VWM, because its amplitude will be lower following the integration of information into one "chunk."

Indeed, several studies manipulated grouping by using Gestalt cues while monitoring the CDA. The results demonstrated that even when the grouping cues were task-irrelevant, VWM was sensitive to these cues and held an integrated representation of the grouped items. Specifically, joint movement ("common fate"), which is a strong Gestalt cue, caused separated items to be grouped in VWM: when two colored squares met and moved together, the color-color conjunction stimuli gradually became integrated in VWM after the joint movement [31]. This study also showed that this integration was not purely perceptual but was based on the movement history, because when the colors only met and did not move together, their representations remained separate, despite resulting in the same final visual input as in the joint movement condition.

A similar study manipulated Gestalt grouping cues of the different parts of a single object [46]. When two halves of a random polygon met and moved together, their integration was immediate. This deviates from what was observed for color-color conjunctions, whose integration took time [31]. Another difference between the two types of stimuli is the need for the joint movement for integration to take place. In a different study, color-color conjunctions that were presented together without movement were partially but not completely integrated [30]. Conversely, two shape-halves that were presented adjacently, to create one whole polygon, were perfectly integrated in a task that did not include movement [46]. These results suggest a dissociation between the integration of different features of a single object, which is fast and presumably mandatory, even for complex items, and the integration of distinct objects into one "chunk," which takes time to complete and requires strong grouping cues. Subsequent research demonstrated that online integration depends on strong grouping cues not only for color-color conjunctions but also for different-dimension stimuli such as orientation-color conjunctions (a tilted bar on top of a colored square; [28]) and that grouping is affected by the global context of the task, as determined by the other conditions included in the experiment [29].

1.2.4 VWM Resetting and Updating

One important aspect of online processing is the ability to change VWM's representations according to changes in the actual items in the environment. Recently, we used the CDA to distinguish between two processes that can modify VWM's representations, namely, updating and resetting. When a represented item changes, the change can either be incorporated into the existing representation in an updating process, or the original representation can be discarded and a new post-change representation established, in a process we termed "resetting" [47]. We argued that for VWM to be able to access the appropriate representation and update it, the representation must be mapped to a unique object and, if this mapping is invalidated, VWM must discard the representation and start over, by resetting. Accordingly, when a single shape moved coherently but then separated into two independently moving parts, we observed a sharp decrease in CDA amplitude, followed by a gradual recovery [47]. Presumably, the shape was originally represented as one object in VWM along with a single mapping to support that representation (as indicated by a CDA amplitude as low as a single shape-half), but following the separation, there were now two independent objects, none of which corresponds to the original object. This caused VWM to remove the integrated object, as indicated by the drop in CDA amplitude, and to encode the shapes as two separate representations (as indicated by a CDA amplitude as high as two separate shape-halves). These results are shown in Fig. 1, where it can also be observed that the CDA is extremely stable regardless of whether the items are visible on the screen (until 1300 ms from trial onset) or are only held in VWM.

In a set of control experiments, we further demonstrated that separation is neither necessary nor sufficient to produce a CDA-drop, meaning to trigger a resetting process [47]. For example, as can be seen in Fig. 2, resetting took place also after object switching, where one relevant object had to be discarded and another one abruptly replaced it. Additionally, when the two shape-halves first moved separately, allowing them to be individuated, and only then met, moved together, and re-separated, the CDA indicated there was no resetting, presumably because two independent mappings could be created to begin with (during the initial separate movement phase). Finally, we confirmed that the CDA-drop is specific to situations involving a loss of object-to-representation mapping, while extremely similar situations result in updating, as long as they allow the mapping to hold [48]. These results are shown in Fig. 3, corroborating the interpretation of the CDA-drop as a unique marker of the resetting process and the mapping invalidation that triggers it.

2 Methods

Because most of the studies that used the CDA relied on classic VWM retention tasks (e.g., change detection), it is important to highlight some guidelines for using it in the less conventional (although gradually growing) way of examining the processing of items within view (*see* **Note 1**). Below, we discuss several issues to consider when designing an online processing CDA experiment. We use our recent CDA studies of updating and resetting as an example.

2.1 General Task Structure

Broadly speaking, there are two different logics for using the CDA to study online processing, dictating two different types of tasks. The first is to use the CDA as a marker for the involvement of VWM

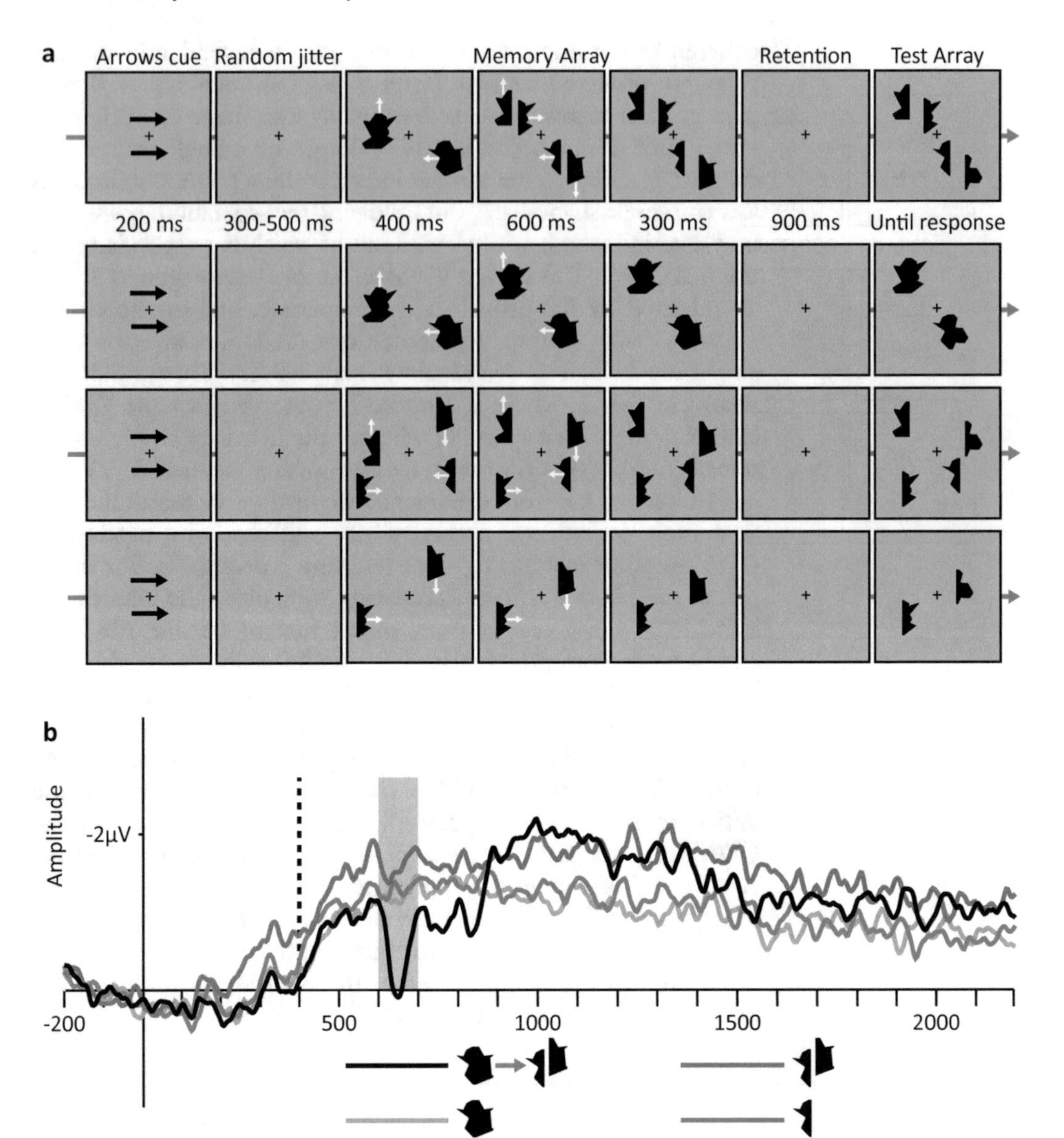

Fig. 1 The paradigm and results of Experiment 2 from Balaban and Luria [47]. (**a**) Examples of trials in the different conditions (white arrows indicate movement directions and were not presented). From top to bottom: separating polygon, integrated polygon, two polygon-halves, and one polygon-half. Note that the task was a shape change detection and the movement was completely irrelevant. (**b**) Grand-averaged CDA waves (averaged across the P7/8, PO3/4, and PO7/8 electrodes), time-locked to memory array presentation. Negative voltage is plotted upward. The vertical dashed line depicts the time of separation. The "drop" time window is depicted by a gray rectangle, and during it, the amplitude of the separating polygon condition significantly dropped

in processes other than classic memory maintenance. For example, one can investigate the involvement of VWM in a visual search process. For this goal, the preferred task is a bilateral version (see below) of a classic task involving the studied process (e.g., visual search).

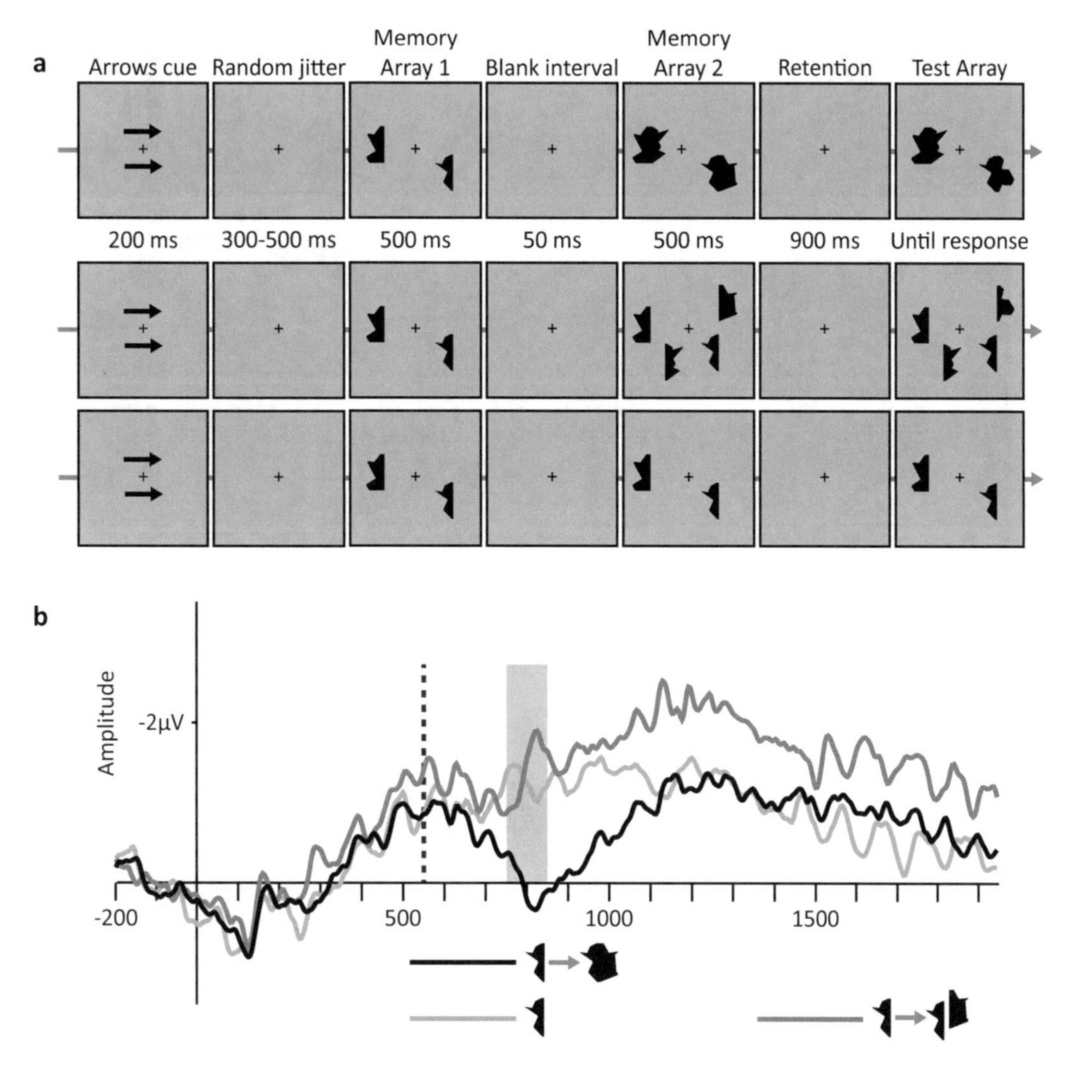

Fig. 2 The paradigm and results of Experiment 3 from Balaban and Luria [47]. (**a**) Examples of trials in the different conditions. From top to bottom: switch, add, and half-polygon repeat. Note that participants were instructed to remember only the shape(s) in the second memory array (making the first one irrelevant) and that the switch and add conditions include the same information, but in different locations. (**b**) Grand-averaged CDA waves (averaged across the P7/8, PO3/4, and PO7/8 electrodes), time-locked to memory array presentation. Negative voltage is plotted upward. The vertical dashed line depicts the time of the presentation of the second memory array. The "drop" time window is depicted by a gray rectangle, and during it, the amplitude of the switch condition, but not the add condition, significantly dropped

The alternative logic is to use the CDA to examine subprocesses of VWM, such as updating, resetting, and other online manipulation processes. If this is the goal, the task is usually a classic VWM task—typically change detection or delayed continuous response—that includes some additional manipulation during the memory array presentation. The VWM task guaranties the involvement of VWM, and the additional manipulation is used to trigger the

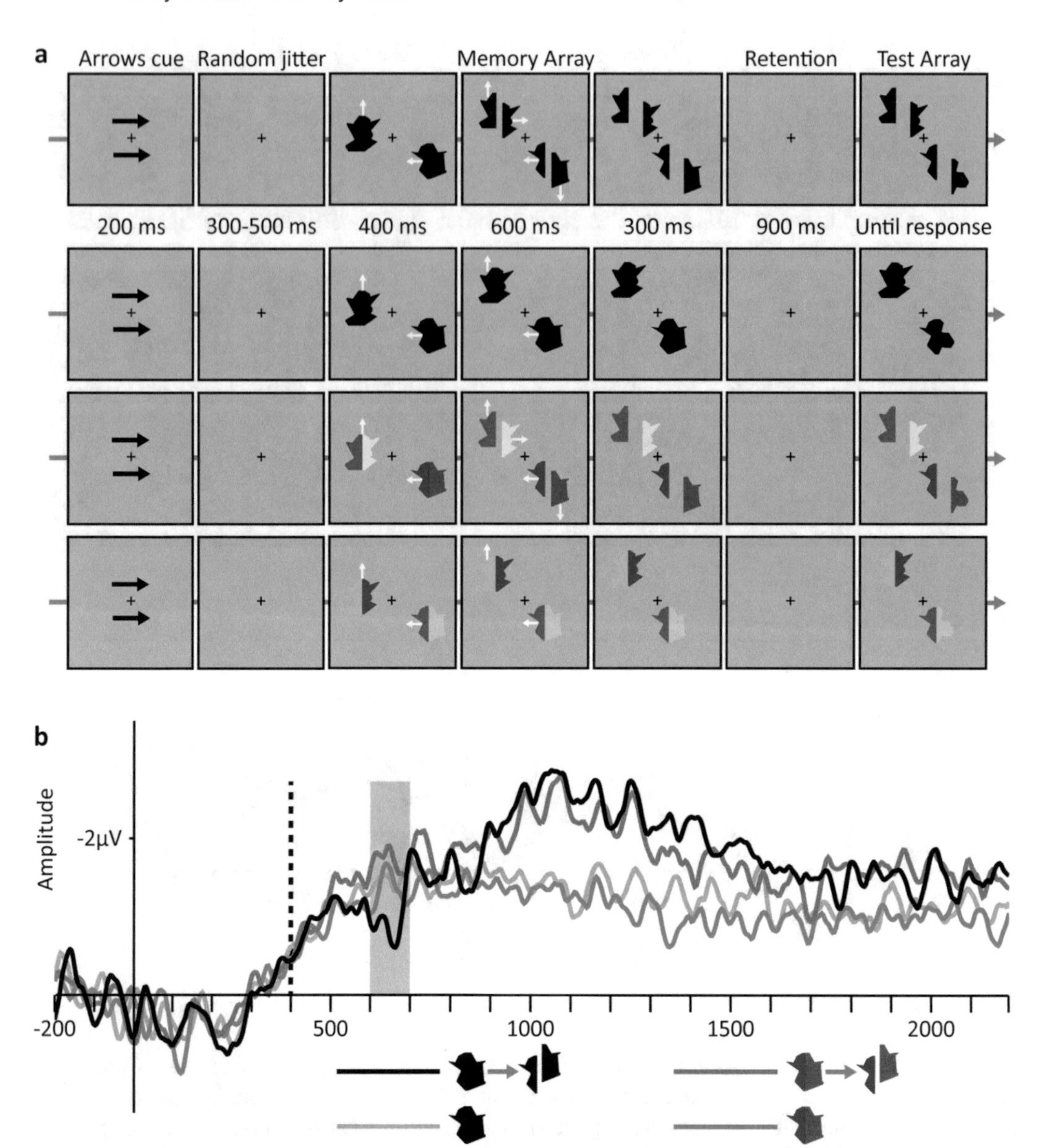

Fig. 3 The paradigm and results of Experiment 1 from Balaban et al. [48]. (**a**) Examples of trials in the different conditions. From top to bottom: black separating polygon, black integrated polygon, bicolored separating polygon, and bicolored integrated polygon. Note that the task was a shape change detection and the colors and movement were completely irrelevant. (**b**) Grand-averaged CDA waves (averaged across the P7/8, PO3/4, and PO7/8 electrodes), time-locked to memory array presentation. Negative voltage is plotted upward. The vertical dashed line depicts the time of separation. The "drop" time window is depicted by a gray rectangle, and during it, the amplitude of the black separating polygon, but not the bicolored separating polygon, significantly dropped

process of interest. An example is using a change detection task with movement that serves as a grouping cue, to study online integration in VWM (e.g., [31]). Note that in this case, the grouping manipulation was task-irrelevant (participants only monitored for a potential color change), to examine whether the items will be integrated despite their joint movement being irrelevant. Indeed, the additional manipulation in this kind of CDA studies can be orthogonal to the memory task, but it can also be a task-relevant manipulation.

Once the task is chosen, it is critical to adapt it to a lateralized display. As mentioned above, the CDA is a difference wave, allowing to control for low-level visual processing. To enable a correct subtraction, each trial should include similar stimuli in the left and right sides of the central fixation point. It is important to equate the left and right sides in terms of factors that are expected to affect the CDA: the number of items, their type (e.g., complex stimuli have higher CDA amplitudes; [34]), and the experimental condition they belong to (e.g., whether they include only relevant items or also distractors; [9]). Conversely, although the two sides should be generally quite similar, it is not necessary to perfectly balance them in terms of low-level features, such as the items' exact brightness, color, size, location, or centrality (*see* **Note 2**). Duplicating the display of one side in the other side is not recommended, because it might lead participants to adopt unwanted strategies (e.g., attending to the wrong side).

There are two main ways to make the task lateralized. The dominant one is to present spatial cues (e.g., arrows above and below the fixation) before the onset of the items in each trial. These cues should be 100% valid, and participants must be aware of that (*see* **Note 3**). If pre-cues are used, it's necessary to include a blank screen (with only a fixation point) between the cues and the task, and as explained below, the duration of this blank screen should be variable. Typically the duration is randomly jittered between 300 and 500 ms, making sure there is enough time to serve as the baseline for the ERP analysis, which is usually the 200 ms immediately before the task. The duration cannot be constant, otherwise the cues will precede the task by a constant lag, and any time-locking to the task locks to the cues as well, making interpretation of the ERPs problematic.

Another possibility is to define the target side by a task-irrelevant feature (usually color), which remains constant throughout the experiment. For example, in a shape task, the cue could be color, such that on each trial, all of the shapes in one side (left or right, randomly determined) have color A and all the shapes in the other side have color B, with half of the participants instructed to attend to color A and half to color B.

When considering the number of trials included in the task, it is important to remember that trials containing artifacts (mostly eye movements and blinks) and/or incorrect responses are excluded from the analysis. We usually include 200–250 trials per condition, leaving a minimum of roughly 100 trials per condition per subject for the most difficult conditions and the noisiest subjects (*see* **Note 4**). Importantly, including a large number of trials allows examining the individual datasets, identifying the effect of interest not only in the grand average, and testing for possible individual differences, as has been successfully done with the CDA in the past (see [25], for a review).

To show the influence of the number of trials on the CDA in online processing tasks, we reanalyzed the data from Experiment 2 in [47], which included 15 blocks of 60 trials (with about 225 trials per condition per subject, before removing artifacts and incorrect responses). We compare the CDA of two conditions in the task (a resetting condition and a baseline condition) for the entire set of blocks with the CDA obtained from dividing the experiment in two halves (blocks 1–7 versus 8–15; about 112 trials per condition per subject, before removing artifacts and incorrect responses) or in four quarters (blocks 1–3, 4–7, 8–11, and 12–15; about 56 trials per condition per subject, before removing artifacts and incorrect responses). As can be seen, using ~100 trials per condition is enough for clean CDA waveforms at the grand average level, as well as for a clear CDA-drop following resetting. Conversely, using ~50 trials produces waveforms that are too noisy, although the CDA-drop can still be observed. Aside from the grand average of all 12 subjects, we also show the individual datasets of two participants: the one with lowest rejection rate (1.4%) and the one with the highest rejection rate (18.7%), to demonstrate the effect of the number of trials at the individual level. The results are shown in Fig. 4, and they suggest that the CDA-drop is still present at the individual level with ~100 trials per condition, but the difference between the different conditions in the CDA becomes quite blurry.

2.2 Avoiding Eye Movements and Blinks

An important issue in visual paradigms generally, and online processing tasks specifically, is the need to control for eye movements and blinks. These artifacts are detrimental not only because of the large direct effect they have on the EEG waveforms but also because of indirect changes they likely cause. This is because an eye movement or a blink, by definition, causes the participant to see different visual information than intended: an eye movement causes the middle of the display to be shifted from the central fixation to another point on the screen, which is critical for the CDA as a difference wave, and a blink means seeing nothing for ~300 ms. This issue is especially important for online processing paradigms, because they typically involve a much longer presentation time than memory paradigms, producing more opportunities for artifacts to occur. Note that mathematically removing these artifacts (e.g., by using ICA) does not solve the problem of these trials being different in terms of their cognitive processing.

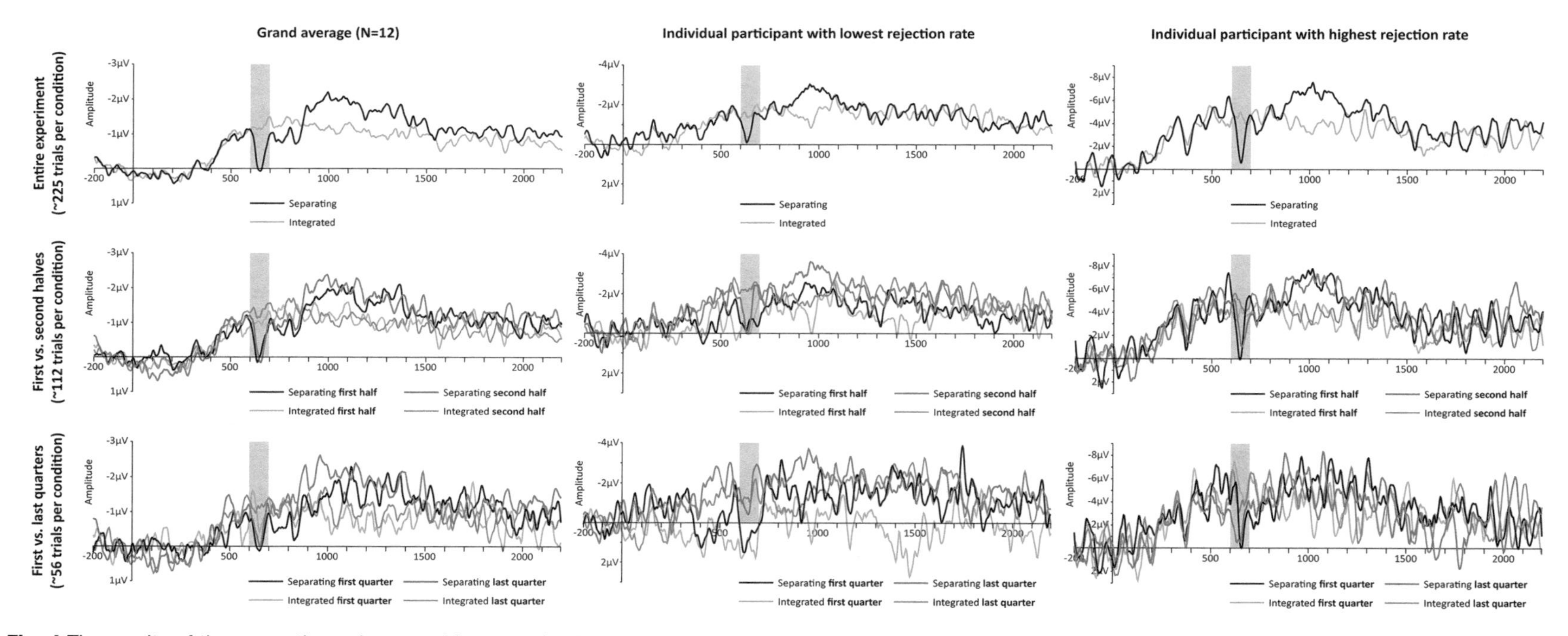

Fig. 4 The results of the separating polygon and integrated polygon conditions of Experiment 2 from Balaban and Luria [47], by the number of trials included. The vertical dashed line depicts the time of separation. The "drop" time window is depicted by a gray rectangle. The left column shows the grand average of all 12 participants, the middle column shows the individual waveforms for the participant with the lowest rejection rate, and the right column shows the individual waveforms for the participant with the highest rejection rate. The top row shows the results for the entire experiment, the middle row shows the results for the first and second halves of the experiment, and the bottom row shows the results for the first and fourth quarters of the experiment. Note that each column has a different scale

Before listing several technical ways to minimize eye movements and blinks, we would like to note that the best way to reduce them is by instructions (*see* **Note 5**). Before the experiment begins, participants must be told that eye movements and blinks are to be avoided, preferably with a short explanation as for the reasons and a demonstration of the effect they have on the raw EEG data. Additionally, the practice phase of the experiment should be used to help participants getting used to not moving their eyes and blinking only between trials (see below), with ongoing feedback from the experimenter. Feedback should also be used throughout the experiments (in real time or at least between blocks), drawing the participant's attention to events of breaking fixation or blinking in the wrong time. With these artifacts, like with any other source of noise, it is always better to not record them in the first place than to struggle removing them during offline processing [49].

Alongside the appropriate task instructions, there are practical tools that can help participants maintain these instructions more easily, mainly by providing them with enough opportunities to rest their eyes throughout the experiment. First, it is critical to include a designated blinking time after each response, by presenting a blank screen for about 1–2 s before the onset of the next trial. Second, breaks should follow every few (no more than 5) minutes of trials. Third, the length of the task itself, including breaks, should be no longer than about an hour and a half. If more trials are needed than can be included in this time frame, the experiment can be divided into two sessions. Fourth, in tasks including movement, which induces more eye movements, if possible, it is better to restrict the movement direction so that items do not move away from the fixation point but only toward it or vertically. Finally, it is always recommended to examine the EOG, to ensure that it is not the source of the effect of interest.

Having highlighted the importance of controlling for eye movements, it is also important to note that the CDA does not reflect these artifacts [33]. Once the proper instructions and task structure are used, the remaining artifacts are minimal and can be removed using standard procedures, such as a moving window peak-to-peak analysis (see [49]). Furthermore, our data suggests that eye movements are not the source for the CDA-drop, even in online paradigms that include movement. Figure 5 shows the horizontal EOG for two experiments, i.e., difference waves of the left EOG electrode minus the right EOG electrode (so that any leftward eye movement creates a negative deflection and any rightward eye movement creates a positive deflection). The results allow comparing situations with and without a CDA-drop (i.e., the switch condition vs. the add condition of Experiment 3 from [47] and the black separating polygon condition vs. the bicolored separating polygon condition of Experiment 1 from [48]) in terms of eye movements. As can be seen in the figure, eye movements are quite minimal across

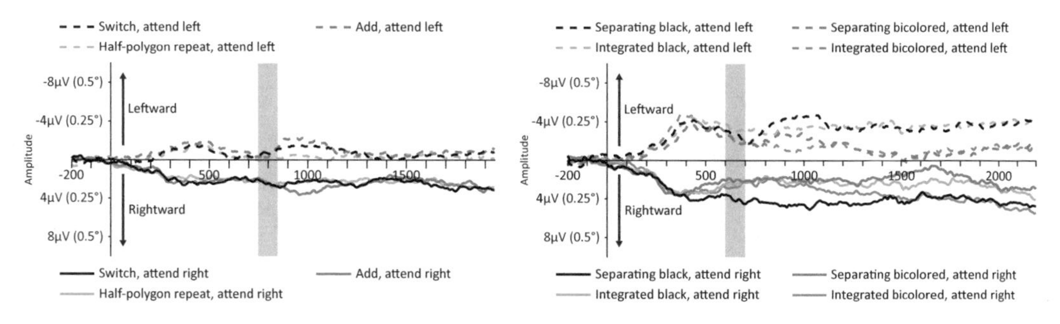

Fig. 5 HEOG waveforms (left minus right) for Experiment 3 from Balaban and Luria [47] (left) and Experiment 1 from Balaban et al. [48] (right), by condition and attended side. The "drop" time window is depicted by a gray rectangle. As a reference point, we used 4 and 8 μV, which correspond to about 0.25° and 0.5° of drifting eye movement, respectively [53]

all conditions, and furthermore, during the relevant time window of the CDA-drop, eye movements are very similar between the different conditions and therefore cannot distinguish between the two CDA patterns.

2.3 Data Analysis

CDA waveforms are typically comprised of only trials that include a correct response (due to the assumption that errors might involve other processes that would complicate the interpretation of the results) and do not include any artifacts: eye movements, blinks, and general noise, all of which are to be removed at the offline processing stage. Activity is time-locked to the onset of the task (e.g., the presentation of the memory array) and baselined to a portion of the blank interval preceding the task. The trials are divided into attend-left and attend-right and averaged by condition. Then, for each electrode pair, the contralateral activity is defined as the average of the right-hemisphere activity in attend-left trials and left-hemisphere activity in attend-right trials, and the ipsilateral activity is defined as the average of the left-hemisphere activity in attend-left trials and right-hemisphere activity in attend-right trials. The CDA is computed as the contralateral minus ipsilateral activity, and mean amplitude of this difference wave is the dependent variable of main interest (*see* **Note 6**). We next discuss the electrodes and timing in which the relevant effects are expected to appear.

The CDA is a parietal-occipital component, mostly observed over the P7/8, PO3/4, and PO7/8 (sometimes referred to as OL/OR) electrode pairs [24]. Usually the strongest activity is in the PO7/8 electrodes, and therefore they are used for statistical analyses. Another common practice is to use the average of all three electrode pairs. While this might attenuate the effects in some cases, it also guarantees that extreme patterns in a single electrode pair will not dominate the interpretation. Figures 6 and 7 present the spatial distribution of the CDA in two of our experiments (using

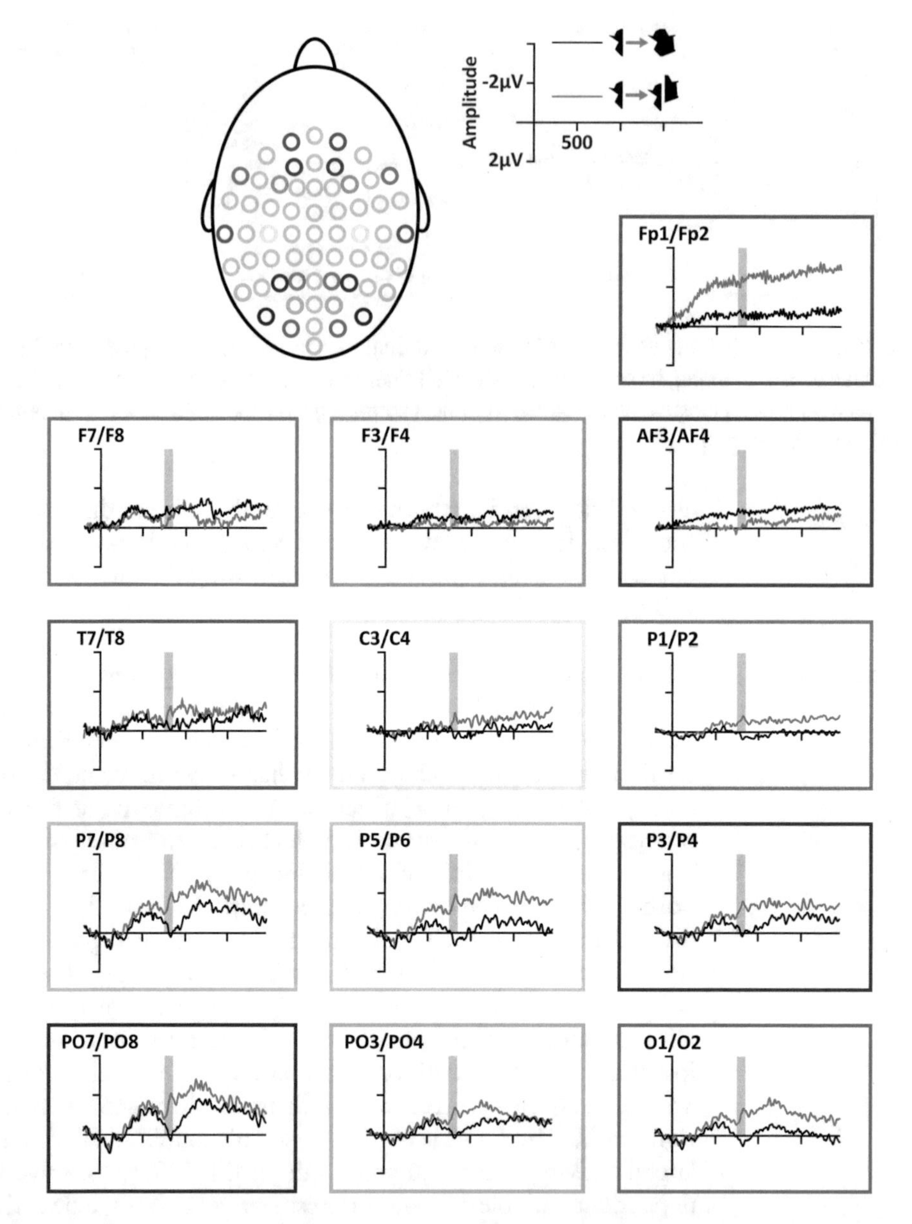

Fig. 6 CDA waveforms for two conditions from Experiment 3 from Balaban and Luria [47], by condition and electrode pair (midline electrodes are not presented). The "drop" time window is depicted by a gray rectangle. The colors of the frames correspond to the colors of the electrodes in the schematic drawing of the head, to show the position of the pair

32 scalp electrodes in a subset of the extended 10–20 system), showing that the spatial distribution is very similar for items' maintenance, updating, and resetting.

The time window used for classic CDA tasks is typically from 300 ms after memory array onset, allowing the CDA time to stabilize, and until the end of the retention interval. However,

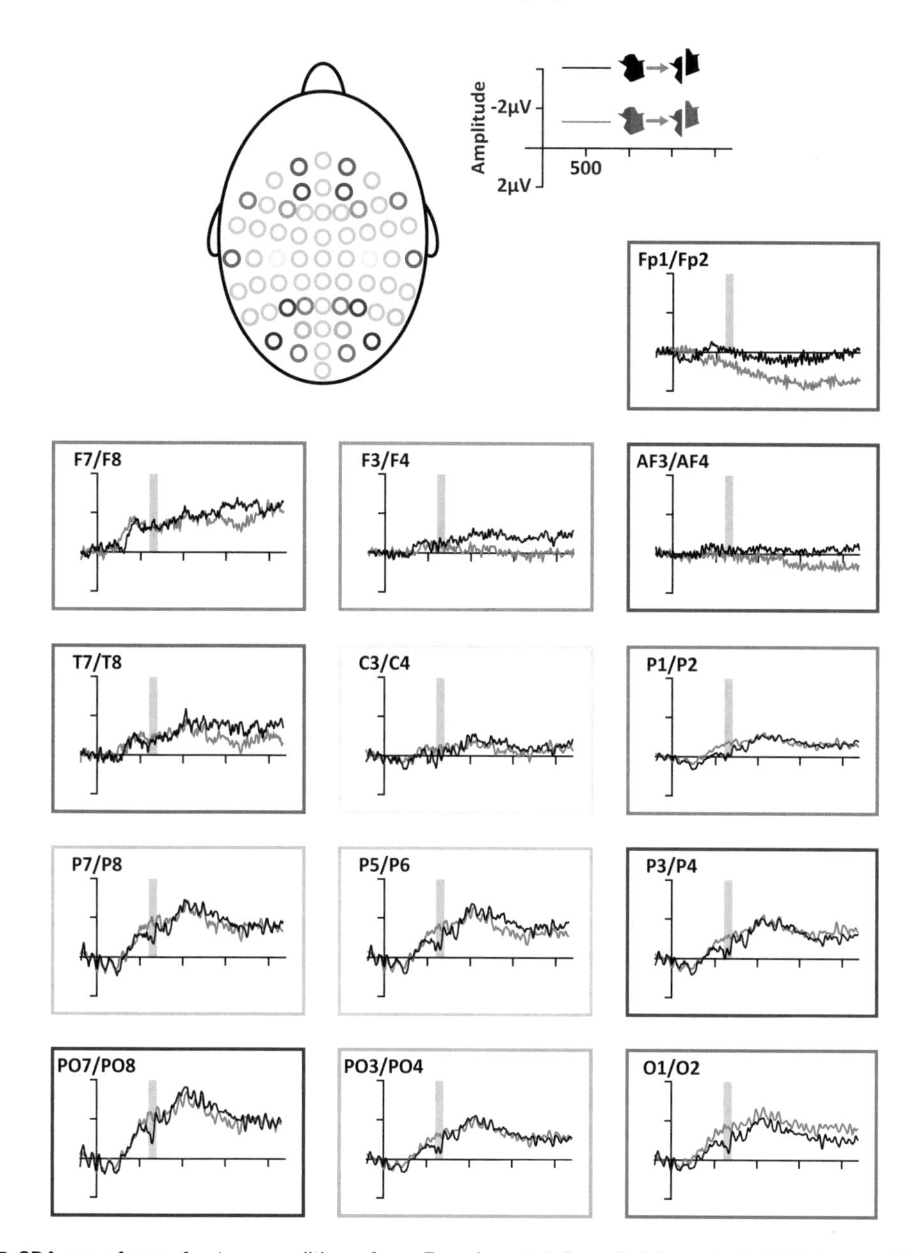

Fig. 7 CDA waveforms for two conditions from Experiment 1 from Balaban et al. [48], by condition and electrode pair (midline electrodes are not presented). The "drop" time window is depicted by a gray rectangle. The colors of the frames correspond to the colors of the electrodes in the schematic drawing of the head, to show the position of the pair

when online processing tasks are used, this is not always suitable. Each task and research question will have slightly different time windows, but several general guidelines can be provided. First, the CDA takes about 200 ms to respond, for example, to initially rise [9], or to drop in VWM resetting [47, 48]. Therefore, time windows should start about 200 ms after the relevant event. Second, if

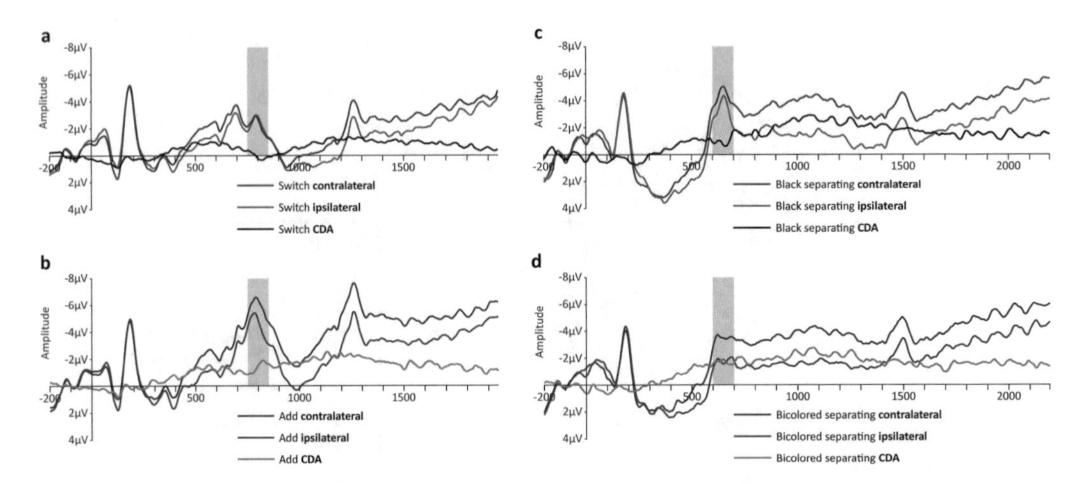

Fig. 8 Contralateral and ipsilateral waveforms for two conditions from Experiment 3 from Balaban and Luria [47] (left) and Experiment 1 from Balaban et al. [48] (right) and the resulting CDA difference wave. The "drop" time window is depicted by a gray rectangle. (**a**) The switch condition of Experiment 3 from Balaban and Luria [47], which resulted in a CDA-drop. (**b**) The add condition of Experiment 3 from Balaban and Luria [47], which did not result in a CDA-drop. (**c**) The black separating polygon condition of Experiment 1 from Balaban et al. [48], which resulted in a CDA-drop. (**d**) The bicolored separating polygon condition of Experiment 1 from Balaban et al. [48], which did not result in a CDA-drop

online dynamics are to be examined, it is usually best to divide the trial into several time windows of interest. For example, in Drew et al.'s [44] MOT paradigm, after an initial tracking period, the targets could change (e.g., adding additional targets). Comparing the CDA before and after this change allowed the researchers to demonstrate the differential ongoing tracking load (e.g., the CDA was higher after more targets were added). Third, each time window should be long enough to minimize the influence of transient fluctuations. Finally, if the resetting process is of interest, the CDA-drop is expected between 200 and 300 ms after the relevant event (e.g., object separation).

2.4 Contralateral and Ipsilateral Waves

As mentioned above, VWM studies usually focus on the CDA as a difference wave. This is mainly because this allows controlling for low-level visual processing, making sure any observed effects are specific to VWM. However, it is always recommended to examine the contralateral and ipsilateral waveforms as well. Figure 8 presents these waves for two of our experiments. Comparing updating and resetting reveals that the CDA-drop originates from an interesting pattern. Namely, while the contralateral wave rises (i.e., becomes more negative) after the critical even (e.g., object separation) in both resetting and updating situations, in updating, it maintains its distance from the ipsilateral wave that rises to the same degree, and thus the CDA is unaffected, while in resetting, the ipsilateral wave rises more than the contralateral wave, and the decreased distance between them is reflected in the CDA-drop. The sharp rise of the contralateral and ipsilateral waves is similar to what happens after the items are first presented, and both effects occur ~200 ms after the relevant event, i.e., in the time window of N1 or N2 components [50], but notably, in most attentional effects, the negative deflection is larger in the contralateral side, while during resetting, we see the increased activity on the ipsilateral side. It might be that the negative deflection reflects the reinstatement of the object-to-representation mappings, similar to their initial creation when the items are first encoded. Alternatively, it could be that there are two superimposed processes that are manifested in the same electrodes, an attentional effect which is the same for the contralateral and ipsilateral sides, as is suggested by updating situations, and another effect of VWM that removes the previous contralateral surplus negativity, thus bringing the contralateral and ipsilateral waves to the same point. This is an interesting path for future research.

3 Notes

1. The CDA emerges when VWM is active, meaning the task has to involve VWM. Specifically, the task has to encourage active *visual* maintenance of the items. If verbal rehearsal of the memoranda could be performed, it should be prevented, e.g., by including verbal suppression. This is especially problematic when using long presentation times with familiar namable stimuli (e.g., colors).
2. If possible, do not present the items at the same locations throughout the task. Instead, randomize locations across trials, or choose the locations on each trial from a large set of locations. Our experience is that using predetermined locations might attenuate the CDA.
3. When explaining the task to participants, some of them (especially those who perform many psychology experiments) become suspicious as for the bilateral nature of the task. They might not say so, but they sometimes do not believe that the uncued side is indeed irrelevant. Because the CDA depends on an effective manipulation of attention, it is critical that participants try the best they can to completely ignore the irrelevant side and not encode it in VWM. Usually the best way to convince them that there is no "trick" and the cues are to be trusted (apart from specifically telling them that) is by explaining the logic of difference waves: a surplus activity of VWM would be present in the contralateral side only if they manage to attend to the relevant side.

4. This large number of trials per condition means the experiment will be quite long and participants tend to get tired. Make sure to encourage them between blocks to reestablish their engagement in the task. Offer them to take breaks that are not too long and not too short, so they can remain focused. Getting them as "on board" as possible is highly recommended and is also the best way to minimize alpha waves that reflect fatigue.
5. Because only one side of the screen is relevant on each trial, a very natural way to perform the task is by moving one's eyes to fixate on the relevant side, which will completely eliminate the CDA and therefore cannot be a legitimate strategy. This points to the fact that the way in which participants should perform the task is often unintuitive for them, and this should be made explicit when explaining the task. The practice phase of the experiment should be long enough to allow participants time to get used to not only the task itself but also its technical constraints, especially not moving their eyes and not blinking. A small percentage of all participants do not succeed in holding fixation without blinking and therefore cannot begin the experiment at all. People that wear contact lenses daily are more prone to this (even when using glasses, as should be done in these experiments), because they are used to blink a lot.
6. Custom scripts for analyzing the CDA using EEGLAB [51] and ERPLAB [52] can be found in our lab website: https://people.socsci.tau.ac.il/mu/royluria/cda-package/.

Acknowledgments

This research was supported by an Israel Science Foundation (grant number 862/17) to R.L. and an Azrieli Fellowship to H.B.

References

1. Baddeley AD, Hitch G (1974) Working memory. Psychol Learn Motiv 8:47–89
2. Cowan N (2001) The magical number 4 in short-term memory: a reconsideration of mental storage capacity. Behav Brain Sci 24 (1):87–114
3. Parra MA et al (2011) Specific deficit of colour-colour short-term memory binding in sporadic and familial Alzheimer's disease. Neuropsychologia 49(7):1943–1952
4. Jost K et al (2011) Are old adults just like low working memory young adults? Filtering efficiency and age differences in visual working memory. Cereb Cortex 21(5):1147–1154
5. Martinussen R et al (2005) A meta-analysis of working memory impairments in children with attention-deficit/hyperactivity disorder. J Am Acad Child Psychiatry 44(4):377–384
6. Johnson MK et al (2013) The relationship between working memory capacity and broad measures of cognitive ability in healthy adults and people with schizophrenia. Neuropsychology 27(2):220–229
7. Cowan N et al (2005) On the capacity of attention: its estimation and its role in working memory and cognitive aptitudes. Cogn Psychol 51(1):42–100
8. Fukuda K et al (2010) Quantity, not quality: the relationship between fluid intelligence and

working memory capacity. Psychon B Rev 17 (5):673–679
9. Vogel EK, McCollough AW, Machizawa MG (2005) Neural measures reveal individual differences in controlling access to working memory. Nature 438(7067):500–503
10. Luck SJ, Vogel EK (2013) Visual working memory capacity: from psychophysics and neurobiology to individual differences. Trends Cogn Sci 17(8):391–400
11. Ma WJ, Husain M, Bays PM (2014) Changing concepts of working memory. Nat Neurosci 17 (3):347–356
12. Brady TF, Konkle T, Alvarez GA (2011) A review of visual memory capacity: beyond individual items and toward structured representations. J Vis 11(5):4
13. Luck SJ, Vogel EK (1997) The capacity of visual working memory for features and conjunctions. Nature 390(6657):279–281
14. Vogel EK, Woodman GF, Luck SJ (2001) Storage of features, conjunctions and objects in visual working memory. J Exp Psychol Human 27(1):92–114
15. Awh E, Barton B, Vogel EK (2007) Visual working memory represents a fixed number of items regardless of complexity. Psychol Sci 18 (7):622–628
16. Wilken P, Ma WJ (2004) A detection theory account of change detection. J Vis 4(12):1–11
17. Zhang W, Luck SJ (2008) Discrete fixed-resolution representations in visual working memory. Nature 453(7192):233–235
18. Fougnie D, Suchow JW, Alvarez GA (2012) Variability in the quality of visual working memory. Nat Commun 3:1229
19. Tsubomi H et al (2013) Neural limits to representing objects still within view. J Neurosci 33 (19):8257–8263
20. McConkie GW, Currie CB (1996) Visual stability across saccades while viewing complex pictures. J Exp Psychol Human 22(3):563
21. Blaser E, Pylyshyn ZW, Holcombe AO (2000) Tracking an object through feature space. Nature 408(6809):196–199
22. Balaban H, Drew T, Luria R (2018) Visual working memory can selectively reset a subset of its representations. Psychon B Rev 25 (5):1877–1883
23. Vogel EK, Machizawa MG (2004) Neural activity predicts individual differences in visual working memory capacity. Nature 428 (6984):748–751
24. McCollough AW, Machizawa MG, Vogel EK (2007) Electrophysiological measures of maintaining representations in visual working memory. Cortex 43(1):77–94
25. Luria R et al (2016) The contralateral delay activity as a neural measure of visual working memory. Neurosci Biobehav R 62:100–108
26. Klaver P et al (1999) An event-related brain potential correlate of visual short-term memory. Neuroreport 10(10):2001–2005
27. Ikkai A, McCollough AW, Vogel EK (2010) Contralateral delay activity provides a neural measure of the number of representations in visual working memory. J Neurophysiol 103 (4):1963–1968
28. Balaban H, Luria R (2016) Integration of distinct objects in visual working memory depends on strong objecthood cues even for different-dimension conjunctions. Cereb Cortex 26(5):2093–2104
29. Balaban H, Luria R (2016) Object representations in visual working memory change according to the task context. Cortex 81:1–13
30. Luria R, Vogel EK (2011) Shape and color conjunction stimuli are represented as bound objects in visual working memory. Neuropsychologia 49(6):1632–1639
31. Luria R, Vogel EK (2014) Come together, right now: dynamic overwriting of an object's history through common fate. J Cogn Neurosci 26(8):1819–1828
32. Feldmann-Wüstefeld T, Vogel EK, Awh E (2018) Contralateral delay activity indexes working memory storage, not the current focus of spatial attention. J Cogn Neurosci 30 (8):1–11
33. Kang MS, Woodman GF (2014) The neurophysiological index of visual working memory maintenance is not due to load dependent eye movements. Neuropsychologia 56:63–72
34. Luria R et al (2010) Visual short-term memory capacity for simple and complex objects. J Cogn Neurosci 22(3):496–512
35. Ye C et al (2014) Visual working memory capacity for color is independent of representation resolution. Plos One 9(3):e91681
36. Woodman GF, Vogel EK, Luck SJ (2001) Visual search remains efficient when visual working memory is full. Psychol Sci 12 (3):219–224
37. Emrich SM et al (2009) Visual search elicits the electrophysiological marker of visual working memory. Plos One 4(11):e8042
38. Luria R, Vogel EK (2011) Visual search demands dictate reliance on working memory storage. J Neurosci 31(16):6199–6207
39. Töllner T et al (2013) Selective manipulation of target identification demands in visual search:

the role of stimulus contrast in CDA activations. J Vis 13(3):23–23
40. Hilimire MR et al (2011) Dynamics of target and distractor processing in visual search: evidence from event-related brain potentials. Neurosci Lett 495(3):196–200
41. Kundu B et al (2013) Strengthened effective connectivity underlies transfer of working memory training to tests of short-term memory and attention. J Neurosci 33 (20):8705–8715
42. Drew T, Vogel EK (2008) Neural measures of individual differences in selecting and tracking multiple moving objects. J Neurosci 28 (16):4183–4191
43. Drew T et al (2011) Delineating the neural signatures of tracking spatial position and working memory during attentive tracking. J Neurosci 31(2):659–668
44. Drew T et al (2012) Neural measures of dynamic changes in attentive tracking load. J Cogn Neurosci 24(2):440–450
45. Drew T et al (2013) Swapping or dropping? Electrophysiological measures of difficulty during multiple object tracking. Cognition 126 (2):213–223
46. Balaban H, Luria R (2015) The number of objects determines visual working memory capacity allocation for complex items. Neuroimage 119:54–62
47. Balaban H, Luria R (2017) Neural and behavioral evidence for an online resetting process in visual working memory. J Neurosci 37(5):1225–1239
48. Balaban H, Drew T, Luria R (2018) Delineating resetting and updating in visual working memory based on the object-to-representation correspondence. Neuropsychologia 113:85–94
49. Luck SJ (2005) An introduction to the event-related potential technique. MIT Press, Cambridge
50. Luck SJ (2014) An introduction to the event-related potential technique. MIT Press, Cambridge
51. Delorme A, Makeig S (2004) EEGLAB: an open source toolbox for analysis of single-trial EEG dynamics including independent component analysis. J Neurosci Methods 134(1):9–21
52. Lopez-Calderon J, Luck SJ (2014) ERPLAB: an open-source toolbox for the analysis of event-related potentials. Front Hum Neurosci 8:213
53. Hillyard SA, Galambos R (1970) Eye movement artifact in the CNV. Electroencephalogr Clin Neurophysiol 28(2):173–182

Neuromethods (2020) 151: 129–156
DOI 10.1007/7657_2019_26
© Springer Science+Business Media, LLC 2019
Published online: 12 September 2019

Multivariate Methods to Track the Spatiotemporal Profile of Feature-Based Attentional Selection Using EEG

Johannes Jacobus Fahrenfort

Abstract

This chapter provides a tutorial style guide to analyzing electroencephalogram (EEG) data contingent on feature-based attentional selection. It is targeted at researchers that currently investigate attentional processes using univariate methods but consider moving to multivariate analyses. The chapter starts by providing examples of classical univariate analysis, in which the EEG signal occurring ipsilateral to the target is subtracted from the signal that occurs in a contralateral electrode (i.e., the classical N2pc, an interhemispheric posterior negativity emerging around 180–200 ms). Next, it shows how the same type of information can also be identified using multivariate pattern analysis (MVPA). MVPA does not restrict one to contrast attentional selection in opposite hemifields but also allows one to assess attentional selection on the vertical meridian, or even within a quadrant of the visual field, opening up new avenues for research. The chapter demonstrates how to visualize topographic maps of attentional selection when using MVPA and shows how to assess timing onsets using the percent-amplitude latency method. Finally, it shows how a forward encoding model enables one to characterize the relationship between a continuous experimental variable (such as attended targets positioned on a circle) and EEG activity. This allows one to construct brain patterns for positions in the visual field that were never attended in the data that was used to create the forward model. This chapter is intended as a practical guide, explaining the methods and providing the scripts that can be used to generate the figures in-line, thus providing a step-by-step cookbook for analyzing neural time series data in the field of feature-based attentional selection.

Keywords Feature-based attention, Attentional selection, EEG, Univariate analysis, N2pc, MVPA, Multivariate pattern analysis, Classification, Decoding, BDM, Forward encoding model, Inverted encoding model, FEM

1 Introduction

Human observers are very good at extracting information from the visual field based on some relevant feature dimension, such as color. For example, when asked to determine whether the red element in a search display is a digit or a letter, they can do so very quickly and with very high accuracy (see Fig. 1). This ability is often referred to as feature-based attentional selection. There is a long history of investigating feature-based attentional selection (from here on referred to as attentional selection) using EEG [1, 2]. Traditionally, attentional selection is investigated in EEG using univariate analyses, in which targets are presented in one of two hemifields [3]. Observers detect

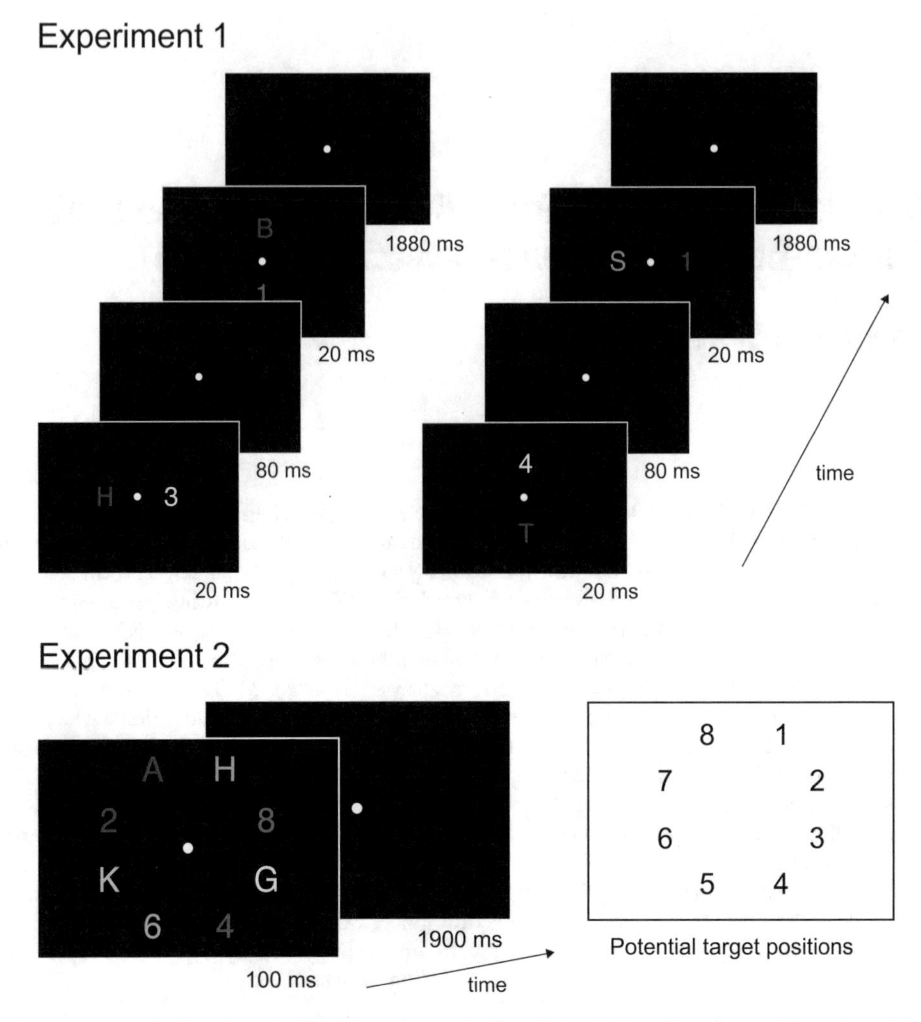

Fig. 1 Task structure in Experiment 1 and Experiment 2. *Top*: Example trial time lines of Experiment 1. There were two types of trials: the first display contained items on the horizontal meridian and the second display items on the vertical meridian, or the first display contained items on the vertical meridian and the second display items on the horizontal meridian. Subjects were required to determine whether a color-defined target was a digit or letter. The target color remained constant within a session (red in this example). Potential target colors were red, green, blue, or yellow (counterbalanced across subjects). Within blocks, subjects either had to detect a target in the first display (D1 blocks) or they had to detect a target in the second display (D2 blocks). We only analyzed task relevant displays. *Bottom*: Task and conditions in Experiment 2. *Bottom left panel*: Trial time line of Experiment 2. Subjects were asked to determine the identity of a colored target (letter or digit); target color could be red, green, or blue (counterbalanced across subjects). *Bottom right panel*: Positions used in Experiment 2, counted clockwise, one position contained the target, the other positions were occupied by distractors

a feature-defined target either on the left or on the right of fixation, and an electrophysiological marker of this selection process is identified by subtracting the event-related potentials (ERPs) on the ipsilateral side of the target from the ERPs occurring on the contralateral side of the target, typically using electrodes PO7 and PO8. This process results in an EEG component that is referred to as the N2pc.

Recently, we have shown how such a signal can also be extracted using multivariate pattern analysis (MVPA) [4]. This chapter compares the traditional method of obtaining electrophysiological markers of attentional selection to those that are obtained through MVPA. First, it explains how to compute the N2pc and provides examples of the N2pc. Next, it compares these to a classification approach in which a neural signature of attentional selection is generated by exploiting the multivariate nature of the EEG signal. Although some aspects of multivariate classification are explained, the focus of the chapter is on illustrating how to execute this analysis oneself. For details regarding multivariate classification, I refer the reader to more specialist texts [5, 6]. The chapter also shows how to properly extract and compare temporal onsets, as well as how to plot topographic maps based on forward-transformed classifier weights.

Finally, the chapter explains how a signature of attentional selection can also be captured using a forward encoding model and how this model can be used to construct cortical activity for conditions that did not occur in the experiment. Potential caveats of the forward encoding approach are briefly discussed. More detailed information is provided with respect to the forward encoding approach than is done for classification, although here too the focus is on practical application. Together, this provides a practical manual for how univariate and multivariate analyses can be carried out in the field of attentional selection, supplying both scripts and some theoretical background along with the analyses. Because of the focus on applicability, it does not follow a standard introduction-methods-results-discussion structure but rather a structure that allows the reader to understand the procedures by reading the explanations along with the results and the scripts that produce these results. Hopefully, this helps the reader to more easily reproduce and apply these analyses to their own data.

The analyses that are presented in this chapter were performed in MATLAB (MathWorks Inc., USA) using the freely available Amsterdam Decoding and Modelling (ADAM) toolbox v1.08-beta [5]. The reader can reproduce the analyses in this chapter in MATLAB by installing the ADAM toolbox and its dependencies from https://github.com/fahrenfort/ADAM/tree/1.08-beta (follow the instructions under "install") and downloading the data and/or results from https://osf.io/r67sc/. To reproduce the results, the reader can either choose to download the preprocessed data from that location (EEGLAB_DATA.zip) and compute the single subject first-level results from scratch or download the first-level results (MVPA_RESULTS.zip) and compute only the group results. Preprocessing of EEG data is not treated or explained in this chapter, as it is not sufficiently relevant to understand the methods that are presented here. As a general remark, I recommend to keep preprocessing to a minimum. The only preprocessing that was applied to these data was (1) epoching the continuous data into trials, (2) performing independent component analysis to remove components reflecting eyeblinks, and (3) removing trials that contain horizontal or vertical eye movements based on the electrooculogram (EOG). No offline high-pass filtering or low-pass filtering was applied to the data, as this can produce unwanted shifts in temporal onsets and produce spurious results; e.g., see [7, 8]. If desired, the preprocessing script that was applied to the data (as well as all the other scripts) can be downloaded from the Open Science Framework (OSF) at the previously noted location (SCRIPTS.zip).

Most of the analyses that are presented in this chapter have also been published in a *Scientific Reports* article that is freely available in the public domain; see [4]. Experimental details that are not reported in this chapter are deemed irrelevant to understanding the methods presented here but can be found in that publication if needed. Further note that small differences between the results from that publication and the results as presented in this chapter are due to the fact that the data in this chapter were analyzed without applying a high-pass filter.

2 Characterizing Attentional Selection Using the N2pc and Multivariate Classification (MVPA)

In two separate experiments, subjects performed the tasks in Fig. 1 (top, Experiment 1, $N = 12$; bottom, Experiment 2, $N = 15$) while EEG was collected. In both experiments, subjects were instructed to fixate a dot in the middle of the screen while reporting the category of a target color (digit or letter) by pressing a response key. In both experiments, the target color was the same throughout the experiment for any given subject, while target colors were counterbalanced across subjects.

In Experiment 1, each trial contained two successively presented stimulus displays. Each display contained two items on opposite sides, one in the target color (e.g., red in Fig. 1) and another one in a nontarget color. In different blocks of the experiment, subjects either had to report the target in the first display or report the target in the second display. One of the two displays on each trial contained a target-nontarget color pair on the horizontal midline (to the left and right of fixation), and the other display contained a target-nontarget color pair on the vertical midline (above and below of fixation). Only task-relevant displays were analyzed. Presentation sequence (vertical stimulus pair preceded by horizontal pair or vice versa) was randomized across trials. The goal of the experiment was to identify a neural signature of attentionally selecting the relevant item. In Experiment 2, each trial contained a single search display comprising eight items in a circular array. The target could appear in any of the eight locations (Fig. 1, bottom). Other than that, the task was identical. In what follows, the chapter describes how to characterize the unfolding of attentional selection using EEG signals acquired during these experiments. The data are analyzed using the ADAM toolbox [5]. At every step, an ADAM script is provided to perform a given analysis, along with a brief explanation of the parameters that are defined for that analysis. Next, the figures are presented that are produced by the analysis.

Typically, EEG data are analyzed by first computing the results at a single subject level. This is also called the first-level analysis. After obtaining first-level results, a statistical analysis is performed at the group level. The ADAM toolbox obtains the first-level results by reading single subject EEG data from disk, performing the relevant univariate and multivariate analyses on these single subject data, and subsequently saving the resulting first-level single subject results to the hard drive. Group-level results are then computed and visualized by reading in the single subject results from disk and subsequently performing and plotting group-level statistics. Below is the script to run the first-level analyses of the first experiment, which can be executed provided that the reader has installed a working copy of MATLAB, the ADAM toolbox and its dependencies, and has downloaded the preprocessed EEGLAB files from OSF. The only thing that requires setting are the proper input directory and output directories. Programming experience is not required to execute these analyses, but the reader should know how to open and execute .m script files in MATLAB and/or running snippets of code from the MATLAB Command Window.

```
%% some general settings regarding experiment 1
filenames = {    'TopDown_1__merged'  'TopDown_2__merged' 'TopDown_3__merged' ...
                 'TopDown_4__merged'  'TopDown_5__merged' 'TopDown_7__merged' ...
                 'TopDown_8__merged'  'TopDown_9__merged' 'TopDown_10__merged' ...
                 'TopDown_12__merged' 'TopDown_14__merged' 'TopDown_16__merged'};
cond_left =     [1 3 11 13]; % these condition codes specify target position
cond_right =    [2 4 12 14];
cond_top =      [5 7 15 17];
cond_bottom =   [6 8 16 18];

% first level analysis
cfg = [];
cfg.datadir = 'C:\EXP1\EEGDATA';
cfg.filenames = filenames;        % specifies the filenames
cfg.erp_baseline = [-.2,-.1];     % baseline period
cfg.resample = 250;               % lower sampling rate to save time
cfg.nfolds = 10;                  % number of folds used in the k-fold
cfg.model = 'BDM';                % 'FEM' for forward encoding model
cfg.channels = 'ALL';             % channel pooling

% classify left versus right
cfg.class_spec{1} = cond_string(cond_left);
cfg.class_spec{2} = cond_string(cond_right);
cfg.outputdir = 'C:\EXP1\RESULTS\LEFTRIGHT';
adam_MVPA_firstlevel(cfg);

% classify top versus bottom
cfg.class_spec{1} = cond_string(cond_top);
cfg.class_spec{2} = cond_string(cond_bottom);
cfg.outputdir = 'C:\EXP1\RESULTS\TOPBOTTOM';
adam_MVPA_firstlevel(cfg);
```

This script performs both univariate and multivariate analysis on the single subject EEG files. It operates by specifying a number of relevant parameters as fields using the cfg variable. Cfg is shorthand for configuration, and the cfg variable is used to pass these parameters to the relevant ADAM function. The actual first-level analyses are executed by the adam_MVPA_firstlevel function (which contains the configuration variable containing the relevant parameters between brackets). It does so by first reading in the epoched EEG data of individual subjects in EEGLAB format (specified in cfg.filenames) from a location on the hard drive (specified in cfg.datadir). Next, it baselines all trials to a window of (−200, −100) (specified in cfg.erp_baseline). Note that that this baseline window was chosen so that the baseline for the second search display would never overlap with the first search display (see Fig. 1, top). The data is also down-sampled to 250 Hz (specified in cfg.resample) prior to classification to expedite the classification analysis.

Next, it performs both a univariate and multivariate analysis. The univariate analysis is executed by computing ERPs for all the electrodes in the EEG data (specified in cfg.channels), separately for targets appearing on the left or on the right of the search display as well as for targets appearing on the top and the bottom of the search display (specified in cfg.class_spec). These two analyses (left versus right and top versus bottom) are executed and saved separately, in a folder called 'LEFTRIGHT' and 'TOPBOTTOM' respectively (specified in cfg.outputdir). We will see later how one can use the ERPs that the function has produced to compute the N2pc.

In addition, the function performs a tenfold (specified in cfg.nfolds) leave-one-out cross-validated multivariate classification analysis using a linear discriminant classifier ('BDM', short for backward decoding model, specified in cfg.model), and it does this across all the EEG electrodes included in the analysis (specified in cfg.channels). As explained, the script above performs two first-level analyses: the first one classifies targets appearing on the left of the search display versus targets appearing on the right of the display, and the second one classifies targets appearing on the top versus targets appearing on the bottom of the search display. The class definitions in cfg.class_spec specify which target positions are classified in the analysis. Classes are specified using the cond_string function, which merely converts integer numerals to comma-separated strings, which is the input format that is required by the ADAM toolbox.

After the first-level analyses are performed, the single subject results are stored at the hard drive location specified in cfg.outputdir, which are later read back in when performing a group-level analysis. Therefore, it is important to specify a meaningful directory name for the location of the first-level results. The directory name should reflect which analysis was performed because (1) this directory needs to be indicated when running the group-level analysis and (2) the name of this directory will be used by ADAM to denote the analysis name in graphs. Note also that one can specify quite a few additional parameters in the cfg variable during first-level analysis. Many of these are not covered here. Without specifying these fields, the analysis is performed using default values. Detailed information about some of the parameters that can be specified when performing an analysis in ADAM, as well as the meaning of some parameters (such as cross-validation) is beyond the scope of this chapter. For more information about these parameters, type "help adam_MVPA_firstlevel" in the MATLAB Command Window. In addition, a freely available open access article is available for those that require more general information regarding MVPA using the ADAM toolbox; see [5].

Next, I describe how one can perform and visualize group-level statistical analyses on these first-level results. As explained in the introduction, the traditional method of identifying attentional selection in experiments like these is using the N2pc. The N2pc is typically computed by subtracting the ERPs on the ipsilateral side of the target from the ERPs occurring on the contralateral side of the target, using electrodes PO7 and PO8 (see the left panel of Fig. 2 for the search display the subject is looking at and the electrodes that need to be subtracted to compute the N2pc). In the first group-level analysis, I illustrate how the ADAM toolbox can extract the N2pc using the ERPs from the first-level analysis, to compute a group-level N2pc for a single hemisphere. The actual

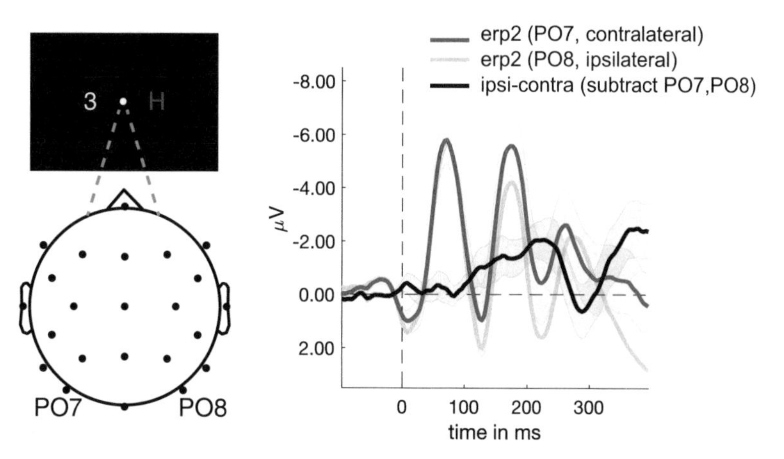

Fig. 2 Illustration of N2pc component. *Left*: Illustration of a subject attentionally selecting the red target on the right in a two-item display in Experiment 1. *Right*: ERP responses on the ipsi- and contralateral electrode to the attended item, as well as the difference between these two. This difference is the univariate N2pc, here for one attended hemifield

script to compute these ERPs and to produce the resulting group-level N2pc is given below.

```
%% extract electrodes P07 and P08, and also subtract them
cfg = [];
cfg.mpcompcor_method = 'none';
cfg.startdir = 'C:\EXP1\RESULTS';
cfg.electrode_def = {{'PO7'},{'PO8'}};
cfg.condition_def = 2;
cfg.timelim = [-100 400];
erpstats = adam_compute_group_ERP(cfg); % select the folder LEFTRIGHT when running this line
cfg.electrode_method = 'subtract';
erpstatsdif = adam_compute_group_ERP(cfg); % select the folder LEFTRIGHT once again

%% plot ERPs
cfg = [];
cfg.acclim = [-8.5 3.5];     % specifies the limits on the y-axis, not required
cfg.acctick = 2;
cfg.singleplot = 'yes';
cfg.line_colors = {[228,30,38]/255 [255,242,0]/255 [0,0,0]};
adam_plot_MVPA(cfg,erpstats,erpstatsdif);
```

The first part of this script loads the single subject data from the RESULTS folder and computes group ERPs when executing the function adam_compute_group_ERP. When this function executes, a folder selection window pops up at the location specified in cfg.startdir, after which one should manually select the LEFTRIGHT directory. Next, it extracts the single subject ERPs from that directory and computes a group-level average of these single subject results. It does so for the second class in the analysis which contained targets presented on the right (specified using cfg.condition_def) from electrode PO7 and PO8 (specified in cfg.electrode_def) within a temporal window of (100, 400) ms (specified in cfg.timelim). No statistical testing is applied (multiple comparison correction, specified as 'none' using cfg.mpcompcor_method). When the adam_compute_group_ERP function is executed, the output of the analysis is stored in a variable called erpstats. Next in the script, the analysis is performed again, now subtracting the ERP from electrode PO8 from PO7 to compute the right-hemispheric N2pc (ipsilateral minus contralateral). This is done by specifying cfg.electrode_method = 'subtract' and the running the same function again, outputting the result in the variable erpstatsdif.

The second part of the script inputs the erpstats and the erpstatsdif result variables into the adam_plot_MVPA function to plot the results. This function produces a graphical depiction of the ERPs that were computed by adam_compute_group_ERP. It plots the separate ERPs from P07 (in red) and PO8 (in yellow) as well as their difference (the N2pc, in black) together in a single figure (cfg.singleplot = 'yes'). Although not required, some additional parameters can be used to further configure the plot. For example, cfg.acclim specifies the limits of the y-axis, and cfg.acctick specifies the tick mark of the y-axis (for more information about plotting parameters, type "help adam_plot_MVPA" in the

MATLAB Command Window). Finally, cfg.line_colors specifies which Red-Green-Blue (RGB) color values to use for the consecutive plots (scaled between 0 and 1). For more information about color specifications in MATLAB type "help colormap" in the MATLAB Command Window. The result of the plotting operation is shown in the right panel of Fig. 2.

However, this only shows the N2pc for a single hemisphere. For the N2pc proper, one should compute the ipsilateral-contralateral difference separately for targets appearing in the left visual field and targets appearing the right visual field and subsequently average those subtractions. This is done using the script below.

```
%% get total N2pc
cfg = [];
cfg.mpcompcor_method = 'cluster_based';
cfg.startdir = 'C:\EXP1\RESULTS';
cfg.electrode_def = {{'PO8'},{'PO7'};{'PO7'},{'PO8'}};
cfg.electrode_method = 'subtract';
cfg.condition_def = [1,2];
cfg.condition_method = 'average';
cfg.timelim = [-100 400];
n2pcstats = adam_compute_group_ERP(cfg); % select the folder LEFTRIGHT when running this line

%% plot N2pc
cfg = [];
cfg.acclim = [-2.5 1];
cfg.singleplot = 'yes';
adam_plot_MVPA(cfg,n2pcstats);
```

The only difference with the earlier script is that this time the ERPs from both class 1 (targets appearing on the left) and from class 2 (targets on the right) are extracted (again specified in cfg.condition_def) and that for each of these, the ipsilateral electrode is subtracted from the contralateral electrode (again specified in cfg.electrode_def). The resulting subtractions are averaged (specified in cfg.condition_method) and tested against zero using a two-sided *t*-test against chance for each time sample. The statistical tests are corrected for multiple comparisons using cluster-based permutation testing (specified using cfg.mpcompcor_method). This method uses group-wise cluster-based permutation testing by taking the sum of the *t*-values for all contiguously significant time points ($p < 0.05$) and computing the number of times this sum is exceeded when computing the maximum cluster-based sum under random permutation [5, 9].

The group average that is computed by adam_compute_group_ERP is stored in variable n2pcstats and subsequently plotted using adam_plot_MVPA. The resulting figure can be found in Fig. 3, left panel. This is the "traditional" N2pc that is often reported in the literature [1, 2]. Note that the figure also contains a vertical dotted line halfway the first peak. This is a measure of the onset of the N2pc component. Note that taking the onset of the cluster itself is an unreliable way of determining the onset latency of an effect [10]. Further note that taking the peak latency is easily

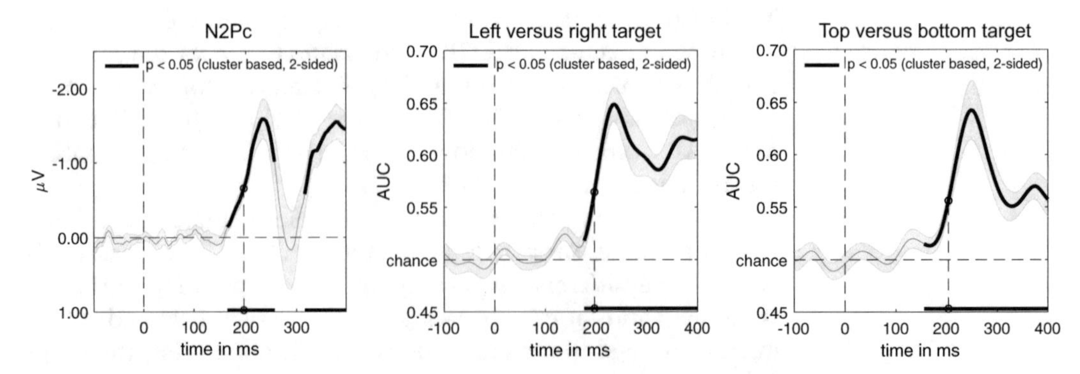

Fig. 3 Average N2pc and classification performance. *Left*: Average N2pc for left and right targets in Experiment 1. Computed using P07 and P08. See main text for details. *Middle and right*: Classification performance of target position for left versus right targets (middle) and classification performance for top versus bottom targets (right). Thick black lines reflect statistical tests that survive cluster-based permutation testing at $p < 0.05$. Shaded areas show $\pm$s.e.m. Onset latency of the 50% amplitude of the peak is indicated by a vertical dotted line. Note the similar temporal evolution between N2pc and classification performance

distorted by neuronal and measurement noise [11, 12]. Instead, a relatively straightforward and reliable way of characterizing temporal onsets is to measure the onset latency as the time when the rising effect of the component has reached 50% of its full amplitude. This is the standard method implemented in the ADAM toolbox, based on freely available code [13]. The onset latency for the N2pc that is estimated this way is stored in the group stats variable, in a field called latencies. Thus, one can access the latency of the N2pc by typing n2pcstats.latencies in the MATLAB command window. The field GA (short for Grand Average) tells us what the onset latency is of the N2pc when computed this way, which for this N2pc is 196 ms. Further below, we will assess whether multivariate measures of attentional selection result in similar onset latencies.

Although the N2pc has been a very successful measure of attentional selection, it also has some prominent shortcomings. The most striking shortcoming is the fact that the N2pc relies on lateral presentation of targets. For example, in Experiment 1 (Fig. 1, top panel), targets can appear both on the horizontal dimension and at the vertical dimension, but using the N2pc, one can only characterize the fingerprint of attentional selection on horizontally lateralized targets. As an alternative to the N2pc, one can use multivariate classification to characterize attentional selection, for example, classifying left versus right targets or classifying top versus bottom targets. These first-level classification analyses were performed when executing the initial script in the beginning of this chapter. The script below computes the associated group-level results of these classification analyses, both for the left versus right targets and for the top versus bottom targets. Finally, it plots these results in two separate graphs.

```
%% get group-level classification performance
cfg = [];
cfg.mpcompcor_method = 'cluster_based';
cfg.startdir = 'C:\EXP1\RESULTS';
cfg.timelim = [-100 400];
cfg.reduce_dims = 'diag';
mvpastats = adam_compute_group_MVPA(cfg); % press OK when the selection dialog pops up

%% plot classification performance over time for the LEFTRIGHT and the TOPBOTTOM dimension
cfg = [];
cfg.acclim = [.45 .7];
cfg.acctick = .05;
cfg.splinefreq = 32;
cfg.line_colors = {[0,0,0] [0,0,0]};
adam_plot_MVPA(cfg,mvpastats);
```

The first part of the script performs the group-level classification analysis on both contrasts: the horizontal dimension (left versus right target contrast) and the vertical dimension (top versus bottom target contrast). The analysis time window is restricted to $(-100, 400)$, specified in cfg.timelim. Classification performance is extracted for the diagonal, so without analyzing temporal generalization (specified in cfg.reduce_dims). This means that the data is trained and tested on the same samples. Details regarding the temporal generalization method are beyond the scope of this chapter, but more information can be found in other sources [5, 14]. The actual group-level analysis is performed by the function adam_compute_group_MVPA. When calling that function, one needs to specify the directory from which the data will be read, which in this case is the same directory as the cfg.startdir, so one can simply press OK after which group analyses from both analyses from contrasts are executed, performing t-testing against chance- level performance and applying cluster-based permutation to correct for multiple comparisons. The results are output in a variable called mvpastats, which has two elements: mvpastats(1) for the left-right classification analysis and mvpastats(2) for the top-bottom classification analysis.

Note that although testing against chance is common in the decoding literature, one caveat when using t-statistics on classification performance is that this does not allow population-level inference, in fact producing fixed effects rather than random effects results; see [15]. The implication is that one cannot formally draw population-level inferences based on such analyses, restricting conclusions to the sample that was tested. For studies that require population-level inference, it would be preferred to either use a completely separate training set (performing the training on different subjects or obtaining training data from a different task) or to replace the t-test with a statistic that explicitly evaluates information prevalence across the sampled subjects again; see [15].

In the second half of the script, the results in the mvpastats variable are plotted side by side, both in black outline. The cfg specifications have been explained before, except the splinefreq field. The splinefreq field causes the timeseries to be smoothed for visualization purposes. Smoothing is achieved by fitting a spline on the performance timeseries after downsampling it to 32 Hz (specified in cfg.splinefreq). The degree of smoothing is controlled by the resampling rate, with lower rates resulting in a smoother graph. The resampled series is centered on peak performance, so that the height of the peak is not affected by the smoothing procedure. This operation is applied for visualization purposes only; all statistical tests are performed on the unsmoothed timeseries.

The actual plotting operation is performed by the adam_plot_MVPA function. The result of the plotting operation is shown in the right two panels of Fig. 3. Note that these graphs no longer show μV on the y-axis, but instead show Area Under the Curve (AUC), a metric that indicates how well two or more classes can be discriminated by the classifier [16]. AUC typically runs between 0.5 (chance performance) and 1.0 (maximum classification accuracy). Interestingly, one can see from Fig. 3 how the two right panels show AUC time courses that are visually similar to the time course of the N2pc in the left panel. For targets on the horizontal meridian, this can be considered somewhat unsurprising, but a similar time course could not have been extracted using the standard N2pc approach for the vertical meridian. Thus, here, we see the first clear advantage of the multivariate classification approach.

As before, temporal onsets are automatically computed, reflecting the point in time where the rising signal reaches 50% of its peak amplitude [13]. As explained before, the exact values of the temporal onsets can be found by inspecting the latencies field of the mvpastats variable. Typing mvpastats.latencies returns two results, one for the left-right dimension (196 ms) and the other for the top-bottom dimension (204 ms). Note that the onset latency for the left-right is identical to the onset latency of the N2pc, and the latency for the top-bottom dimension is highly similar. One can test whether two onset latencies are different using jackknifing. Jackknifing is the practice of repeatedly computing an average while leaving out one subject, until each subject has been left out once. This way, you get the same number of observations as you have subjects in the dataset, but each observation is a group average with one of the subjects left out. This is useful when the single subject results are too noisy to determine a peak for every single subject (and thus to compute the 50% amplitude latency onset). The resulting jackknifed values can be used in a regular *t*-test or ANOVA, as long the resulting *t*- or *F*-values are corrected for jackknifing [17, 18]. This correction is applied automatically in the *t*-test function jackT from the latency package [13], which is included in the ADAM toolbox. Thus, to test whether the N2pc has a different onset latency from the left-right classification timeseries, one can simply type:

```
jackT(n2pcstats.latencies.jackknife, mvpastats(1).latencies.jackknife)
```

in the MATLAB command window. As a result, the function jackT will run a corrected t-test based on the jackknife latency onsets in the latencies fields of the n2pcstats and left vs. right classification in the mvpastats variables. Unsurprisingly, the result shows that the onset latencies between N2pc and classification are not significantly different; $t(11) = 0$, $p = 1$, providing converging evidence that the N2pc and the classification results tap into the same underlying signals. Similarly, one can also test whether the onsets between classifying left vs. right and classifying top vs. bottom are different by typing:

```
jackT(mvpastats(1).latencies.jackknife, mvpastats(2).latencies.jackknife)
```

Here too, the onset latencies are not significantly different between left vs. right and top vs. bottom; $t(11) = -1.62$, $p = 0.13$.

Multivariate classification seems the superior approach compared to the N2pc, as it allows one to characterize attentional selection both on the horizontal and on the vertical meridian. Moreover, it precludes one from having to perform a priori electrode selection. However, one might object that the downside of the approach is that it is hard to ascertain the source of the performance metric in the brain. Although the classifier produces training weights for the electrodes for every time sample, these classifier weights cannot be directly interpreted as neural sources [19]. Luckily, there are alternatives. The easiest way of characterizing the underlying cortical activity is to transform the classifier weights to forward weights by multiplying them with the data covariance matrix. Because the weights obtained from linear discriminant analysis contain the difference between the two compared sets normalized by the covariance matrix, this operation creates activation patterns that return the mass-univariate difference between the compared conditions, but which unlike classifier weights, are interpretable as neural sources [19]. Forward-transformed weights are equivalent to the univariate difference between conditions, except that they are derived from the classification analysis itself, providing a sanity check that the classification analysis results in a meaningful pattern of results. The ADAM toolbox automatically computes forward-transformed weights during the first level and stores these in the output variable during group-level analysis. The script to plot the forward-transformed weights for the two classification analyses (right versus left and top versus bottom) is given below.

```
%% plot topographic maps
cfg = [];
cfg.plotweights_or_pattern = 'covpattern';
cfg.timelim = [240 250];
cfg.weightlim = [-1.8 1.8];
cfg.mpcompcor_method = 'cluster_based';
adam_plot_BDM_weights(cfg,mvpastats);
```

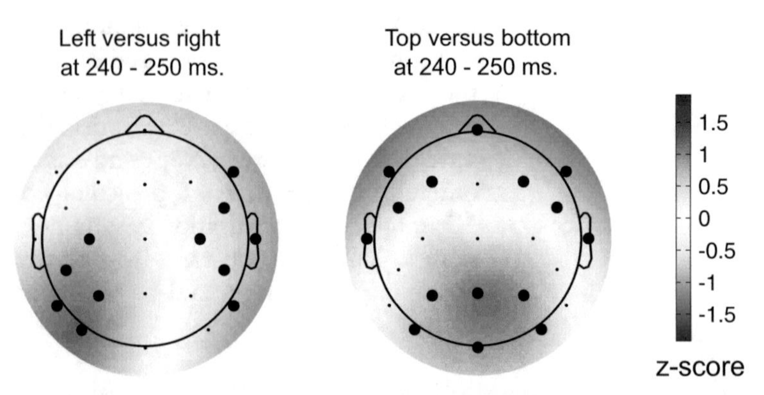

Fig. 4 Activation patterns associated with peak decoding accuracy (240–250 ms) in Experiment 1. Derived from the product of the weight vectors and the covariance matrix, normalized across space (see main text). *Left*: The pattern associated with left versus right decoding. Note the clearly lateralized distribution. This lateralized pattern shows the distribution of neural activity underlying successful discrimination between targets appearing on the left and the right of fixation and is equivalent to the mass-univariate difference between left and right targets. *Right*: The pattern associated with top versus bottom decoding, now showing a posterior-anterior distribution. Thick electrode dots belong to clusters having $p < 0.05$ under cluster-based permutation testing

The adam_plot_BDM_weights function computes two topographic plots from the mvpastats data, which can be found in Fig. 4. The cfg variable first specifies that the function should plot the pattern based on the covariance matrix (indicated in cfg.plotweights_or_pattern). Further, it averages the plot in the temporal window between 240 and 250 ms (specified in cfg.timelim), approximately corresponding to the peak of the performance timeseries. Each of the electrodes is tested against zero using a *t*-test, after which a cluster-based permutation test is executed based on the adjacency of neighboring electrodes to correct for multiple comparisons. Electrodes that survive the cluster-based permutation test are indicated as thick dots on topographic map. Note that the adam_plot_BDM_weights function spatially normalizes the pattern for every subject prior to computing the group average, so that the amplitude at every electrode is expressed as a *Z*-score across electrodes.

To further highlight the advantage of multivariate classification to characterize attentional selection, we now move our attention to the second experiment. In this experiment, targets were not presented on the horizontal or on the vertical meridian, but rather in a circular array (see Fig. 1, bottom). Here, we first ask the question whether it is possible to characterize attentional selection without even crossing the meridian, so, for example, within a quadrant.

For example, we may ask whether one can dissociate attentional selection of targets within each of the quadrants of the visual field, such as between targets on position 1 and targets on position 2, between position 3 and 4, etc. (see Fig. 1, right bottom). The script to execute the first level decoding analyses to extract these analyses is given below.

```
%% general information about experiment 2
filenames = { 'DecExp3_1_R'  'DecExp3_2_G'  'DecExp3_3_B'  'DecExp3_4_R'  'DecExp3_5_G' ...
              'DecExp3_6_B'  'DecExp3_7_R'  'DecExp3_8_G'  'DecExp3_9_B'  'DecExp3_10_R' ...
              'DecExp3_11_G' 'DecExp3_12_B' 'DecExp3_13_R' 'DecExp3_14_G' 'DecExp3_15_B' };
for c=1:8
    pos{c} = [ 10+c 20+c ];
end

%% settings for first level quadrant analysis of experiment 2
cfg = [];
cfg.datadir = 'C:\EXP2\EEGDATA';
cfg.filenames = filenames;
cfg.erp_baseline = [-.1,0];
cfg.resample = 250;
cfg.nfolds = 10;
cfg.model = 'BDM';
cfg.channels = 'ALL';

% classify attentional selection in the upper right quadrant
clear class_spec;
class_spec{1} = cond_string(pos{1});
class_spec{2} = cond_string(pos{2});
cfg.class_spec = class_spec;
cfg.outputdir = 'C:\EXP2\RESULTS\QUADRANT\1_2';
adam_MVPA_firstlevel(cfg);

% classify attentional selection in the bottom right quadrant
clear class_spec;
class_spec{1} = cond_string(pos{3});
class_spec{2} = cond_string(pos{4});
cfg.class_spec = class_spec;
cfg.outputdir = 'C:\EXP2\RESULTS\QUADRANT\3_4';
adam_MVPA_firstlevel(cfg);

% classify attentional selection in the bottom left quadrant
clear class_spec;
class_spec{1} = cond_string(pos{5});
class_spec{2} = cond_string(pos{6});
cfg.class_spec = class_spec;
cfg.outputdir = 'C:\EXP2\RESULTS\QUADRANT\5_6';
adam_MVPA_firstlevel(cfg);

% classify attentional selection in the upper left quadrant
clear class_spec;
class_spec{1} = cond_string(pos{7});
class_spec{2} = cond_string(pos{8});
cfg.class_spec = class_spec;
cfg.outputdir ='C:\EXP2\RESULTS\QUADRANT\7_8';
adam_MVPA_firstlevel(cfg);
```

The above script performs analyses analogous to the analysis that was performed in Experiment 1, but now within the four quadrants of Experiment 2. The group-level analyses can be executed and plotted using the script below.

```
%% get group-level classification performance within each quadrant quadrant, experiment 2
cfg = [];
cfg.startdir = 'C:\EXP2\RESULTS';
cfg.mpcompcor_method = 'cluster_based';
cfg.timelim = [-100 400];
cfg.reduce_dims = 'diag';
cfg.channelpool = 'ALL';
cfg.plotmodel = 'BDM';
mvpastats_quadrant = adam_compute_group_MVPA(cfg); % select the folder QUADRANT

%% plot classification performance over time for each of the four quadrants
cfg = [];
cfg.acclim = [.46 .7];
cfg.acctick = .05;
cfg.splinefreq = 32;
cfg.line_colors = {[0,0,0],[0,0,0],[0,0,0],[0,0,0]};
cfg.plotorder = {'7_8' '1_2' '5_6' '3_4' };
cfg.nolatency = true;
adam_plot_MVPA(cfg,mvpastats_quadrant);
```

The first part of the script once again reads in the first level results and computes the group-level results. When the function adam_compute_group_MVPA is executed, a dialog appears. Select "QUADRANT" to read in the analyses from the four different quadrants. These will be stored in the variable mvpastats_quadrant, the contents of which can subsequently be plotted using adam_plot_MVPA. The cfg that is used to plot the results is much the same as before, with two minor additions. A field plot_order was added to control the order in which the analyses are plotted. The names in plot_order are taken directly from the folder names that are used to store the different first-level analyses. Further, a field nolatency is used to preclude the plotting of latency information. Although the graphs in Fig. 5 clearly show that it is possible to classify which item was attentionally selected within each quadrant, eyeballing the data already suggests that classifier performance is not reliable enough over time to estimate consistent onset latencies (e.g., see the top left panel in Fig. 5). For this reason, I chose not to plot latency information in this graph.

Indeed, to reliably determine onset latency, the analysis would require more data. The search display has eight target positions, so to increase the signal-to-noise ratio, one can perform an 8-way classification analysis, inputting each of the eight positions as classes into the classification analysis. This should provide the best possible estimate in Experiment 2 of the time course of attentional selection across the display and is achieved using the script below (keeping the same filename and condition definitions as in the quadrant analysis above).

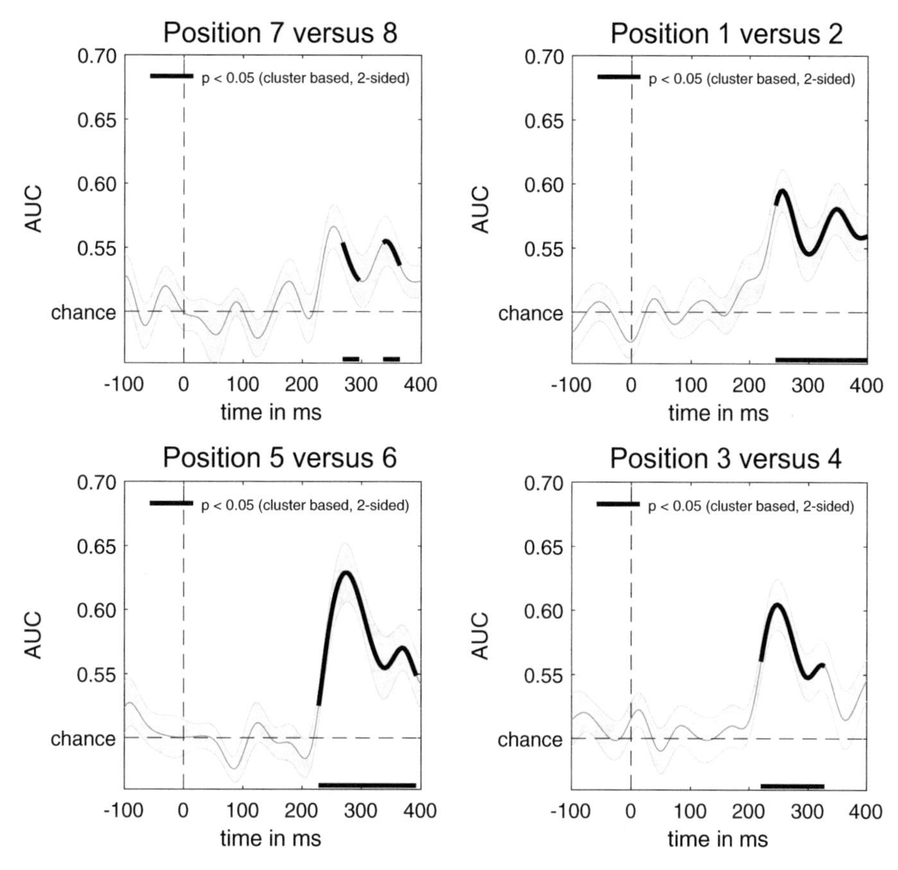

Fig. 5 Per quadrant decoding accuracy of target position in Experiment 2. Thick black lines reflect statistical tests that survive cluster-based permutation testing at $p < 0.05$. Shaded areas are $\pm$s.e.m

```
%% settings for first level analysis of all positions in experiment 2
cfg = [];
cfg.datadir = 'C:\EXP2\EEGDATA';
cfg.filenames = filenames;
cfg.model = 'FEM,BDM';
cfg.resample = 250;
cfg.channels = 'ALL';
cfg.erp_baseline = [-.1,0];
cfg.sigma_basis_set = 0;

% classify attentional selection across all eight target positions
clear class_spec;
for c = 1:8
    class_spec{c} = cond_string(pos{c});
end
cfg.class_spec = class_spec;
cfg.outputdir = 'C:\EXP2\RESULTS\ALLPOS';
adam_MVPA_firstlevel(cfg);
```

Note that the above script not only runs a decoding analysis ('BDM' specified in cfg.model) but also a forward encoding analysis ('FEM' specified in cfg.model). We will return to this analysis in the next section. But before we do so, we first plot the result of the 8-way decoding analysis and determine the concomitant onset latency from that analysis. Below is the script to compute the group results and plot the 8-way classification analysis.

```
%% get group-level 8-way classification performance, experiment 2
cfg = [];
cfg.startdir = 'C:\EXP2\RESULTS';
cfg.mpcompcor_method = 'cluster_based';
cfg.timelim = [-100 400];
cfg.reduce_dims = 'diag';
cfg.channelpool = 'ALL';
cfg.plotmodel = 'BDM';
mvpastats_8way = adam_compute_group_MVPA(cfg); % select the folder ALLPOS

%% plot 8-way decoding
cfg = [];
cfg.acclim = [.45 .75];
cfg.acctick = .05;
cfg.splinefreq = 32;
adam_plot_MVPA(cfg,mvpastats_8way);
```

When the adam_compute_group_MVPA function executes, a folder selection window pops up in which one should manually select the ALLPOS folder. This folder contains the first-level results for all 8-way classification of the eight target positions. As before, this will compute the group results and assign this outcome to a variable (mvpastats_8way), subsequently plotting this outcome using the adam_plot_MVPA function. The resulting plot can be found in Fig. 6 below, which now also highlights the 50% peak amplitude onset latency using a vertical dotted line. The numerical value associated with the 50% peak amplitude onset latency in this plot can be found by typing mvpastats_8way.latencies.GA, which returns 220 ms. Interestingly, this onset latency seems slightly later

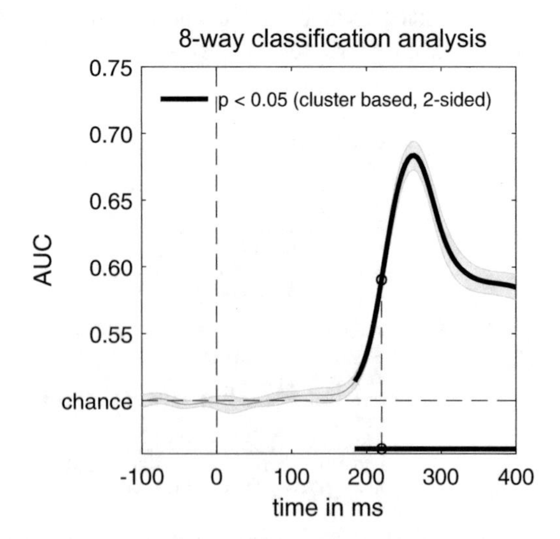

Fig. 6 8-way classification accuracy of target position in Experiment 2. Note again the similarity to the temporal evolution of the N2pc (Fig. 3, left) and decoding performance in Experiment 1 (Fig. 3, middle and right), although plausibly having a slightly later onset because of the increase in the number of items on the screen. Thick black lines reflect statistical tests that survive cluster-based permutation testing at $p < 0.05$. The shaded area is $\pm$s.e.m

than the onset latencies that were identified in Experiment 1, possibly because of the larger number of items in the display. Determining the true cause of this apparent latency difference is beyond the scope of the current chapter and would require further experimentation. I suffice to point out here that one can investigate such onset latency differences using the methods that are explained in this chapter.

3 Characterizing the N2pc Using Forward Encoding

So far, this chapter has covered classification approaches to characterize attentional selection. In this section, I discuss a complementary multivariate approach, which is to use a forward encoding model to establish a continuous relationship between an experimental variable of interest and cortical activity [4, 20, 21]. This approach has been further extended using inverted encoding models that estimate model responses from the data in so-called channel tuning functions (CTFs); e.g., see [22–25]. Below, I first describe the general approach that is taken in these models, and I discuss some caveats of the method [26–28]. Next, I provide details and script that applies a forward encoding model (FEM) to the data obtained from Experiment 2 and show how it can be used to reconstruct cortical activity for conditions that did not occur during the experiment.

The principal goal of forward encoding models is to characterize a direct link between a continuous stimulus parameter space and the cortical responses that are measured (here through EEG). The advantage of this approach is that one can predict (reconstruct) cortical activations for novel stimulus values that were never presented during the experiment or stimulus parameter estimates for novel brain data for which no condition labels were acquired [20]. In addition, a number of studies have suggested that one can use inverted encoding models to estimate the model response (referred to as channel tuning function, or CTF), to assay of how broad-scale cortical activity is tuned to a continuous experimental variable, somewhat akin to neural tuning functions at the level of single neurons [29].

Despite their initial promise, it has recently been shown that the width of a CTF not only reflects the degree of tuning to the parameter space but also the signal-to-noise ratio of the data on which the model is fitted [27]. Moreover, it has been shown that such a fitting procedure recovers arbitrary starting parameters of the model, rather than a recovering a CTF that reflects the actual relationship between the experimental variable and cortical activity [26]. With these caveats in mind, one might still use the simplest form of these models (a delta function) to construct a CTF to provide insight into the degree to which neighboring experimental parameter values produce overlapping cortical activations.

Here, we generate a FEM of the data in the second experiment (Fig. 2, bottom). To do so, we employ a procedure previously described by Brouwer and Heeger [20] using the same tenfold cross validation scheme as in the previously described classification analyses. In this procedure, the training set is used to estimate the response in each of eight hypothetical position "channels" (corresponding to the eight target positions on the screen). The nomenclature "channels" here should not be confused with MEG or EEG sensors; EEG sensors are referred to as electrodes in the current chapter. To provide an initial estimate of the channel responses, a preliminary "basis set" is used to estimate the weights that specify the relationship between the observed multivariate signal and the channel responses. Typically, authors have used a basis set in the form of a Gaussian or a sinusoid raised to a power, but as explained above, it has recently been shown that CTF estimation using this method can recover any arbitrary basis set, rather than assaying the true relationship between the continuous experimental variable under investigation and the measured multivariate activity [26].

For this reason, I recommend here to only use the simplest form of the basis set, containing a 1 for the corresponding target position and a 0 for all other positions, so that the shape of any resulting CTF cannot reflect the initial basis set, but must be caused by the data itself. A binary on-off basis set like this is sometimes also referred to as a delta function. Here, we use eight basis sets (one for each target position), each shifted by one position compared to its neighbor, to construct a regression matrix C1. C1 has the form $k \times n1$, in which k is the number of position channels (1 to 8) and n1 is the number of trials in the training set. Next, we estimate the response amplitude to each of the eight hypothetical position channels by performing an ordinary least squares regression of the C1 matrix onto the B1 matrix from the EEG training set. B1 contained EEG data of the form $m \times n1$, in which m is number of electrodes and n1 is the number of trials in the training set. This regression yields a weight matrix W in which each electrode obtains a regression coefficient (a "weight") for each hypothetical channel. The weight matrix W has the form $m \times k$, in which m is the number of electrodes and k is the number of position channels.

Next, the model is inverted by performing ordinary least squares regression of these weights onto the B2 matrix from the EEG testing set to produce the estimated channel responses for each trial. B2 has the form $m \times n2$, in which m is number of electrodes and n2 is the number of trials in the testing set. The resulting estimated channel responses are contained in matrix C2, having the form $k \times n2$, in which k are the observed channel responses and n2 are the trials in the testing set. This procedure is repeated for all folds in the train-test procedure, until all data has been tested once. Next, the channel responses are averaged across trials in the testing set, separately for each of the eight trial types that correspond to each of the eight target positions on the screen.

The channel responses from this testing phase in combination with the associated weights contain the validated and invertible one-to-one relationship between a particular attended location in the search display and the multivariate EEG response. The script to perform the above procedure was executed when specifying 'FEM' during the first-level analyses of Experiment 2. The (averaged) C2 channel responses constitute a CTF per condition. These can be shifted to a common center, so that the channel responses for each of the eight target positions are aligned and averaged to obtain a canonical CTF. Mathematical (less verbal) descriptions as well as more graphical depictions of this train-test estimation procedure have been provided elsewhere, e.g., [20–28]. The script below extracts and plots the CTFs from the first-level analyses.

```
%% compute FEM in experiment 2
cfg = [];
cfg.startdir = 'C:\EXP2\RESULTS';
cfg.mpcompcor_method = 'cluster_based';
cfg.plotmodel = 'FEM';
cfg.channelpool = 'ALL';
cfg.timelim = [-100,1000];
cfg.reduce_dims = 'diag';
femstats = adam_compute_group_MVPA(cfg); % select the folder ALLPOS

%% plot CTF at 260-270 ms
cfg = [];
cfg.plotfield = 'CTFpercond';
cfg.shiftindiv = true;
cfg.weightlim = [-.2 .6];
cfg.CTFtime = [260 270];
cfg.BLtime = [-100 0];
CTF = adam_plot_CTF(cfg,femstats);
```

First, it extracts the group-level channel responses and corresponding weights using adam_compute_group_MVPA (cfg. plotmodel as 'FEM') and stores these results in a stats variable called femstats. Next, it uses the function adam_plot_CTF to plot the CTF for each condition (specified in cfg.plotfield as 'CTFpercond') in the period between 260 and 270 ms for the CTF (specified in cfg.CTFtime, corresponding the peak classification accuracy in Fig. 6) as well as the CTF in the baseline between −100 and 0 ms (specified in cfg.BLtime). The resulting figure is shown in the top left panel of Fig. 7. These CTFs are taken from the condition-specific averages of C2, but note that these responses are shifted to a common center, so that the channel responses for each of the eight target positions are aligned. This is specified by indicating ctf. shiftindiv = true (see labels under the x-axis of the figure to see how they were shifted). One can also plot the average of these shifted condition-specific CTFs to show the canonical CTF. This can be

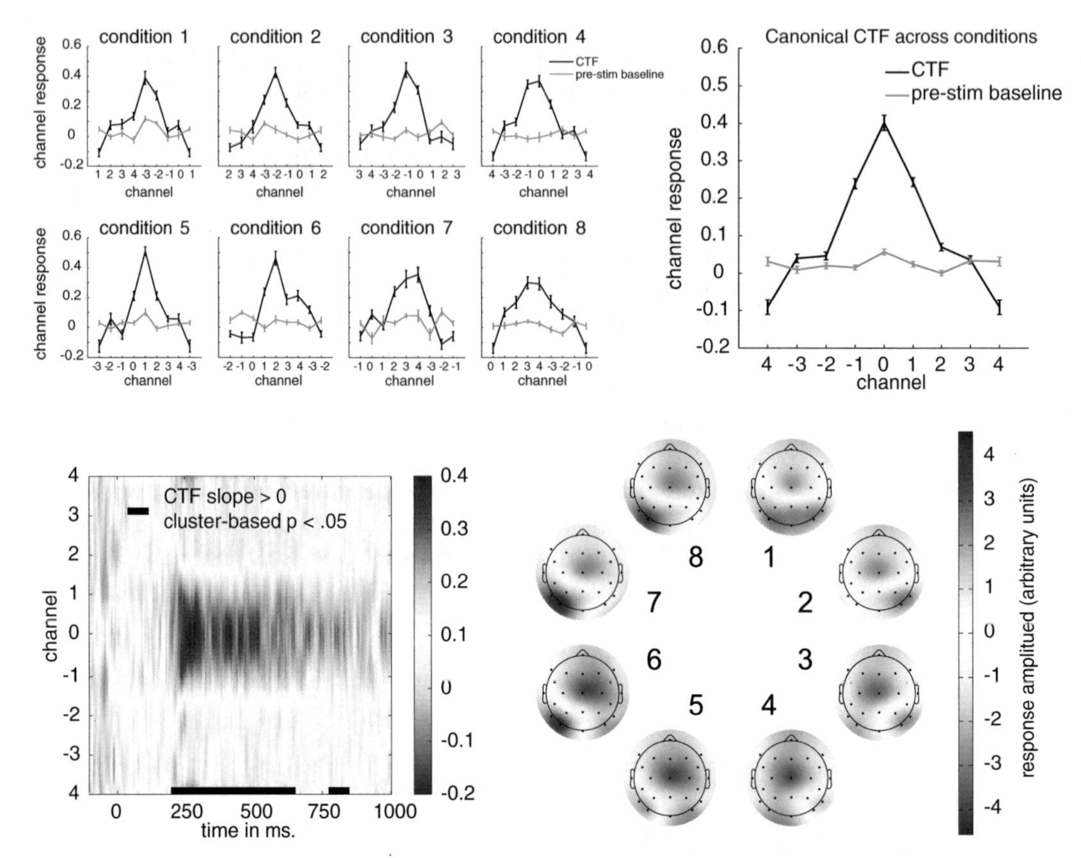

Fig. 7 Channel tuning functions. *Top left*: CTFs for individual conditions show that the CTF is not driven by particular target positions. *Top right*: CTF for the 260–270 ms window and CTF during baseline (−100 to 0 ms) obtained by shifting the individual condition CTFs to align to the same channel. *Bottom left*: CTF development over time in which color reflects channel responses. The black line near the time axis shows the time windows where the slope of the CTF is significantly different from 0 ($p < 0.05$, cluster-based permutation test). *Bottom right*: Topographic weight plots for each condition in the 260–270 ms time window. Weights from forward models are directly interpretable in terms of neural sources. These plots therefore show how neural activity changes as a function of variability in attended target position

done by running the same script as above, the only difference being that cfg.plotfield should be changed to 'CTF' prior to running the script, like this:

```
cfg.plotfield = 'CTF';
```

Figure 7 top right shows the averaged empirical CTF across conditions. Importantly, we used a basis set that did not make any assumptions about the shape of the CTF beforehand (the delta function), so we can be sure that the CTF reflects the relationship between an experimental parameter of interest (the hypothetically attended location channel, on the x-axis) and the strength of the multivariate response (on the y-axis). What this

CTF shows is that there is some degree of overlap between multivariate responses for neighboring attended locations, with the caveat that the exact strength of this overlap as quantified by the CTF is also affected by the signal-to-noise ratio [27]. Note further that in the top of Fig. 7, we plotted the CTFs during peak classification performance, between 260 and 270 ms. However, the estimation procedure was done for every time sample, yielding a CTF over time. The CTF over time can also be plotted, using the script below.

```
%% plot CTF over time
cfg = [];
cfg.plotfield = 'CTF';
cfg.reduce_dims = 'diag';
cfg.colorlim = [-.2, .4];
adam_plot_CTF(cfg,femstats);
```

The script is the same as before, the only difference being that the we no longer specify CTFtime, so that the function plots the CTF over the entire time interval rather than averaging over a time window, now using color to denote the strength of the channel response for every time point, and plotting the channels on the y-axis. The resulting plot can be found in Fig. 7, left bottom. Aside from plotting CTFs, one might also be interested in knowing the topographic distribution of these responses. Fortunately, the weights resulting from a forward encoding model can be interpreted directly as a neural source [19]. The script to plot the weights for each of the eight target positions is given below.

```
%% plot FEM weights, experiment 2
cfg = [];
cfg.timelim = [260 270];
cfg.mpcompcor_method = 'none';
cfg.normalized = false;
adam_plot_FEM_weights(cfg,femstats);
```

This produces the topographic weight maps for each of the eight target positions. This is shown in Fig. 7, bottom right. Note that no statistics are applied (cfg.mpcompcor_method as 'none'), and the plots are not spatially normalized (cfg. normalized = false).

Finally, we establish how a forward encoding model can be used to reconstruct cortical activity for experimental stimulus values that were used to generate the model. The top left of Fig. 8 shows the target display from Experiment 2, containing four target positions that were not present in the experiment (top, bottom, left, and right position). The aim of the following section is to reconstruct cortical activity associated with these positions using the specified forward model. The first step in this reconstruction is to construct CTFs (channel responses) that would have occurred when these positions would have

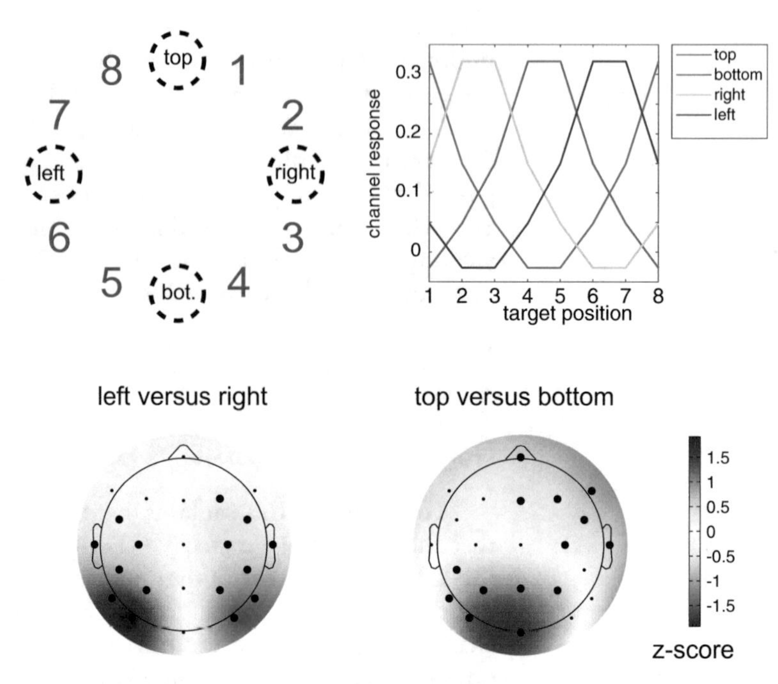

Fig. 8 Reconstructing the neural signature of attentional capture for target positions that were never presented during Experiment 2. *Top left*: The target positions that are reconstructed: top, bottom, left, and right. *Top right*: The constructed channel responses that are associated with these positions using the CTF from Fig. 7 (see main text for details). Top is in between target position 8 and 1, so the channel response to top is constructed by averaging the hypothetical channel response to position 8 and position 1. Similarly, right is created from averaging channel responses to 2 and 3, etc. Any position on the circle can be constructed using a weighted average of channel responses. Left, right, bottom, and top weights were reconstructed using the product of the constructed channel responses and the weight matrix at 260–270 ms. *Bottom left*: The left versus right pattern was generated by subtracting the left from the right pattern. *Bottom right*: The top versus bottom pattern was created by subtracting the bottom pattern from the top pattern. Note the similarity with the left-right and top-bottom patterns from Experiment 1. Patterns are normalized across electrodes. Thick electrode dots survived cluster-based permutation testing under $p < 0.05$

occurred in the experiment. For the top position, this would be the interpolated CTF between location 8 and 1. For the right position, this would be the interpolated CTF between location 2 and 3 and so forth. These CTFs are generated by taking the canonical CTF and using it to generate these interpolated CTFs. The CTF values during peak classification response were returned when plotting that CTF above (CTF = adam_plot_CTF(cfg,fem-stats) in the script). These values are used below to create the channel-specific CTFs.

```
%% create new CTFs from the canonical CTF
basis_set = mean(CTF.indivCTFmean); % a basis set created from the canonical CTF

% mirror
basis_set = (basis_set + basis_set(end:-1:1))/2;
basis_set = basis_set(2:9); % remove duplicate end point

% generate channel responses for each condition (shifted)
old_chan_responses = nan(numel(basis_set),numel(basis_set));
for c=1:numel(basis_set)
    old_chan_responses(:,c) = circshift(shiftdim(basis_set),-floor(numel(basis_set)/2)+c);
end
% interpolate new channel response sets for positions between 1 and 8, between 4
% and 5 (the vertical ones) and between 2 and 3 and 6 and 7 (the horizontal ones)
new_chan_responses = nan(numel(basis_set),4);
new_chan_responses(:,1) = mean([ old_chan_responses(:,8) old_chan_responses(:,1) ],2);
new_chan_responses(:,2) = mean([ old_chan_responses(:,4) old_chan_responses(:,5) ],2);
new_chan_responses(:,3) = mean([ old_chan_responses(:,2) old_chan_responses(:,3) ],2);
new_chan_responses(:,4) = mean([ old_chan_responses(:,6) old_chan_responses(:,7) ],2);

% plot the result
figure; plot(new_chan_responses);
```

These channel responses for top, bottom, left, and right are plotted in Fig. 8, top right. Next, these interpolated channel responses are used to generate new weight matrices for these positions. This is done in the first section of the script below.

```
%% reconstruct patterns, respectively for top, bottom, right and left
for cSubj = 1:size(femstats.weights.indivWeights,1)              % subject loop
    for cT = 1:size(femstats.weights.indivWeights,2)              % time loop
        W = squeeze(femstats.weights.indivWeights(cSubj,cT,:,:)); % extract channel weights
        indivWeights(cSubj,cT,:,:) = W*new_chan_responses;        % new weights
    end
end

%% subtract channel weights on the horizontal and vertical meridian
newIndivWeights(:,:,:,1) = indivWeights(:,:,:,4) - indivWeights(:,:,:,3); % bottom from top
newIndivWeights(:,:,:,2) = indivWeights(:,:,:,1) - indivWeights(:,:,:,2); % left from right

%% insert new subtractions into femstats for plotting
constructed_femstats = femstats;                                % copy what we had
constructed_femstats.weights.indivWeights = newIndivWeights; % inject new weights

% plot reconstructed patterns
cfg = [];
cfg.timelim = [260 270];
cfg.weightlim = [-1.8 1.8];
cfg.mpcompcor_method = 'cluster_based';
adam_plot_FEM_weights(cfg,constructed_femstats);
```

The second section of this script subtracts the bottom from the top and the left from the right position, to get the distribution associated with top versus bottom and left versus right. Finally, these new weight matrices are injected into a femstats variable for plotting using the adam_plot_FEM_weights function that we used before. The resulting topographic maps are plotted in Fig. 8, bottom. If these topographic plots look familiar, that is no coincidence. They seem to nicely correspond to the topographic maps that were obtained in the first experiment (Fig. 4). However, a crucial difference between Figs. 4 and 8, bottom, is that the topographies in Fig. 4 were derived from actual data in Experiment 1, whereas the topographies in Fig. 8 were constructed from the forward encoding model and do not correspond to actual data that was collected during Experiment 2. The correlation between the topographic maps from Experiment 1 and Experiment 2 is extremely high ($r = 0.85$, $p < 10^{-6}$ for the horizontal meridian and $r = 0.86$, $p < 10^{-6}$ for the veridical meridian), thus providing converging evidence that the forward encoding model is able to successfully construct cortical activation maps for data that was not actually present in the data that was used to generate the forward encoding model.

4 Conclusion

This chapter compared univariate to multivariate methods when analyzing EEG data obtained during tasks in which subjects need to use feature-based attention to select items in a display. This shows that multivariate classification is superior to traditional univariate analysis when characterizing the spatiotemporal profile of attentional selection. Experiment 1 shows how multivariate classification analyses enable one to not only assess attentional selection on the horizontal meridian but also on the vertical meridian. Experiment 2 shows that one can use classification analyses to assess the time course of attentional selection within quadrants of the visual field and that one can even look at the time course of attentional selection across a large number of attended positions. Further, Experiment 2 shows how one can use a forward modeling approach to construct spatiotemporal responses for locations that were never attended during the experiment. The chapter has also demonstrated how one can assess onset latencies, as well as how one can plot spatiotemporal maps of both decoding and forward encoding analyses. Together, this should provide a useful introduction for those in the field of feature-based attentional selection that want to move from traditional univariate analysis to multivariate analysis.

Acknowledgments

I would like to thank Anna Grubert, Martin Eimer, and Chris Olivers for allowing me to freely use and share the data from these experiments, as well as the analysis plans that we used on these data. This chapter would not have existed without them.

References

1. Eimer M (1996) The N2pc component as an indicator of attentional selectivity. Electroencephalogr Clin Neurophysiol 99(3):225–234. https://doi.org/10.1016/0013-4694(96)95711-9
2. Luck SJ, Hillyard SA (1994) Electrophysiological correlates of feature analysis during visual search. Psychophysiology 31(3):291–308
3. Woodman GF (2010) A brief introduction to the use of event-related potentials (ERPs) in studies of perception and attention. Atten Percept Psychophys 72(8):2031–2046. https://doi.org/10.3758/APP.72.8.2031
4. Fahrenfort JJ, Grubert A, Olivers CNL, Eimer M (2017) Multivariate EEG analyses support high-resolution tracking of feature-based attentional selection. Sci Rep 7(1):1886. https://doi.org/10.1038/s41598-017-01911-0
5. Fahrenfort JJ, van Driel J, van Gaal S, Olivers CNL (2018) From ERPs to MVPA using the Amsterdam decoding and modeling toolbox (ADAM). Front Neurosci 12. https://doi.org/10.3389/fnins.2018.00368
6. Grootswagers T, Wardle SG, Carlson TA (2017) Decoding dynamic brain patterns from evoked responses: a tutorial on multivariate pattern analysis applied to time series neuroimaging data. J Cogn Neurosci 29(4):677–697. https://doi.org/10.1162/jocn_a_01068
7. van Driel J, Olivers CNL, Fahrenfort JJ (2019) High-pass filtering artifacts in multivariate classification of neural time series data. bioRxiv. https://doi.org/10.1101/530220
8. VanRullen R (2011) Four common conceptual fallacies in mapping the time course of recognition. Front Psychol 2:365. https://doi.org/10.3389/fpsyg.2011.00365
9. Maris E, Oostenveld R (2007) Nonparametric statistical testing of EEG- and MEG-data. J Neurosci Methods 164(1):177–190. https://doi.org/10.1016/J.Jneumeth.2007.03.024
10. Sassenhagen J, Draschkow D (2019) Cluster-based permutation tests of MEG/EEG data do not establish significance of effect latency or location. Psychophysiology 35(2):e13335. https://doi.org/10.1111/psyp.13335
11. Kiesel A, Miller J, Jolicoeur P, Brisson B (2008) Measurement of ERP latency differences: a comparison of single-participant and jackknife-based scoring methods. Psychophysiology 45(2):250–274. https://doi.org/10.1111/j.1469-8986.2007.00618.x
12. Luck SJ (2014) An introduction to the event-related potential technique. MIT Press, Cambridge, MA. https://doi.org/10.1086/506120
13. Liesefeld HR (2018) Estimating the timing of cognitive operations with MEG/EEG latency measures: a primer, a brief tutorial, and an implementation of various methods. Front Neurosci 12:765. https://doi.org/10.3389/fnins.2018.00765
14. King JR, Dehaene S (2014) Characterizing the dynamics of mental representations: the temporal generalization method. Trends Cogn Sci 18(4):203–210. https://doi.org/10.1016/j.tics.2014.01.002
15. Allefeld C, Görgen K, Haynes J-D (2016) Valid population inference for information-based imaging: from the second-level t-test to prevalence inference. Neuroimage 141:378–392. https://doi.org/10.1016/j.neuroimage.2016.07.040
16. Hand DJ, Till RJ (2001) A simple generalisation of the area under the ROC curve for multiple class classification problems. Mach Learn 45(2):171–186. https://doi.org/10.1023/A:1010920819831
17. Miller J, Patterson T, Ulrich R (1998) Jackknife-based method for measuring LRP onset latency differences. Psychophysiology 35(1):99–115
18. Ulrich R, Miller J (2001) Using the jackknife-based scoring method for measuring LRP onset effects in factorial designs. Psychophysiology 38(5):816–827. https://doi.org/10.1111/1469-8986.3850816
19. Haufe S, Meinecke F, Goergen K, Daehne S, Haynes J-D, Blankertz B, Biessgmann F (2014) On the interpretation of weight vectors of linear models in multivariate neuroimaging. Neuroimage 87:96–110. https://doi.org/10.1016/j.neuroimage.2013.10.067
20. Brouwer GJ, Heeger DJ (2009) Decoding and reconstructing color from responses in human visual cortex. J Neurosci 29(44):13992–14003. https://doi.org/10.1523/JNEUROSCI.3577-09.2009
21. Garcia JO, Srinivasan R, Serences JT (2013) Near-real-time feature-selective modulations in human cortex. Curr Biol 23(6):515–522. https://doi.org/10.1016/j.cub.2013.02.013
22. Ester EF, Sprague TC, Serences JT (2015) Parietal and frontal cortex encode stimulus-specific mnemonic representations during

visual working memory. Neuron 87 (4):893–905. https://doi.org/10.1016/j.neuron.2015.07.013

23. Foster JJ, Sutterer DW, Serences JT, Vogel EK, Awh E (2017) Alpha-band oscillations enable spatially and temporally resolved tracking of covert spatial attention. Psychol Sci 28 (7):929–941. https://doi.org/10.1177/0956797617699167
24. Foster JJ, Sutterer DW, Serences JT, Vogel EK, Awh E (2016) The topography of alpha-band activity tracks the content of spatial working memory. J Neurophysiol 115(1):168–177. https://doi.org/10.1152/jn.00860.2015
25. Samaha J, Sprague TC, Postle BR (2016) Decoding and reconstructing the focus of spatial attention from the topography of alpha-band oscillations. J Cogn Neurosci 28 (8):1090–1097. https://doi.org/10.1162/jocn_a_00955
26. Gardner JL, Liu T (2019) Inverted encoding models reconstruct an arbitrary model response, not the stimulus. eNeuro 6(2). pii: ENEURO.0363-18.2019. https://doi.org/10.1523/ENEURO.0363-18.2019
27. Liu T, Cable D, Gardner JL (2018) Inverted encoding models of human population response conflate noise and neural tuning width. J Neurosci 38(2):398–408. https://doi.org/10.1523/JNEUROSCI.2453-17.2017
28. Sprague TC, Adam KCS, Foster JJ, Rahmati M, Sutterer DW, Vo VA (2018) Inverted encoding models assay population-level stimulus representations, not single-unit neural tuning. eNeuro 5(3). pii: ENEURO.0098-18.2018. https://doi.org/10.1523/eneuro.0098-18.2018
29. Hubel DH, Wiesel TN (1962) Receptive fields, binocular interaction and functional architecture in the cat's visual cortex. J Physiol 160:106–154

Neuromethods (2020) 151: 157–176
DOI 10.1007/7657_2019_28
© Springer Science+Business Media, LLC 2019
Published online: 29 August 2019

How to Perceive Object Permanence in Our Visual Environment: The Multiple Object Tracking Paradigm

Christian Merkel, Jens-Max Hopf, and Mircea Ariel Schoenfeld

Abstract

The ability to simultaneously maintain multiple representations through motion is an essential feature of the visual system. The multiple object tracking paradigm (MOT) has been devised in order to develop an understanding of how the visual system retains the correspondence between visual objects and their neural representation across time. A multitude of potential mechanisms maintaining this correspondence have been proposed, each being either supported or challenged by several studies. In order to provide a background for developing MOT paradigms focusing on current MOT literature, we will discuss design strategies for creating object tracking environments and present methods to quantify tracking performance under different task conditions. Finally, methods to measure resource deployment for spatial locations during tracking will be presented that will allow for inferences about potential underlying tracking mechanisms.

Keywords Multiple object tracking, Correspondence problem, Attentional spotlight, Spatial probe

1 Introduction: The Correspondence Problem

Imagine driving through the crowded streets of a city. Through the windshield you can see a number of other cars moving along side, in front or toward you. Pedestrians are crossing the street. All those objects have the potential to influence your behavior, since each of the aforementioned object trajectories can intersect with yours. Your visual system is therefore tasked with the maintenance of each of those objects' positions through space and time in order to ensure appropriate behavior. The situation becomes even more complicated since new relevant objects can enter your field of vision while others may leave it. Some objects might just be occluded for a while by other objects but might remain to be highly relevant. Take, for example, a car approaching in the rearview mirror, disappearing in the car's blind spot just to reappear right next to you moments later. It is crucial to realize that this is still the same car in contrast to an entirely new vehicle appearing out of thin air next to you. The perception of object permanence in the real world is, although intuitively effortless, still one of the insufficiently understood features of the visual system. Studying object permanence with standard visual attention paradigms poses a challenge because the majority of stimuli in such paradigms are typically kept spatially invariable. Early on, studies have shown that features of an object can be searched for [1, 2], integrated [3, 4], or temporally stored [5, 6] based on their mostly invariant spatial position. However, experimental designs employing stationary objects are artificial since in everyday life the visual system must continuously resolve object permanencies in a highly dynamic environment. Real-world visual motion challenges the construction, maintenance, and release of parallel existing associations between object identities and their spatial positions, as they change dynamically through time. This notion summarizes the central problem of correspondence that our visual system has to solve.

2 The Multiple Object Tracking Paradigm

In order to investigate how we perceive the persistence of several visual objects despite the undergoing transitions of their spatiotemporal properties, Pylyshyn devised the multiple object tracking (MOT) paradigm in 1988 [7]. The fundamental setup of MOT experiments consists of an even number of visually identical stimuli or items (Fig. 1a). Half of these items are designated as relevant targets, usually by flashing them a couple of times at the beginning of the trial. The remaining objects are distractors. Next, all objects randomly move across space for a couple of seconds until the motion seizes and the belonging of one or more items to the target set has to be determined by the subject. Usually subjects are required to either respond to one single probed item (Fig. 1a(1)) or have to select all items believed to be targets manually (Fig. 1a (2)). Surprisingly, most subjects can track up to four items simultaneously with an accuracy around 85% in this type of experiment. Since all items are visually indistinguishable during motion, object identities for the target set items have to be maintained based on the spatial positions that are subject to change. Intuitively, this task could be accomplished using spatial visual attention by continuously updating the spatial information of each relevant object (target). However, classical studies of visual selective attention suggest that the relevant process in question operates with an indivisible spatial focus at a single item level. Although still a matter of great debate [8–10], the idea of an indivisibility of attentional resources has been advanced for the spatial [11, 12], featural [13, 14], and object-based domains [15, 16]. Thus, given the resource limitations of selective attention, the question emerges: how are subjects able to maintain the spatiotemporal information of about four different items with such a high accuracy?

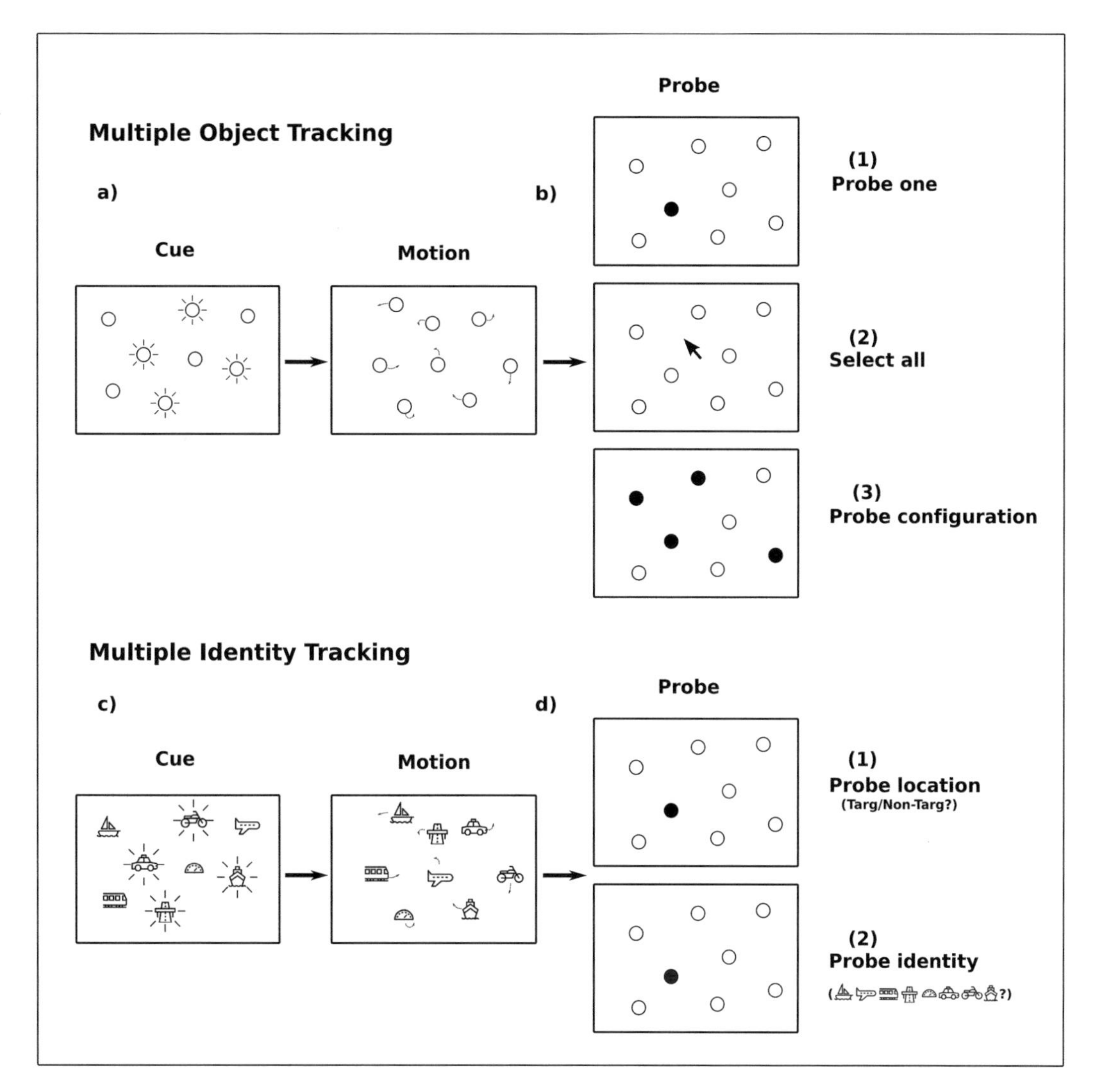

Fig. 1 The multiple object tracking paradigm. (**a**) A subject is presented with a number of visually identical items of which half are usually designated relevant during a cueing phase. Subsequently, all items move within the aperture randomly for some time (a few seconds) until motion seizes. (**b**) Tracking performance can be quantified by different probing methods: (1) The subject discriminates a single probe as being a target or nontarget item. (2) All targets have to be selected by the subject with a cursor. (3) The subject determines whether a set of probes corresponds with all the targets fully or not. Multiple identity tracking. (**c**) Subjects are required to track a number of individually labeled items. (**d**) After motion, all objects are masked, and the subjects are asked whether a probed item has been tracked or not (similar to the single probe in the MOT task). Furthermore the exact identity of the tracked item has to be reported

3 Theoretical Framework: Solving Correspondence in the Temporal Domain

In his seminal paper, Pylyshyn [17] tested the idea of a rotating attentional spotlight, updating all relevant locations of the objects in succession. Assuming the spatial displacement of a particular

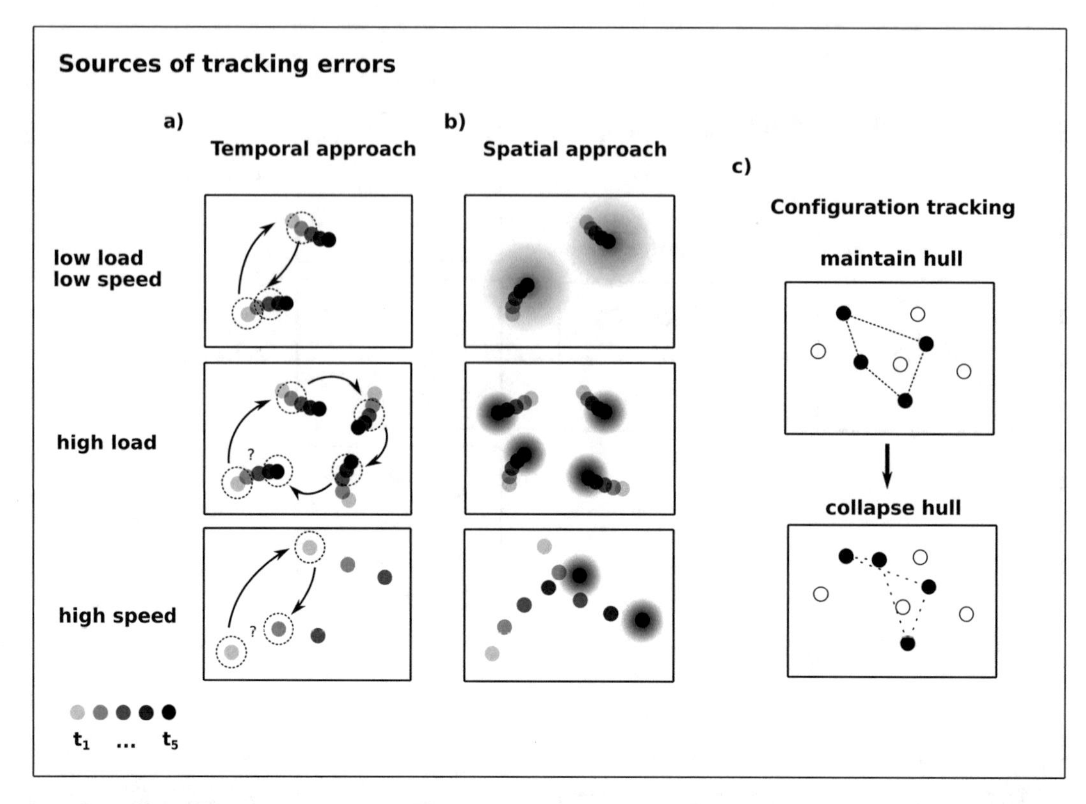

Fig. 2 Sources of tracking errors. (**a**) Solving tracking in the temporal domain: Theoretical frameworks incorporating a single attentional spotlight in solving the correspondence problem during tracking require a higher temporal resolution in more demanding settings. Equal sampling rates of item maintenance result in more frequent inter-item confusions for more relevant (second row) or faster items (third row). Eventually, the distance travelled of one item during a sampling cycle exceeds inter-item proximities, resulting in swapping of target locations. (**b**) Solving tracking in the spatial domain: A higher spatial resolution is required for the multifocal approach whenever the number of relevant items (second row) or the item speed (third row) increases in order to resolve smaller inter-item proximities. (**c**) Configuration tracking: The maintenance of the relevant targets as a singular configurational representation is disrupted primarily by collapse of its convex hull. Events of collapse occur whenever a single item crosses the imaginary edge of the entire hull

object is small enough throughout maintenance intervals, the correspondence between two nearby locations could be bridged, and the perception of object permanence should prevail (Fig. 2a). However Pylyshyn concluded that for his experiments, the average duration of the attentional shift [18, 19] would afford spatial displacements at or beyond the order of inter-item distances and therefore create confusions between relevant and irrelevant items and thus lead to even poorer-than-expected tracking performance [7, 20, 21]. Although the attentional spotlight is able to accomplish smooth pursuits of fast single moving objects, a setup with multiple items requires discrete shifts of the focus between the objects, involving multiple engaging and disengaging processes, which last some hundred milliseconds [22, 23], and are therefore definitively too long to resolve of the correspondence problem during multiple object tracking. However, the idea of a serial tracking process, cycling through the relevant items, has also been revisited [24, 25] under the assumption of constant switching times between items irrespective of their spatial proximity. More target items or higher tracking speed would however increase the temporal proximity of sampling intervals, resulting in a poorer tracking performance (Fig. 2a).

4 Theoretical Framework: Tracking Through Spatiotemporal Pointers

In order to provide an explanation for the initial behavioral data, suggesting that subjects maintain the positional information of multiple items without much effort, the concept of spatiotemporal pointers has been introduced [17, 26, 27]. These pointers are thought to attach to individual objects at a pre-attentive stage and are therefore not affected by the temporal and spatial resource limitations of spatial attention. The drawback of solving the correspondence problem at this early stage is that no properties besides spatiotemporal features can be referenced to by these pointers (also called FINSTs (fingers of instantiations)). Any subsequent access to nonspatial properties of a relevant object to determine whether it is a target or distractor requires visual spatial attention and is therefore resource dependent. The concept of spatiotemporal pointers has been subsequently refined to account for more complex observed behavioral patterns. Pointer associations are thought to degrade over time, explaining tracking accuracy declines for longer tracking periods [20, 27]. Nonetheless, the general existence of a universal number of initially available spatial pointers of around 4–5, similar to slot models of visual short-term memory [6, 28], has been maintained for a long time in the literature. A very similar concept for solving the correspondence problem has been introduced by Kahneman [29]. Object permanence is hereby maintained by storing spatiotemporal references in separate so-called object files. The appearance of a new object induces the creation of an object file, representing the objects spatiotemporal description together with other loosely filed nonspatial features of the object [29, 30]. However Kahneman as well realized the problem that nonspatial features like shape, color [20, 31, 32], and even motion direction [33] would be represented only very poorly as unbound features within the individual object file.

5 Theoretical Framework: Solving Correspondence in the Spatial Domain

The individual performance in an object tracking task was discovered to depend on a wide-ranging set of stimulus parameters.

Especially the inter-item proximity appears to be a crucial factor determining the subjects tracking ability [34, 35]. The spacing between items directly varies with the overall number of relevant items [34] as well as the mean tracking speed [34, 36, 37]. Suggesting the relevance of the spatiotemporal setup in the observed variations in tracking performance led to the development of a multifocal attentional model for tracking multiple objects [34, 38]. The main assumption is that the attentional spotlight can be spatially distributed among multiple objects. The global attentional resource can hereby be flexibly assigned toward different objects according to the current task demands. Similar to the zoom lens model [12], a low tracking demand, defined by low velocities and high inter-item distances [36, 39], allows for the assignment of low-resolution foci among several relevant objects in order to sufficiently resolve the individual items spatially. As soon as the task demands increase, more resources in the form of higher spatial resolution have to be assigned to fewer objects [40, 41] to avoid spatial confusion, which in turn leads to a decrease in tracking ability (Fig. 2b). A growing body of data suggest that the central resource can be flexibly assigned only toward multiple items within one hemifield. This would suggest two attentional processes operating independently within each hemisphere [42]. This setup allows for twice as many objects being able to be tracked within two apertures located rather in both hemifields than in only one hemifield. In comparing performances for within and between hemifield tracking, impairment due to hemifield independence seems to increase with the overall number of relevant targets, having little to no effect at all when tracking only two items [43]. Furthermore, it has been suggested recently that additional processes are required to "hand over" representations between those two central resources whenever they cross the vertical meridian.

6 Theoretical Framework: Tracking Configurations

An entirely different idea about maintaining multiple objects has been presented by Yantis [21]. He designed tracking tasks under the assumption that multiple items can form perceptual groups guided by different organization principles (Gestalt). Spatial interactions between the items are hereby an important source of information. In his tasks, the set of relevant targets would form a convex polygon hull throughout the motion phase [21]. During the crucial condition, this polygon would collapse, by one target crossing an imaginary boundary of the hull connecting all target items (Fig. 2c). Performance turned out to be worse during those trials compared when the hull of the polygon remained intact. In those cases, the relevant target items are believed to be grouped into a single higher-order object, to which attentional resources can be assigned to. Additionally, tracking has been shown to improve when subjects are given the instruction to interpret the target set as one single morphing polygon. In fact, there is mounting evidence that the correspondence problem during tracking is partly solved by an object-based mechanism operating on the configuration of the target set, rather than on multiple single locations. In this way targets that share motion features can be tracked more accurately [44]. Similarly, whenever target and distractor items are linked by visual cues, tracking is highly impaired, suggesting a spread of attentional resources across a group of objects [30]. Object-based accounts of object tracking would also easily explain the loss of item identity during maintenance, discussed below. To further investigate the role of grouping strategies as well as of attention during tracking, a different paradigm was developed. As an alternative to either probing a single item or selecting all tracked items, the probes consisted of different item configurations [45–47]. Subjects were asked to track four out of eight items. After the motion stopped, four items were always probed (Fig. 1a(3)). These four probes were different in their congruity with the target set (they matched none, one, two, three, or all four target items). The reasoning for this task is that any potential grouping advantage can only be measured by probing the configurational information itself. Interestingly, during these tasks, subjects exhibited higher performance for the detection of fully congruent probe sets compared to partly congruent or incongruent configurations. This finding provides strong support for grouping-based accounts of tracking. Tracking the relevant targets as a higher-order object requires the maintenance of a continuously changing, spatial representation within a reference space. This reference seems to be detached from a simple retinal representation. Spatially transforming the whole tracking aperture, while maintaining relative relations within the aperture between the items, similar to an air traffic control environment, does not impair tracking [37]. Local changes of spatiotemporal properties of single items do not interfere with the high-order object representation.

7 Do We Track Non-spatiotemporal Information? Multiple Identity Tracking

All accounts developing a theoretical framework for tracking multiple items discussed thus far actually focus solely on tracking the location of those items, i.e., the spatiotemporal features of the relevant items independent of any further visual properties. In fact, initial accounts of multiple object tracking that sought to solve the correspondence problem merely imply the existence of a discrete reference between the unique identity of an object and its position [17, 29], drawing no distinction between its spatial and nonspatial features during tracking. Through its identity, any feature of an object should be accessible. However several studies failed to provide sufficient evidence to support such a "discrete reference principle." For example, subjects in a tracking task were not only asked to track a number of target items but also their unique label assigned to them before the movement. Even those subjects who were very accurate in reporting whether a probed item was a target or a distractor had a very poor performance in reporting the label of the probed target [31]. Furthermore, changes of nonspatial features during tracking are only detected when they elicit salient transients [32, 48]. To further investigate how nonspatial features are maintained over time in a dynamic environment, the so-called multiple identity tracking task (MIT) was developed [20, 49] (Fig. 1b). Crucially, in this class of paradigms, all relevant objects are distinguishable throughout the tracking by one or more nonspatial features. Once the motion seizes, all objects are masked, and one of the two questions are asked. Using a single probe, one can ask whether that probed item was a target or not (similar to the original MOT task) (Fig. 1b(1)). Furthermore, one can inquire about the identity of the probe (Fig. 1b(2)). The first question provides information about the association between a location and the membership to the group of target objects. The second question probes the retention of specific object features linked to the object at that location. The general observation in these experiments is that the capacity for maintaining multiple nonspatial features is substantially lower than maintaining multiple locations [31, 49], which cannot be explained simply by a low VSTM capacity. Based on such different performance measures for object tracking and identity tracking tasks, it has been suggested that the maintenance of the location of an object and its identity might be mediated by different cognitive processes [50, 51]. According to Oksama, location information for a number of objects can be maintained by a pre-attentive process similar to an indexing mechanism. Any additional binding between a location and a nonspatial identity however would have to be continuously updated by relocating attentional resources. Importantly, the later serial, demanding task of updating specific object identities does not seem to interfere with the location tracking [31, 50]. Further support for the existence of separate processes underlying the tracking of spatial and nonspatial properties of multiple objects has been provided by eye-tracking recordings during object tracking tasks. Under free viewing conditions, subjects unconsciously tend to fixate the centroid of the group of relevant moving targets during classical object tracking tasks [52–54], which strongly argues for a parallel tracking strategy. It is important to note here however that in general no systematic differences in tracking performances have been observed between studies enforcing central fixation and those which allow free viewing. However, fixating during multiple identity tracking seems to have a crucial significance. The gaze appears more volatile during identity tracking [55] and shifts continuously across relevant objects in order to keep their nonspatial features updated [56].

8 Trajectory Extrapolation as Tracking Strategy?

Overall, the evidence from MOT and MIT paradigms clearly suggests a primary relevance for location information over nonspatial features in the perception of object permanence. Interestingly, a number of studies examining the retention of further spatial information like motion speed and direction during object tracking provide contradictory evidence regarding the effortless maintenance of spatial information. At least two reports show that subjects are well aware of the last direction target objects are heading before motion seizes, either by directly reporting the angle of the last bearing of a probed target [57] or by showing a response tuning toward the motion direction when the positions of the tracked targets have to be reproduced [41]. Intuitively, spatial information about the motion vector of the relevant objects would be useful during tracking, permitting the extrapolation of the subsequent movement direction. However, a number of studies were not able to confirm the idea of motion trajectory prediction. Introducing short gaps during motion in which all objects disappear distorts subject's perception of persistence even when the motion continues at positions consistent with the extrapolated trajectory [58, 59]. In fact, correspondence of a reappearing object appears to be solved solely based on its proximity to locations of disappearance rather than by extrapolation of movement trajectories [60]. Probabilistic models of human observers suggest that contrary to intuition, extrapolation, especially in crowded displays, is rarely employed, mainly because of a large number of potential confusions between targets and nontargets [61, 62]. A key factor to solve the problem is the number of relevant objects. Observers tracking one single object consistently report its position shortly after disappearance as being slightly ahead along the last motion vector [63], both in direction [64] and speed [65]. In these cases attention seems to pursuit the target slightly ahead along its motion vector [66]. A case for the involvement of extrapolation during multiple object tracking has been made by Scholl [67] in showing its relevance during moments of occlusion. The tracking performance does not decline when single targets are being briefly occluded by visible or non-visible barriers. On the contrary, additional resources might be applied to a target briefly out of sight to facilitate maintenance. This so-called "high-beam" effect leads to an improved detection of visual probes on locations of currently occluded targets, relative to non-occluded targets [68]. However, whether trajectories are extrapolated during occlusion or not seems to be determined by the type of occlusion. Subjects are more likely to drop a target item after a sharp discontinuation compared to when the item was gradually occluded [67]. The process of spontaneous disappearance and subsequent reappearance hereby drives the perception of two different objects, whereas a natural manner of occlusion rather provides the perception of a single object with two successive locations [69, 70]. Further support for this idea is provided by a study, in which the persisting perception of the last motion vector was reflected by an enhanced BOLD signal within the visual area MT after the continuous transition of an object and during occlusion [71]. The extrapolation of a motion vector not only appears to be limited to two objects but also be impaired by a random trajectory. The degree of single direction changes can be quantified accurately by subjects only for about two straight moving objects [33, 72]. Likewise, the aforementioned studies supporting the idea of extrapolation in a multiple object tracking setting [41, 57] both used linear tracking algorithms with straight trajectories and predictable direction changes after collision with other objects or the apertures boundary. Furthermore, irrespective of the nature of the trajectories, the extrapolation of movement trajectories does not seem to influence tracking performance for more than two items [61, 73].

9 Constructing Motion Trajectories

The growing experience with the object tracking paradigm lead to the development of a variety of algorithms for determining the trajectories of the tracked objects. Given the potential effect of predictable motion vectors on the subjects tracking ability, the application of a specific algorithm to investigate certain aspects of tracking has to be considered carefully. A common design strategy is to assign motion vectors with random direction but equal length to each object at the beginning of each trial. Typical speeds (length of vectors) range from $3°/s$ to $10°/s$. Using these parameters, a tracking capacity between 3 and 5 objects can be observed [34, 36]. The easiest but also most predictable trajectory design is to keep the motion direction constant during the trial and avoid contacting the outer boundary by a simple reversal of the x or y component of the motion vector whenever a horizontal or vertical boundary is reached, respectively [41, 73]. To further avoid occlusions of objects, such reversals can randomly be introduced, whenever two objects reach a minimum distance. Usually all objects additionally maintain a minimum distance to fixation in order to avoid any “capture” of the attentional spotlight by any of the objects. The group of Scholl and Pylyshyn used a strategy of introducing a fixed number of possible motion vectors that would differ in direction and length, by combining x and y components that can vary in value between -3 and 3. An object could thus assume

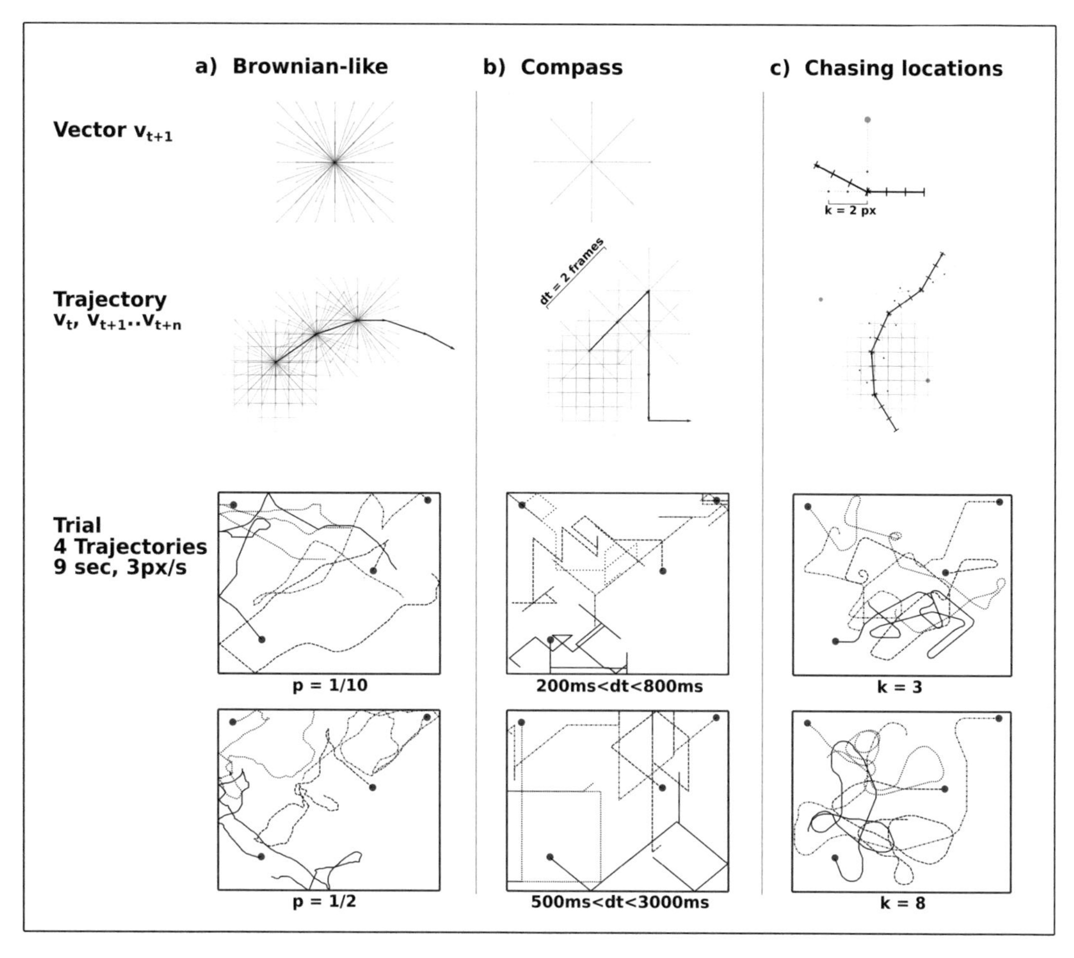

Fig. 3 Calculating trajectories—examples of motion patterns for four identical starting locations using three different algorithms. (**a**) Brownian motion: A set of 49 motion vectors is predefined. At each frame the previous vector has a chance of *p* to change the value of one of its components by 1. Changing the parameter of *p* leads to more or less erratic motion patterns. (**b**) Compass: The motion vector can assume one of the eight compass directions. The vector changes to a new random direction after a random interval *dt*. (**c**) Chasing dot: The motion vector at each frame is a normalized weighted sum of the previous vector and a directional vector between the item and a location within the aperture changing position randomly. The weight *k* determines the "curvature" of the motion pattern

49 different combinations of heading and speed (0–9°/s). At each frame each of the *x* and *y* component would have a low chance (around 10%) of changing its value by 1, introducing random, non-erratic changes in direction and speed for each object during motion (Fig. 3a). This approach is usually described as producing a Brownian-like motion in the literature. Whenever changes in motion vectors are needed due to imminent collisions or occlusions, they are randomly introduced. Oksama used a similar strategy, applying the eight compass directions as motion vectors with equal length. Individual vectors for each object change values

randomly every 200–800 ms (Fig. 3b). More advanced strategies to deal with potential infractions with other objects and boundaries include the introduction of repulsion fields, which alter the direction of an object as a function of the proximity to every other object [74, 75]. Especially the undershoot of a minimum distance to the outer boundary is usually resolved by a simple repulsion, making the imminent direction of motion at these instances fairly predictable (natural reflection of an object from a surface). In an attempt to introduce multiple truly independent, non-predictable trajectories, Merkel et al. suggested an algorithm updating the direction of a motion vector by adding a weighted target vector of an imagined random target location within the boundaries of the aperture [47]. Each object would thus essentially "chase" a dot jumping to new random locations in random intervals (10–500 ms) (Fig. 3c). A large number of trajectories stored offline would be combined in a subsequent step to form a trial of eight objects that maintains preset constraints in each frame and move fully independent of each other and the boundary. This approach has been used to attach two trajectory histories with a shared starting location and reversed starting directions to one object trajectory in order to be able to control for location at a particular frame during tracking [46].

10 Quantifying Tracking Performance: Capacity Estimates

Most tracking accounts share the concept of a bijective correspondence between single objects and referencing units of a central resource (attentional spotlights, spatiotemporal pointers, or object files). (Tracking higher-order representations can be viewed as an exception here.) A fundamental process in these concepts is the quantification of the referencing units that directly translate into the potential number of targets that can simultaneously be tracked. This tracking capacity can vary across subjects as well as across different tracking task designs, but it is defined by the underlying model and can be experimentally measured using behavioral data. In detail, most tracking tasks share the following sequence of events. Subjects are presented with a number of visually identical items, of which some are cued as relevant targets. The task is to subsequently keep track of the target items throughout the motion phase. The following retrieval phase usually falls into two major categories. Subjects either have to select all tracked items or they have to indicate a single probed item as being part of the targets or distractors (Fig. 1a(1, 2)). In the former task, after movement halts, a cursor appears with which all to believe targets have to be marked. In this class of tracking tasks, tracking accuracy p is defined as the ratio between correctly selected objects and the absolute amount of target objects across all trials, hence the hit rate. Any selection of a nontarget item would constitute an error. Initially, Scholl [30] determined p in its simplest form as all correct selections among the target items and the correct random guesses for the remaining selections among the targets. Since there are as many targets as distractors, the guess rate is 0.5, therefore: $p = m/n + ((n - m)/n)/2$ (1). m denoted the number of effectively tracked items (ENOT) [20, 30] or tracking capacity [74], whereas n is the number of targets. (1) can be reduced to $p = (m/n + 1)/2$ (2). This term can be solved for m and allows for an estimate of the individual tracking capacity of a subject based on the performance: $m = n(2p - 1)$ (3). This relation has been used for determining a lower bound for p assuming subjects are unable to keep track of multiple objects [30, 76]. Note that in this case, m equals one instead of zero, meaning that subjects would always be able to track one single item. Hence the lower bound of p in a task of tracking four items would be 0.625 instead of 0.5. The pitfall of (1) is that this relation systematically underestimates m, since the guessing term assumes drawing a random item from the same initial n targets and n nontargets, (1) therefore constituting a sampling with replacement. However in the actual task, after choosing m correct targets, the target-distractor ratio does shift toward the distractors, making it more likely to randomly choose more distractors than targets. Hulleman [77] rectified this issue by introducing an estimate for p assuming sampling without replacement. In this case, still assuming an equal number of targets and distractors, $p = (m + (n - m)^2/(2n - m))/n$ (4). Solving for m, the tracking capacity can be deduced from the subjects performance through $m = n(2p - 1)/p$ (5) [34, 49]. In this case the lower performance bound for tracking only one item among four targets would be estimated at 0.571 instead of 0.625.

For tasks in which subjects are requested to identify a single probe as target or distractor during the retrieval phase, the underlying reasoning for determining performance accuracy p based on the number of effectively tracked items m is very similar. In these tasks p would be the ratio between correctly identified targets (hits) plus distractors (correct rejections) and the total number of trials presented. Assuming a best guess strategy [21], subjects would respond to a probe which has not been tracked, based on the perceived probability for that probe to be either a target or a distractor. Whenever at least one target was tracked and none of the tracked targets are probed, those probabilities would always tilt toward responding "nontarget," since out of the remaining non-tracked items, a majority would be distractors. In 50% of all trials, a target is probed; thus the amount of hits (target probe being one of the tracked items) equals to $(m/n)/2$. The same assumptions lead to the conclusion of 100% correct rejections for the trials including a probed distractor, thus 1/2. Overall the relation of performance and number of effectively tracked items for subjects using the most conservative best guess can be expressed as $p = (m/n)/2 + 1/2$ (6). Interestingly, (6) is equivalent to the performance for selection tasks assuming sampling with replacement (2). In this case the capacity estimates for tasks involving a single probe do not differ from those using the selection method. Introducing random guessing into (6) and therefore circumventing the artificial assumption of a 100% rejection rate whenever the probe was not tracked lead to a relation between performance and capacity of $p = (m/n + ((n - m)/n)/2)/2 + 1/4$ (7). Note that in this case, even a capacity meeting the task demand $m = n$ would lead to a maximum performance rate of 75% only. This is due to the fact that responses toward probed distractors would remain fully random, even though all targets are being tracked. This problem might be addressed by introducing a separate capacity measure for distractors [77], although one has to reflect on which underlying tracking model any applied capacity estimate is based on.

11 Quantifying Allocated Resources: Using a Location Probe

Determining tracking capacities is useful to identify stimulus parameters that either promote or disturb successful tracking. However, in order to tackle the question of *how* exactly multiple items can be maintained simultaneously, the tracking process itself has to be analyzed. The most commonly used strategy is to introduce an additional probe detection task during the motion phase of each tracking trial. Probe detection has been utilized early on in psychophysics in order to sample the spatial distribution of attentional resources within a static visual display [11]. The same principle can be applied for the dynamic situation of a multiple object tracking task. For a short period of time (about 50–100 ms) during motion, a small dot appears on a specific spatial location relative to other dynamic visual stimuli (Fig. 4). The detection performance for this probe allows for inferences about the allocation of attentional resources toward this location at this particular moment in time. Several facilitatory as well as inhibitory processes are believed to influence the complex spatiotemporal distribution of attentional resources during tracking. This complicates the interpretation of probe detection results greatly as indicated by a variety of contradicting results. Nevertheless, one of the more consistent patterns observed is an improved detection rate for probes appearing on targets versus on distractors during tracking [27, 68, 78–80]. A major point of dispute relates to the potential enhancement for targets and inhibition of distractors contributing to this effect. A slightly higher [78] or similar [79] detection rate for probes appearing on empty locations compared to the location of a target might suggest a primary distractor suppression mechanism involved in object tracking. However, probes appearing closer to visual stimuli might be subject to masking, thus be more difficult to

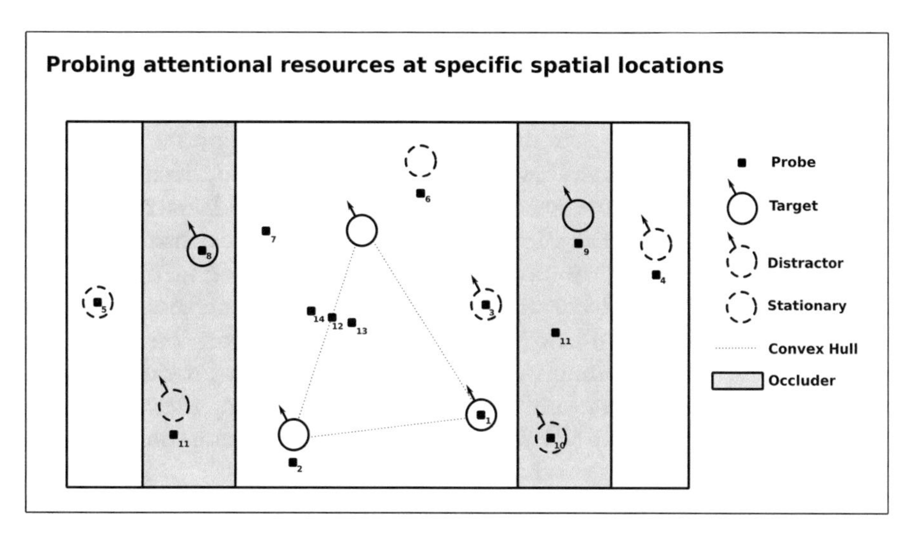

Fig. 4 Probing attentional resources at specific spatial locations. In a variety of studies collecting behavioral and electrophysiological data, probes have been introduced at (1) and around (2) the moving target, moving distractor (3, 4) as well as further task-irrelevant stationary items (5, 6). Behavioral and electrophysiological responses toward those probes are usually compared to a baseline probe located at an empty location (7). Spatial attention has been probed on and close to momentarily occluded targets (8, 9) and distractors (10, 11) as well. Most consistently described are target enhancement and distractor inhibition effects. Moreover a suppressive center-surround configuration has been found for target as well as distractor locations. Most intriguing is the "high-beam" effect elevating the processing of momentarily occluded objects. Probes have also been introduced at (12) and around (13, 14) the convex hull of the configuration of relevant items

detect in general [68, 79]. Further studies suggest a suppressive center-surround profile of enhancement or inhibition for targets and distractors, respectively [68, 81], with detection rates being highest close to an item compared to directly on the item. However, using elongated objects this effect seems to disappear with attentional enhancement being largest at the center of the object [82]. Suggesting a certain degree of extrapolation contributing to multiple object tracking, one study showed probe detection rates for probes near a target being biased toward locations within the future trajectory of the target [81]. Additionally, probes appearing on target locations, which are currently occluded, can be detected more accurately than probes on non-occluded targets [68].

The allocation of resources toward a location can additionally be studied by measuring the electrophysiological response elicited by the probe appearing at that location. The probe itself can remain task-irrelevant which allows for predictions about the spatial distribution of attention without the subjects' awareness of the probe [83–85]. Spatial attention is expected to enhance the amplitude of the P1 and N1 components around 100–180 ms after the appearance of the probe. The first study utilizing a task-irrelevant probe to elicit an evoked electrophysiological response in an object tracking task found enhanced P1 and N1 responses for target locations, while those components did not differ for distractors, empty locations, or stationary items [86]. Interestingly, these results primarily suggest a process of target enhancement during tracking, contrary to earlier behavioral probe detection studies. Further studies found an enhancement specifically for targets over distractors and empty (neither target nor distractor there) locations during the P1 time range [87, 88] or an enhancement for targets and empty locations over distractors during the N1 time range [89]. Electrophysiological studies thus suggest a combination of target enhancement and distractor suppression mechanisms both contributing to multiple object tracking. Providing evidence for a configuration-based tracking mechanism, Merkel [46] found an enhanced P1 effect for probes appearing on the convex hull between two relevant targets.

Comparing this multitude of results is complicated by the diversity of experimental designs including differences in the number of relevant targets used, different motion patterns, and probe positioning. Measuring electrophysiological responses toward visual probes requires a high degree of matching visual stimulation parameters across conditions (probe target vs. probe distractor) at the time of probe occurrence. One has to keep in mind that early electrophysiological components originate in visual networks that are retinotopically organized. Thus, amplitudes elicited by visual probes not only differ in relation to their relative location to target or distractor stimuli but also depending on their absolute spatial location within the visual field.

12 Summary

In order to explain object persistence for multiple objects in motion, a number of temporal, spatial, and objects-based accounts have been put forward. All have their specific merits but also limitations, and typically they are not successful at explaining all currently available behavioral and electrophysiological findings. Multifocal approaches fail to recognize the advantage of tracking a set of relevant items as a single morphing object through space, which is supported by electrophysiological evidence of enhanced processing of solely imaginary parts (e.g., imagined connection between two target items) of this higher-order object. On the other hand, pure object-based accounts of tracking are less successful in accounting for parametric variations of tracking performance due to changes of physical stimulus features. A major problem to reconcile between the different tracking theories is the huge variety of stimulus parameters that can influence tracking performance. These include the actual number of relevant tracking items, item speed, spatial proximity, occlusion, gestalt principles governing the imagined shape of the configuration of relevant targets, hemifield independence, and trajectory predictability. All these features have a potential impact on our ability to track multiple objects, thereby highlighting the importance but also the complexity of a well-controlled experimental design for studying potential tracking mechanisms.

References

1. Wolfe JM (1994) Guided Search 2.0 A revised model of visual search. Psychon Bull Rev 1 (2):202–238. https://doi.org/10.3758/BF03200774
2. Treisman AM, Gelade G (1980) A feature-integration theory of attention. Cogn Psychol 12(1):97–136
3. Treisman AM (1996) The binding problem. Curr Opin Neurobiol 6(2):171–178
4. Driver J, Davis G, Russell C et al (2001) Segmentation, attention and phenomenal visual objects. Cognition 80(1-2):61–95
5. Jonides J, Smith EE, Koeppe RA et al (1993) Spatial working memory in humans as revealed by PET. Nature 363(6430):623–625
6. Luck SJ, Vogel EK (1997) The capacity of visual working memory for features and conjunctions. Nature 390:279–281
7. Pylyshyn Z, Storm RW (1988) Tracking multiple independent targets: evidence for a parallel tracking mechanism. Spat Vis 3:1–19
8. Muller MM, Malinowski P, Gruber T et al (2003) Sustained division of the attentional spotlight. Nature 424(6946):309–312. https://doi.org/10.1038/nature01812
9. McMains SA, Somers DC (2004) Multiple spotlights of attentional selection in human visual cortex. Neuron 42(4):677–686
10. Cave KR, Bichot NP (1999) Visuo-spatial attention: beyond a spotlight model. Psychon Bull Rev 6:204–223
11. Posner MI (1980) Orienting of attention. Q J Exp Psychol 32:3–25
12. Eriksen CW, James JDS (1986) Visual attention within and around the field of focal attention: a zoom lens model. Percept Psychophys 40(4):225–240
13. Schoenfeld MA, Hopf JM, Merkel C et al (2014) Object-based attention involves the sequential activation of feature-specific cortical modules. Nat Neurosci 17(4):619–624. https://doi.org/10.1038/nn.3656
14. Huang L, Treisman A, Pashler H (2007) Characterizing the limits of human visual awareness. Science 317(5839):823–825
15. VanRullen R (2013) Visual attention: a rhythmic process? Curr Biol 23(24):R1110–R1112. https://doi.org/10.1016/j.cub.2013.11.006
16. Scholl BJ (2001) Objects and attention: the state of the art. Cognition 80(1–2):1–46
17. Pylyshyn Z (1989) The role of location indexes in spatial perception: a sketch of the FINST spatial-index model. Cognition 32:65–97
18. Posner MI, Walker JA, Friedrich FA et al (1987) How do the parietal lobes direct covert attention? Neuropsychologia 25(1A):135–145
19. Eriksen CW, Murphy TD (1987) Movement of attentional focus across the visual field: a critical look at the evidence. Percept Psychophys 42 (3):299–305
20. Oksama L, Hyona J (2004) Is multiple object tracking carried out automatically by an early vision mechanism independent of higher-order cognition? An individual difference approach. Vis Cogn 11(5):631–671
21. Yantis S (1992) Multielement visual tracking: attention and perceptual organization. Cogn Psychol 24(3):295–340
22. Carlson TA, Hogendoorn H, Verstraten FA (2006) The speed of visual attention: what time is it? J Vis 6(12):1406–1411. https://doi.org/10.1167/6.12.6
23. Horowitz TS, Holcombe AO, Wolfe JM et al (2004) Attentional pursuit is faster than attentional saccade. J Vis 4(7):585–603. https://doi.org/10.1167/4.7.6
24. Hogendoorn H, Carlson TA, Verstraten FA (2007) The time course of attentive tracking. J Vis 7(14):2.1–2.10. https://doi.org/10.1167/7.14.2
25. Holcombe AO, Chen WY (2013) Splitting attention reduces temporal resolution from 7 Hz for tracking one object to <3 Hz when tracking three. J Vis 13(1):12. https://doi.org/10.1167/13.1.12
26. Pylyshyn Z (2001) Visual indexes, preconceptual objects, and situated vision. Cognition 80:127–158
27. Sears CR, Pylyshyn ZW (2000) Multiple object tracking and attentional processing. Can J Exp Psychol 54(1):1–14

28. Cowan N (2001) The magical number 4 in short-term memory: a reconsideration of mental storage capacity. Behav Brain Sci 24 (1):87–114. discussion 114-185
29. Kahneman D, Treisman A, Gibbs BJ (1992) The reviewing of object files: object-specific integration of information. Cogn Psychol 24 (2):175–219
30. Scholl BJ, Pylyshyn ZW, Feldman J (2001) What is a visual object? Evidence from target merging in multiple object tracking. Cognition 80(1–2):159–177
31. Pylyshyn Z (2004) Some puzzling findings in multiple object tracking: I. Tracking without keeping track of object identities. Vis Cogn 11(7):801–822
32. Saiki J (2003) Feature binding in object-file representations of multiple moving items. J Vis 3(1):6–21. https://doi.org/10.1167/3.1.2
33. Tripathy SP, Barrett BT (2004) Severe loss of positional information when detecting deviations in multiple trajectories. J Vis 4 (12):1020–1043. https://doi.org/10.1167/4.12.4
34. Alvarez GA, Franconeri SL (2007) How many objects can you track? Evidence for a resource-limited attentive tracking mechanism. J Vis 7 (13):14.11–14.10
35. Shim WM, Alvarez GA, Jiang YV (2008) Spatial separation between targets constrains maintenance of attention on multiple objects. Psychon Bull Rev 15(2):390–397
36. Franconeri SL, Jonathan SV, Scimeca JM (2010) Tracking multiple objects is limited only by object spacing, not by speed, time, or capacity. Psychol Sci 21(7):920–925. https://doi.org/10.1177/0956797610373935
37. Liu G, Austen EL, Booth KS et al (2005) Multiple-object tracking is based on scene, not retinal, coordinates. J Exp Psychol Hum Percept Perform 31(2):235–247. https://doi.org/10.1037/0096-1523.31.2.235
38. Cavanagh P, Alvarez GA (2005) Tracking multiple targets with multifocal attention. Trends Cogn Sci 9:349–354
39. Franconeri SL, Alvarez GA, Cavanagh P (2013) Flexible cognitive resources: competitive content maps for attention and memory. Trends Cogn Sci 17(3):134–141. https://doi.org/10.1016/j.tics.2013.01.010
40. Chen WY, Howe PD, Holcombe AO (2013) Resource demands of object tracking and differential allocation of the resource. Atten Percept Psychophys 75(4):710–725. https://doi.org/10.3758/s13414-013-0425-1
41. Iordanescu L, Grabowecky M, Suzuki S (2009) Demand-based dynamic distribution of attention and monitoring of velocities during multiple-object tracking. J Vis 9(4):1.1–1.12. https://doi.org/10.1167/9.4.1
42. Alvarez GA, Cavanagh P (2005) Independent resources for attentional tracking in the left and right visual fields. Psychol Sci 16:637–643
43. Stormer VS, Alvarez GA, Cavanagh P (2014) Within-hemifield competition in early visual areas limits the ability to track multiple objects with attention. J Neurosci 34 (35):11526–11533. https://doi.org/10.1523/JNEUROSCI.0980-14.2014
44. Howe PD, Cohen MA, Pinto Y et al (2010) Distinguishing between parallel and serial accounts of multiple object tracking. J Vis 10 (8):11. https://doi.org/10.1167/10.8.11
45. Merkel C, Hopf JM, Heinze HJ et al (2015) Neural correlates of multiple object tracking strategies. Neuroimage 118:63–73. https://doi.org/10.1016/j.neuroimage.2015.06.005
46. Merkel C, Hopf JM, Schoenfeld MA (2017) Spatio-temporal dynamics of attentional selection stages during multiple object tracking. Neuroimage 146:484–491. https://doi.org/10.1016/j.neuroimage.2016.10.046
47. Merkel C, Stoppel CM, Hillyard SA et al (2014) Spatio-temporal patterns of brain activity distinguish strategies of multiple-object tracking. J Cogn Neurosci 26(1):28–40. https://doi.org/10.1162/jocn_a_00455
48. Bahrami B (2003) Object property encoding and change blindness in multiple object tracking. Vis Cogn 10:949–963
49. Horowitz TS, Klieger SB, Fencsik DE et al (2007) Tracking unique objects. Percept Psychophys 69(2):172–184
50. Botterill K, Allen R, McGeorge P (2011) Multiple-object tracking: the binding of spatial location and featural identity. Exp Psychol 58 (3):196–200. https://doi.org/10.1027/1618-3169/a000085
51. Oksama L, Hyona J (2008) Dynamic binding of identity and location information: a serial model of multiple identity tracking. Cogn Psychol 56(4):237–283. https://doi.org/10.1016/j.cogpsych.2007.03.001
52. Fehd HM, Seiffert AE (2008) Eye movements during multiple object tracking: where do participants look? Cognition 108(1):201–209. https://doi.org/10.1016/j.cognition.2007.11.008
53. Fehd HM, Seiffert AE (2010) Looking at the center of the targets helps multiple object tracking. J Vis 10(4):19.11–19.13. https://doi.org/10.1167/10.4.19

54. Zelinsky GJ, Neider MB (2008) An eye movement analysis of multiple object tracking in a realistic environment. Vis Cogn 16 (5):553–566
55. Oksama L, Hyona J (2016) Position tracking and identity tracking are separate systems: evidence from eye movements. Cognition 146:393–409. https://doi.org/10.1016/j.cognition.2015.10.016
56. Li J, Oksama L, Hyona J (2019) Model of multiple identity tracking (MOMIT) 2.0: resolving the serial vs. parallel controversy in tracking. Cognition 182:260–274. https://doi.org/10.1016/j.cognition.2018.10.016
57. Horowitz TS, Cohen MA (2010) Direction information in multiple object tracking is limited by a graded resource. Atten Percept Psychophys 72(7):1765–1775. https://doi.org/10.3758/APP.72.7.1765
58. Fencsik DE, Klieger SB, Horowitz TS (2007) The role of location and motion information in the tracking and recovery of moving objects. Percept Psychophys 69(4):567–577
59. Keane BP, Pylyshyn ZW (2006) Is motion extrapolation employed in multiple object tracking? Tracking as a low-level, non-predictive function. Cogn Psychol 52(4):346–368
60. Franconeri SL, Pylyshyn ZW, Scholl BJ (2012) A simple proximity heuristic allows tracking of multiple objects through occlusion. Atten Percept Psychophys 74(4):691–702. https://doi.org/10.3758/s13414-011-0265-9
61. Zhong SH, Ma Z, Wilson C et al (2014) Why do people appear not to extrapolate trajectories during multiple object tracking? A computational investigation. J Vis 14(12):12. https://doi.org/10.1167/14.12.12
62. Vul E, Frank M, Alvarez GA et al (2009) Explaining human multiple object tracking as resource-constrained approximate inference in a dynamic probabilistic model. Adv Neural Inf Process Syst 22:1955–1963
63. Freyd JJ, Finke RA (1984) Representational momentum. J Exp Psychol Learn Mem Cogn 10:126–132
64. Freyd JJ, Finke RA (1985) A velocity effect for representational momentum. Bull Psychon Soc 23:443–446
65. Finke RA, Shyi GC (1988) Mental extrapolation and representational momentum for complex implied motions. J Exp Psychol Learn Mem Cogn 14(1):112–120
66. Verghese P, McKee SP (2002) Predicting future motion. J Vis 2(5):413–423. https://doi.org/10.1167/2.5.5
67. Scholl BJ, Pylyshyn ZW (1999) Tracking multiple items through occlusion: clues to visual objecthood. Cogn Psychol 38(2):259–290
68. Flombaum JI, Scholl BJ, Pylyshyn ZW (2008) Attentional resources in visual tracking through occlusion: the high-beams effect. Cognition 107(3):904–931. https://doi.org/10.1016/j.cognition.2007.12.015
69. Watamaniuk SN, McKee SP (1995) Seeing motion behind occluders. Nature 377 (6551):729–730. https://doi.org/10.1038/377729a0
70. Yantis S (1995) Perceived continuity of occluded visual objects. Psychol Sci 6:182–186
71. Olson IR, Gatenby JC, Leung HC et al (2004) Neuronal representation of occluded objects in the human brain. Neuropsychologia 42 (1):95–104
72. Tripathy SP, Narasimhan S, Barrett BT (2007) On the effective number of tracked trajectories in normal human vision. J Vis 7(6):2. https://doi.org/10.1167/7.6.2
73. Howe PD, Holcombe AO (2012) Motion information is sometimes used as an aid to the visual tracking of objects. J Vis 12(13):10. https://doi.org/10.1167/12.13.10
74. Bettencourt KC, Somers DC (2009) Effects of target enhancement and distractor suppression on multiple object tracking capacity. J Vis 9 (7):9. https://doi.org/10.1167/9.7.9
75. Intriligator J, Cavanagh P (2001) The spatial resolution of visual attention. Cogn Psychol 43 (3):171–216. https://doi.org/10.1006/cogp.2001.0755
76. VanMarle K, Scholl BJ (2003) Attentive tracking of objects versus substances. Psychol Sci 14 (5):498–504. https://doi.org/10.1111/1467-9280.03451
77. Hulleman J (2005) The mathematics of multiple object tracking: from proportions correct to number of objects tracked. Vis Res 45 (17):2298–2309. https://doi.org/10.1016/j.visres.2005.02.016
78. Pylyshyn ZW, Annan V Jr (2006) Dynamics of target selection in multiple object tracking (MOT). Spat Vis 19(6):485–504
79. Pylyshyn ZW, Haladjian HH, King CE et al (2008) Selective nontarget inhibition in multiple object tracking. Vis Cogn 16 (8):1011–1021. https://doi.org/10.1080/13506280802247486
80. Tran A, Hoffman JE (2016) Visual attention is required for multiple object tracking. J Exp Psychol Hum Percept Perform 42 (12):2103–2114. https://doi.org/10.1037/xhp0000262

81. Atsma J, Koning A, van Lier R (2012) Multiple object tracking: anticipatory attention doesn't "bounce". J Vis 12(13):1. https://doi.org/10.1167/12.13.1
82. Alvarez GA, Scholl BJ (2005) How does attention select and track spatially extended objects? New effects of attentional concentration and amplification. J Exp Psychol Gen 134 (4):461–476
83. Heinze HJ, Mangun GR, Burchert W et al (1994) Combined spatial and temporal imaging of brain activity during visual selective attention in humans. Nature 372(8):543–546
84. Hillyard SA, Anllo-Vento L (1998) Event-related brain potentials in the study of visual selective attention. Proc Natl Acad Sci 95 (3):781–787
85. Hopf J-M, Boehler CN, Luck SJ et al (2006) Direct neurophysiological evidence for spatial suppression surrounding the focus of attention in vision. Proc Natl Acad Sci 103 (4):1053–1058
86. Drew T, McCollough AW, Horowitz TS et al (2009) Attentional enhancement during multiple-object tracking. Psychon Bull Rev 16 (2):411–417. https://doi.org/10.3758/PBR.16.2.411
87. Stormer VS, Li SC, Heekeren HR et al (2013) Normal aging delays and compromises early multifocal visual attention during object tracking. J Cogn Neurosci 25(2):188–202. https://doi.org/10.1162/jocn_a_00303
88. Sternshein H, Agam Y, Sekuler R (2011) EEG correlates of attentional load during multiple object tracking. PLoS One 6(7):e22660. https://doi.org/10.1371/journal.pone.0022660
89. Doran MM, Hoffman JE (2010) The role of visual attention in multiple object tracking: evidence from ERPs. Atten Percept Psychophys 72(1):33–52. https://doi.org/10.3758/APP.72.1.33

Neuromethods (2020) 151: 177–205
DOI 10.1007/7657_2019_24
© Springer Science+Business Media, LLC 2019
Published online: 11 May 2019

Combining Transcranial Direct Current Stimulation and Electrophysiology to Understand the Memory Representations that Guide Attention

Shrey Grover and Robert M. G. Reinhart

Abstract

How attention is used during visual search is intricately associated with memory. A considerable body of work has demonstrated that representations in both working memory and long-term memory can guide attention in a variety of different circumstances. Neural evidence of such memory-mediated attentional guidance has been elegantly shown using noninvasive electrophysiological measurements of human brain activity. Recently, with the rising popularity of noninvasive brain stimulation techniques, such as transcranial direct current stimulation (tDCS), researchers have been able to gain insight into the causal mechanisms of memory-guided attention. Here, we review our current understanding of the role of memory representations in guiding attention and how tDCS can be used to characterize the mechanisms and establish causal relationships. We further discuss the translational implications of using tDCS to alleviate memory-based attentional deficits in psychiatric disorders, such as schizophrenia.

Keywords Memory-guided attention, Attention, Working memory, Long-term memory, tDCS, CDA, Anterior P1, Schizophrenia, N2pc

1 Introduction

The human brain is constantly interacting with a complex environment. At any given moment, our senses are bombarded with numerous streams of information. Yet, for the most part, we function seamlessly. We identify what is relevant, disregard what is not, and store information in the "backs of our minds" for later use. The ability to selectively attend to targets within a field of distracting information is essential for goal-directed behavior. Whether it is food in the grasslands, an explosive among luggage, or an abnormal lesion in a mammogram, our attention plays a vital role in determining what is perceived, remembered, and processed for later action. But, what guides our perceptual attention toward some inputs in the environment and away from others?

Searching for something is easier when we know what we are looking for. When we hold a representation of a search target in memory, we are able to find it better. For example, remembering that the keychain attached to our car key is blue and shiny will help us see it more easily in a bowl of similar-looking keys. The internal representations of search targets, which can be stored in different memory systems of the brain, have been of interest to scientists for nearly a century, but only recently have we begun to examine the nature of these memory representations and how they guide our perceptual attention and behavior.

Advances in neuroscience have considerably improved our knowledge about how information stored in working memory or previously laid down in long-term memory can guide future action. New behavioral paradigms combined with noninvasive neuromonitoring methods, such as electroencephalography (EEG) with its exquisite temporal resolution, have provided us with a window into the neural dynamics of attentional guidance processes as they rapidly unfold over time. The opportunity to visualize the dynamics of the memory mechanisms guiding attention has opened up an entirely new realm of theoretically important, empirical hypotheses about the nature of the cognitive mechanisms guiding attention. In addition to electrophysiological measurements of brain activity, the field of memory-guided attention may benefit from a new class of neuroscience tools that offer the possibility of directly modulating neural activity in a noninvasive and reversible manner. Such tools promise the unique opportunity to establish causal relations between the brain, cognitive processes, and behavior and to more rigorously address questions about the nature of goal-driven behavior.

In this chapter, we focus on the neuromodulation method of transcranial direct current stimulation (tDCS). We discuss how tDCS has been used in conjunction with electrophysiological measurements of human brain activity to examine the memory systems underlying the control of attention during the analysis of complex visual scenes. We begin with a brief overview of the major theories of memory-guided attention. Then we discuss the electrophysiological signatures in human EEG that provide the correlative evidence for these theories. This is followed by a brief discussion on tDCS and its purported mechanisms of action. Next, we consider in some detail how tDCS has helped clarify the dynamic interplay among working memory and long-term memory in guiding visual search. Finally, we discuss the translational potential of tDCS and its ability to temporarily reverse attentional impairments by modulating faulty memory representations in schizophrenia.

2 Memory-Guided Attention: Theory and Electrophysiological Evidence

Intuitively, we know that we can pay attention to only a small part of what we see at any given time. Since there are often important consequences for what we pay attention to, identifying relevant inputs from the environment is critical. Current theories describe attention as an interplay between factors that make an input stand out against others (the bottom-up, stimulus-driven capture of attention) and factors that allow the volitional focus on inputs relevant for current task demands (top-down control and guidance of attention) [1–3]. Here, we discuss some of these top-down factors that guide our attention.

Theoretical accounts of attention have sought to explain how top-down factors influence the selection of one input over others. As Desimone and Duncan [4] suggest, "objects in the visual field compete for representation, analysis, or control." In an influential computational theory of visual attention [5, 6], Bundesen and colleagues propose how this competition could be solved in favor of one input over others. This critical role is accomplished by the memory systems of the brain. Representations of a target object in these memory systems can provide "templates" to the attentional system describing what to search for in the environment. In this role, memory systems provide top-down control of information to the attention systems. These templates are eventually used to categorize inputs as relevant targets or irrelevant distractors [5, 6]. Maintenance of an internal template to match inputs in order to identify them as targets has been a prevailing feature in many theoretical and descriptive accounts of visual search (e.g. [4, 7–9]). In the brain, these representations are thought to provide a biasing signal that feeds the neurons engaged in visual perception, increasing their sensitivity for detecting target features of interest [4, 6, 9].

The question that naturally follows is how are these templates maintained within our cognitive makeup? The predominant view implicates working memory in this process. When searching for a target object in a cluttered scene, an active representation of the target is maintained in working memory. These representations provide biasing signals to perceptual areas. To experimentally validate this hypothesis, variations of the memory-guided visual search paradigm have been used across laboratories (see Fig. 1a for an example). In these tasks, participants are required to search for a target in an array of non-target distractors. However, before the search array is presented, participants are cued with the identity of the upcoming search target. During the time interval between cue and search array presentation, the cued target is thought to be actively maintained in working memory. That is, because participants are completely informed beforehand about what the target looks like, they can maintain a representation of the target in

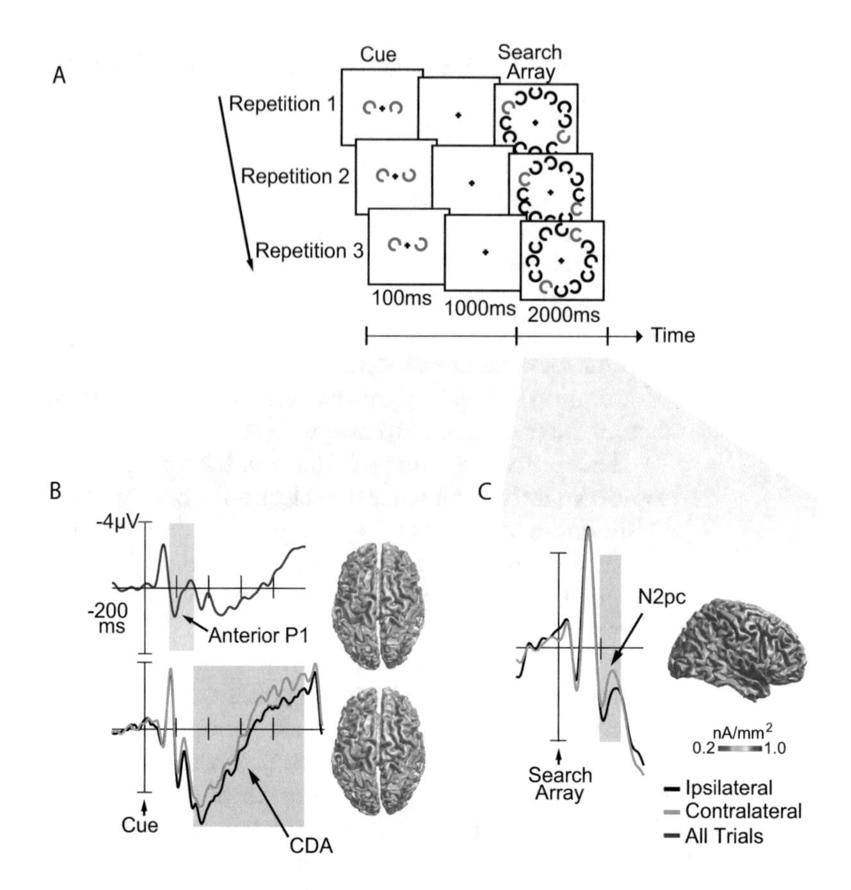

Fig. 1 Memory-guided visual search paradigm and the ERP components used to study working memory, long-term memory, and attentional focus. (**a**) Task-relevant cue (red) is presented indicating the target that should be searched for in the search array. A task-irrelevant cue (blue) is also presented to balance visual input across hemifields. Following a delay period, the search array is presented, and the participant indicates whether the target is present in the array or not. The cue could remain fixed across runs of three to seven trials. (**b**) During the delay period between presentation of cue and search array, memory representations of the cue can be observed using electrophysiology. Working memory involvement is observed using the CDA (bottom panel), while long-term memory can be indexed using the anterior P1 (top panel). Both representations have distinct spatial and temporal profiles, with anterior P1 appearing early relative to the CDA (gray shaded areas), and the former having a bilateral frontal topography while the latter is localized to the contralateral posterior sites. (**c**) Both memory representations improve attentional focus which is measured using the N2-posterior-contralateral potential. The N2pc is evident after the presentation of the search array suggesting a completely independent temporal profile

working memory to provide a biasing signal to facilitate target detection and speed behavioral responses. Thus, working memory can be an appropriate store for maintaining such templates.

Indeed, extensive work in cognitive psychology has found reliable evidence for the role of working memory representations in guiding attention. When a representation is held in working memory, participants are quicker to attend to objects that match the representation than those which do not [10–17]. This tuning of attention is often measured as changes in reaction times or how long it takes to react to a target when some representation is within memory versus when it is not. If a representation is maintained in working memory, attention is likely to be drawn to the matching objects. However, a parallel body of work has also suggested that simply keeping a representation in working memory is not enough to automatically attract attention. Such representations do not always capture attention and can even be strategically used to guide attention away from potential distractors [18–23]. Further, working memory representations can differentially affect target selection based on current task goals [16, 17], task difficulty and demands [24–27], magnitude of rewards [28], and state of the representation within working memory [29, 30]. Such evidence agrees with more recent additions to the theories of visual attention which propose that upstream executive control mechanisms can provide directions on how working memory representations should guide attention [6, 31]. Together, these studies have further expanded the initial proposition that representations in working memory guide attention, by suggesting that the exact manifestation could differ across different environmental conditions.

With recognition of this complexity, the need for supplementing behavioral evidence with established neural measures becomes necessary. The original theoretical models made explicit proposals on neuronal modulation by memory contents [4]. This strengthens the need to determine neural correlates of working memory in attention. The first set of neural evidence in favor of working memory-guided attention came from studies in nonhuman primate electrophysiology. Chelazzi and colleagues observed sustained firing in the inferotemporal cortex during maintenance of the cued target representation prior to visual search [32, 33]. Firing activity even after the offset of visual input suggests that an internal representation of the cue is maintained to eventually select the appropriate target and suppress attention to distractors. Thus, at least in nonhuman primates, information maintenance in working memory could be reflected in sustained firing patterns in the brain. This provides a potential neural measure to examine the contribution of working memory in nonhuman primates, but a noninvasive measure is more appropriate for studying the phenomenon in humans.

Supporting evidence from human studies was made possible with the discovery of an event-related potential (ERP) observed when a representation is maintained in working memory. ERPs are systematic voltage deflections evident in averaged EEG segments time-locked with respect to stimulus or response events. When storing an object presented on one half of the visual field, the average voltage at electrodes on the back of the head (posterior), opposite to the side where the cue was presented (contralateral), becomes more negative compared to the corresponding electrodes on the same side of the target (ipsilateral) [34]. Subtracting the potentials from the contralateral and ipsilateral sites provides a measure called the "contralateral delay activity," or CDA, as it is observed during the delay period of working memory tasks, mirroring the sustained activity pattern of inferotemporal neurons in primates (see the bottom panel of Fig. 1b for an example). The effect emerges approximately 300 ms after the presentation of the object. A large body of work has demonstrated the CDA to be a reliable signature of information maintenance in working memory. The CDA tracks the quantity of objects represented in working memory, increasing in amplitude up to each individual's visual working memory capacity [35–37]. It is sensitive to the quality or precision of the stored object representations [38] and to various features of stored objects, such as color [35, 39], orientation [36, 39], or shape [37, 40]. The CDA does not only track retention of previously displayed items but also reflects online processing of concurrently visible items as discussed in a preceding chapter by Balaban and Luria [41]. The CDA occurs even when the information stored in memory needs to be compared with objects that can appear anywhere in the visual field [28, 42–47], making it an ideal tool for studying the role of working memory in attentional guidance during visual search in humans.

To test the role of working memory in guiding attention using the CDA, variations of the cued visual search paradigm are typically used. In this paradigm, on each trial, a target-cue array is followed by a complex search array of objects (Fig. 1a). The cue informs participants of the identity of the target prior to search. Thus, the cue can be maintained in working memory to guide attention during search. Importantly, EEG can be acquired as participants perform this task, and the voltage fluctuations in response to the cue and search array can be later analyzed. Consistent with the view that attentional templates are maintained in visual working memory, the cue-elicited CDA has been observed (Fig. 1b, bottom panel) to continue until search is performed and to correlate with later search performance, with higher CDA amplitudes predictive of faster search reaction times [42, 43]. Immediately following the search array, a second ERP component can be measured called the N2-posterior-contralateral or N2pc (Fig. 1c), a well-validated and reliable index of the deployment of covert visual attention [48–51]. The N2pc has been measured in a number of studies in conjunction with the CDA to provide further evidence that working memory mechanisms can direct perceptual attention under various circumstances [28, 44, 45, 52–54]. The results from these studies show that ERPs can be used to directly measure the visual working memory representations that proactively drive the attention-demanding search process, when targets switch from moment to moment, as is typically the case in the real world.

While our environment is often dynamic and search demands frequently change, we also come across situations that are similar to earlier experiences. If such experiences are repeatedly and consistently encountered, we learn and adapt our behavior to perform required actions more efficiently, skillfully, and automatically. Researchers studying learning and skill acquisition have proposed theoretical models to explain how behavior becomes automatized with increasing experience [55–59]. In Logan's instance theory of attention and memory [56], each encounter with a given object leaves a trace or an "instance" in memory. As more instances are accumulated over time, they increasingly influence the choice among various behavioral options such that one option becomes more biased, leading to faster selection of that choice. Similar speeding of performance due to attentional tuning has long been observed in the field of visual search when search targets repeat trial after trial [60–63]. Thus, it is possible that even the search process can be automatized if the search environment remains consistent over time. In a way, we can *learn* to guide our attention automatically in relatively stable environments.

We know that visual working memory representations benefit attentional search in dynamic environments. But, are working memory representations equally important when performing search in consistent environments? When working memory is filled to capacity by information irrelevant for the search process, one might expect visual search to worsen. This is because the limited working memory resources can no longer be leveraged to maintain active target representations to guide the search process. In dynamic environments where search demands change from moment to moment, Woodman and colleagues [64] observed precisely this effect. In their study, participants performed an orthogonal demanding task along with memory-guided visual search. The second task demanded working memory resources, leaving little available for maintaining an active cue representation for the search. Consequently, if the cues changed from trial to trial, participants became worse at maintaining those cue representations, and their search performance suffered. However, if the cue remained consistent for a series of consecutive three to seven trials (called "runs"; Fig. 1a), the lack of working memory resources had no effect on search performance. That is, search can be performed equally well regardless of whether working memory is available or not as long as one is searching for the same target over and over again. Using electrophysiology, Carlisle and colleagues [43] observed that the CDA amplitude in response to the cues systematically decreases if the cue identity remains fixed during a run of trials, even though search performance progressively improves. This result again suggests that the need for maintaining a representation in working memory reduces over time in consistent environments. Since search performance is unaffected, these data suggest that visual search can be aided by some source other than working memory.

Despite the prevailing theoretical view that working memory representations of target objects provide top-down control of attention as we search for these objects in cluttered visual scenes [4–7, 29], a more current alternative view, though not mutually exclusive, is that long-term memory representations play a critical role in guiding attention [45, 56, 65–74]. The incorporation of long-term memory into the mechanisms that tune visual attention has the advantages of both freeing up limited resources from working memory and leveraging the virtually unlimited capacity of long-term memory [75, 76], which enables the storage of a large number of items without incurring costs on behavior [65]. This perspective has strong real-world significance, as the guidance of attention during everyday activities often relies on past experiences without explicit cues. Thus, when environments are stable, attention may be aided by long-term memory in lieu of the working memory system, which is more important when environments are dynamic or unpredictable.

Electrophysiology has again been tremendously useful to address the hypothesis that long-term memory is taking over the role of attentional guidance from working memory in consistent environments. To examine how search behavior becomes rapidly automatized, researchers have employed an additional ERP measure indexing long-term memory called the anterior P1, in addition to the CDA component indexing visual working memory. The anterior P1 or P170 is a product of the long-term recognition memory literature using visual memoranda [77–79], in which modulation of this early frontal positivity has been demonstrated to predict the magnitude of behavioral priming effects across protracted memory intervals (i.e., hundreds of stimuli in the past) [80]. Although the anterior P1 is a positive component, it becomes increasingly more negative as encounters with a stimulus accumulate traces in long-term memory [28, 45, 79] (see the top panel of Fig. 1b for an example). As a result, the anterior P1 is generally thought to reflect the successful buildup of information in long-term memory on the basis of stimulus familiarity [79–82]. Given the different spatial and temporal profiles of the anterior P1 (frontal; 150–200 ms after stimulus onset) and the CDA (posterior; 300 ms onward after stimulus onset), both ERP components can be independently measured, thereby giving simultaneous measures of working memory and long-term memory.

If the hypothesis is correct that long-term memory takes over the control of attention from working memory during the tuning of visual attention, then the automation of attentional performance should derive from a decline in attentional templates in working memory, indexed by decreasing CDA amplitudes, and a rise in attentional templates in long-term memory, indexed by increasing anterior P1 negativity. Indeed, these patterns of CDA and anterior P1 effects were observed by Woodman and colleagues [45] using data from Carlisle et al. [43]. Using the cued visual search task with runs of trials in which the same target was repeatedly cued, participants showed an increase in negative anterior P1 amplitudes, a decrease in CDA amplitudes, and faster reaction times, as they continued to search for the same target object over runs of trials. The results suggest that the tuning of attention was accomplished by the transitioning of target representations guiding attention from working memory to long-term memory. These data provide a prime example of how control of attention can be construed as a dynamic process among multiple memory systems in line with theoretical proposals emphasizing a role for mechanisms of long-term memory [45, 56, 65–74] and executive control [31, 70] in the tuning of perceptual attention.

Subsequent studies have further investigated the dynamic interplay between the memory systems that direct attention when participants analyze complex scenes for target objects. First, the co-occurrence of mounting anterior P1 negativity and declining CDA amplitudes during attentional tuning has been replicated in several studies using simple and more complex stimuli [47, 53]. Second, we now also know that anterior P1 amplitudes continue to change over longer time scales of learning, while the trends of CDA amplitude modulation do not change over time. This suggests that the two memory systems are at least partially independent with non-overlapping roles in guiding attention [53]. Third, it has been found that even after working memory relinquishes control when the targets of search become increasingly familiar, its resources can be recruited under situations of heightened cognitive control [28, 52]. Although the precise role of the anterior P1 is under debate (see [47]), and its modulations have yet to be as thoroughly investigated as the CDA, the linkage between long-term memory and the anterior P1 seems encouraging. Further, the negatively correlated changes in anterior P1 and CDA suggest a transition of representations from working memory to long-term memory, with the role of attentional guidance switching between the two.

We have gained some insight into the memory mechanisms that control attention using theories from cognitive psychology and various electrophysiological tools. The literature outlined here has shown that the strength of working memory representations, as indexed by the CDA, is predictive of attentional behavior and how repetitive visual search comes to rely less on working memory, potentially transferring control to long-term memory in order to guide attention. However, in the field of human memory-guided attention, the body of evidence to date has been largely correlative in nature. That is, our inferences derive from relating changes in behavior to changes in neural activity according to experimental manipulations of cognitive demands. It is true that visual search performance has been associated with changes in the CDA and anterior P1, but we do not yet know whether these changes are mechanistically causing behavioral enhancements. Further, the nature of the anterior P1 and its relationship to long-term memory attentional control has yet to be sufficiently characterized. Thus, stronger evidence is needed to determine how these electrophysiological signals interrelate and affect search performance.

3 Transcranial Direct Current Stimulation

Transcranial direct current stimulation (tDCS) has gained increasing popularity among scientists in recent years as a noninvasive neuromodulation technique that can provide an opportunity to obtain causal insight into how human brain activity relates to cognitive processing and behavior, with implications for the development of therapeutic interventions for clinical populations [83]. The approach is safe, well tolerated, inexpensive, portable, and straightforward to administer [84], which are partly the reasons for its growing popularity. TDCS involves the noninvasive application of a low-intensity and sustained direct electrical current with the goal of modulating neuronal excitability in a causal and reversible manner. Conventionally, two relatively large electrodes (5 × 5 cm) are used with an appropriate electrolyte buffer (e.g., conductive gel, paste, or saline) applied between electrodes and the surface of the scalp. The electrode from which current enters the body is designated the anode, and the electrode from which current exits the body is designated the cathode. Classic in vivo animal studies have shown that neuronal excitability tends to increase nearest the inward current flow of the anode and decrease nearest the outward current flow of the cathode [85–89]. More current work showing anodal/cathodal stimulation leads to somatic depolarization/hyperpolarization, and increase/decrease in firing rate [90, 91] further supports the sliding-scale rationale that the anode excites and the cathode inhibits, which, although oversimplified, remains a reasonable first approximation for introducing how differences in tDCS polarity generally influence the brain. Thus, application of a weak current has the potential to modify how excitable a cortical tissue of interest is in a polarity-dependent manner, allowing researchers to obtain bidirectional control over brain activity and behavior [92]. Using computational models that estimate how the applied current should flow through the brain, we can fine tune stimulation parameters, such as the location and intensity values of the anodal and cathodal electrodes, to maximize the effect on a given brain region while minimizing impact on nearby regions [93]. Further, improved neuromodulation methods, such as high-definition tDCS (HD-tDCS), which use an array of smaller electrodes (e.g., 10 mm in diameter) in carefully designed montages have been shown to stimulate one brain region or a network of regions with higher spatial resolution and longer-lasting effects compared to conventional tDCS [94].

The applied direct current interacts with the physiology of neurons and other cells in the brain [95–97]. Unlike other brain stimulation techniques, such as transcranial magnetic stimulation (TMS), tDCS does not cause direct spiking or firing of neurons. A conventional tDCS experiment involves applying between 1 and 2 mA current using a regular 9-V battery. Most of the applied current is filtered out by the skull, scalp, and intermediate layers which protect the brain [98]. The remaining current produces a weak electrical potential gradient typically less than 1 V/m [99, 100], which, by itself, is not sufficient to cause neurons to directly fire [90, 98]. Instead, the beneficial effects of tDCS that outlast the stimulation duration happen through gradual, cumulative changes in neuronal membrane potential [88]. Although this implies that effects of tDCS take minutes to develop and are not as immediate as compared to TMS effects, tDCS has been shown to exert a surprising degree of temporal precision by influencing specific electrophysiological mechanisms that are as brief as 100 ms during a 1–5 s flow of information processing [101].

The low-intensity electrical current used in typical tDCS studies confers benefits to participant comfort and safety and to the effectiveness of sham control procedures. TDCS is most commonly associated with only minor side effects, such as slight itching sensations under electrodes observed, as the current ramps up and initially spreads over the scalp [84, 102]. In order to control for these side effects, an "active" stimulation session is often coupled with a sham stimulation session (analogous to a placebo). Participants are generally kept blind to the stimulation condition they undergo in a given session. Of note, the experimenter can also be kept blind to the stimulation condition which greatly facilitates the use of tDCS for randomized controlled designs. The most common type of sham design involves brief periods of ramping up of the current before switching it off completely. This ramp-up period simulates the itching sensations experienced in the actual "active" stimulation condition. After the ramp-up, the current is ramped down and switched off for a period of time that matches the duration of the total stimulation in the active condition. As no current is being delivered, few neurophysiological effects are expected (however, see [103] for a review of mixed results). Toward the end of this period, the current is ramped up and down again to simulate the itching sensations observed at the end of an active stimulation session. Alternate designs involve continuous delivery of a current at the same site as in active stimulation but at much weaker intensity or application of a current with the same intensity but at a different site [103–107]. Newer designs that provide the same current intensity but use a montage that effectively shunts the current within the scalp preventing it from reaching the brain have also been proposed [108]. Determining optimal sham controls is still an active area of tDCS research, but the ease of designing sham controls and the tolerability of the side effects during stimulation make tDCS an attractive tool for neuromodulation.

Complex biophysical mechanisms potentially underlie the effects observed with tDCS. As the current moves through the brain tissue, it affects the membrane potential of neurons in its path. This changes the flux of ions across the neuronal membranes, changing the neurons' likelihood of firing [84, 109–111]. Prolonged changes in the membrane potential can modify the efficacy of neurotransmitter receptors, such as NMDA, which leads to changes in the overall excitability of the tissue [111]. As excitability changes, small modulations are observed in the firing rates which get amplified due to the intrinsic connectivity across neurons in the cortical networks, leading to the macroscopic effects observed in behavior [91]. Much work remains to be done on the biophysical basis of tDCS, but we know that a variety of neurotransmitter systems are potentially implicated including GABA, glutamate, dopamine, serotonin, and acetylcholine [112–114] suggesting that tDCS broadly affects a variety of networks. Thus, tDCS is potentially a powerful tool to modulate intrinsic brain activity through both electrophysiological and chemical changes.

Various avenues of research are currently in progress to maximize the utility of tDCS. For developing better montages and fine-tuning stimulation parameters, the biophysical mechanisms have to be precisely elucidated. In this regard, examination of in vivo and in vitro models of applied electric field on neural tissue is a major area of investigation [115]. Special emphasis is also being given to merge tDCS with neuroimaging techniques such as fMRI and EEG, with other stimulation methods such as TMS, and with structural connectivity measurements obtained from diffusion imaging. Doing so provides a window to observe the macroscopic effects in neuronal functioning and broad, network-based changes in brain states due to stimulation [116–121]. A major body of work uses tDCS to modulate brain networks that are hypothesized to underlie a specific cognitive function such as attentional allocation, learning, and memory [120, 122]. Another active area of work examines how tDCS, with its neuromodulatory abilities, can be used as a therapeutic intervention to correct for deficits in neuropsychiatric disease [123]. In the following sections, we build on these areas of research, discussing recent multimodal evidence obtained via tDCS in combination with EEG about memory-guided attention in healthy population and in people with schizophrenia.

4 Visual Attention Improved in Healthy Young Adults by Modulating Long-Term Memory Electrophysiology with tDCS

The ability to manipulate neural processes using transcranial stimulation has important implications for studying cognition, in general, and memory-guided attention, in particular. We have seen that modulation of electrophysiological signals indexing working memory and long-term memory is associated with enhanced attentional performance. Transcranial stimulation allows us to examine this relationship more stringently by modifying the neural processes directly. We can dissect how multiple control mechanisms converge, diverge, or compete to direct attention by studying which mechanisms are most differentially affected when we externally modulate attention.

A growing body of work shows evidence that tDCS can be used to transiently improve attentional performance (see [124, 125] for a review). Given that both dorsal and ventral attentional networks span a variety of brain areas [126], researchers have targeted various nodes of these networks to enhance attention, including dorsolateral prefrontal cortex (DLPFC) [127, 128], inferior frontal cortex [105], and posterior parietal cortex (PPC) [129, 130]. However, few studies have taken advantage of such converging evidence of attentional enhancement with brain stimulation to study top-down attentional guidance.

An earlier study using TMS tested how attention is captured by items that either match active working memory representations or match items that are perceptually familiar due to prior exposure [131]. The authors observed that activating visual cortex with TMS differentially affected the two forms of attentional capture. This suggests that the two representations influence attention differently – an important finding that supports multiple sources of attentional control. In another recent study, TMS was performed over DLPFC and PPC, and it was found that activating these regions leads to greater attentional capture by working memory representations [132]. These studies add to growing causal evidence that memory representations lead to different attentional states which can be demarcated in the brain. However, concurrent measurements of working memory and long-term memory would be necessary in order to examine the dynamics of their interactions.

Reinhart and Woodman [53] tested whether representations in working memory and long-term memory can be modified by tDCS to guide attention. In an earlier study, they observed that performing anodal stimulation over the medial frontal cortex improved target detection [92]. Medial frontal cortex is not commonly thought of as a region involved in the canonical attentional network of frontoparietal structures. However, medial frontal cortex is not a unitary structure. It is a broad categorization of various functional areas such as the anterior cingulate cortex (ACC), the supplementary motor area (SMA), and the pre-supplementary motor area (pre-SMA). The cingulate cortex is considered a part of a wider cingulo-opercular network that shows functional connectivity with the attentional network during perceptual search [133]. This area also performs core cognitive control functions, including specifying task sets or rules, and the making and monitoring of choices [134–140]. Due to this performance monitoring role, the ACC has also been considered a part of an anterior attention system influencing recruitment of attention [141]. The pre-SMA and SMA have also been shown to perform performance monitoring [142, 143], indicating their potential involvement in attentional control. Therefore, attentional control could be modified by broadly targeting the medial frontal cortex which could improve detection performance [92].

The positioning of the medial frontal cortex makes it relatively inaccessible for tDCS to directly target. However, using current flow models, the authors [53] were able to confirm that by placing the anode over the FCz site in the international 10–20 convention and placing the cathode over the right cheek, current could be maximally distributed over the pre-SMA, the SMA, and the fronto-medial surface of the brain including the ACC. Therefore, such a stimulation montage that broadly targets various nodes within the medial frontal cortex could influence the control of which top-down signals need to be sent to the downstream visual areas to guide attention.

Based on this evidence, Reinhart and Woodman [53] targeted medial frontal cortex as the site of anodal stimulation during memory-guided visual search. The same paradigm was previously used to test both working memory and long-term memory contributions by repeatedly looking for the same target for a series of trials [43]. Specifically, participants searched for a target among a bilateral array of distractors (Fig. 1a). The target identity was informed prior to search onset and would remain constant for three to seven consecutive trials. Earlier studies observed reduction in the CDA amplitudes and increasing anterior P1 negativity across these short bursts of learning trials, suggesting that the control of attention shifted from working memory to long-term memory [28, 45]. By combining tDCS with these electrophysiological signals, the authors examined whether attentional enhancements brought about by medial-frontal stimulation were accompanied by changes in the CDA or anterior P1 or both. It is possible that exciting the medial frontal cortex leads to improved working memory representations providing stronger downstream attentional control. If that is the case, then the reduction in the CDA amplitudes might disappear or even reverse. Alternatively, if the behavioral enhancement happens due to increasing contribution from long-term memory, anterior P1 amplitudes would become exceedingly negative, without much change in the CDA.

As expected, attentional tuning was observed as participants repeatedly searched for the same target across the short runs of trials in the baseline (sham) condition. Stimulation of the medial frontal cortex enhanced this effect even further. In fact, attentional performance became so efficient that the time taken to search for the target reached floor levels within the first two repetitions. This effect was supported by an electrophysiological correlate of attentional focus. With stimulation, the N2pc component in response to the search target reached its highest amplitudes within the first two repetitions, indicating that modulation of medial frontal cortex could significantly enhance attentional allocation. However, the crucial question was whether stimulation improved the working memory representation making the search process more efficient or whether the stimulation caused participants to rely on a stronger long-term memory representation.

Comparing the CDA amplitudes between sham and stimulation conditions can test whether working memory representations changed as a consequence of stimulation. In the sham condition, working memory indexed by the CDA was found to reduce with repeated target search in line with previous studies. However, there was no evidence that stimulation affected this rate of change of the CDA with repetition. The authors tested this effect in two experiments where participants were asked to remember and look for simple shapes (Landolt-Cs) or real-world objects. Neither experiment showed evidence that stimulation affected the natural modulation of the CDA with repetition. This suggests that the medial-frontal stimulation-induced improvements in visual search were not due to improved representations in working memory.

Although evidence using the CDA indicated that working memory representations remained unaffected following tDCS, long-term memory representations may have been driving the attentional benefits. As the participants engaged in multiple instances of searching for the same target, these instances could contribute to a greater buildup of target representation in long-term memory [55, 56]. Thus, behavioral improvements can happen due to changes in long-term memory instead of relying on active working memory. Since anterior P1 can index the buildup in the long-term memory, its amplitude might become more negative with repetitive search during stimulation. This is precisely what the authors observed. With stimulation, it took less repetitions for anterior P1 to reach its maximum amplitudes (highest negativity) than in the sham condition. In fact, anterior P1 amplitudes peaked within the first two repetitions, paralleling behavioral and N2pc observations. The change in the amplitude of anterior P1 across repetitions correlated with the change in reaction times across repetitions during stimulation across all participants. Further, these results were consistent across search for simple shapes and real-world objects. These results suggest that when looking for the same target repeatedly, exciting the medial frontal cortex caused long-term memory representations to rapidly take over the control of attention without affecting working memory.

Similar evidence of enhanced long-term memory aiding attention was evident at even more protracted time scales. The runs of three to seven trials where the search target remained constant were repeated throughout the experiment. We have already discussed how reaction times progressively decreased within each run, with simultaneous decreases in the CDA and increases in anterior P1 negativity. As runs are repeated, offering more opportunity to learn the identity of a previously encountered target, learning effects could accumulate throughout the experiment. If learning-induced changes in anterior P1 underlie improvements in attention, then learning at even longer time scales such as the time course of the experiment should be reflected in anterior P1 amplitudes and behavioral performance. In order to examine this, the authors looked at performance and ERP amplitudes for the same runs during the first third, middle third, and last third of the experiment. The authors observed that even without stimulation, reaction times at the beginning of the runs progressively reduced across the early, middle, and late phases of the experiment. Anterior P1 amplitudes similarly became more negative potentially reflecting greater local consolidation of the target item in long-term memory over the 3-h time course of the experiment. The pattern of CDA amplitudes within a run, however, did not vary across the duration of the experiment. Thus, visual search improved with repetition throughout the experiment not because of a better use of working memory representations but potentially through increasing consolidation in long-term memory. Anodal stimulation over the medial frontal cortex led to even faster reaction times and greater attentional tuning over the course of the experiment. Anterior P1 amplitudes became even more negative even though the CDA amplitudes remained unaffected. These results imply that visual search efficiency could be improved with stimulation and were potentially brought about by greater reliance on long-term memory. Of note, stimulation was performed for only 20 min which was still enough to improve representations in long-term memory and enhance attentional performance for nearly 3 h. Therefore, stimulation could provide behavioral benefits that outlast its duration if the appropriate neuronal networks are targeted.

To our knowledge, this is the only study so far testing simultaneous contributions of working memory and long-term memory in attentional guidance using tDCS. The study provides additional evidence that the contributions from one type of memory system could be preferentially enhanced without affecting the other, suggesting the presence of independent memory guidance mechanisms that likely produce different downstream effects in the sensory regions, as proposed by Soto and colleagues [131].

The study was informative in various aspects. First, the study suggests that stimulation-induced benefits in attentional allocation can happen due to rapid recruitment of long-term memory for guiding attention. Second, it lends further support to the idea of multiple memory systems guiding attention [70] since stimulation enhanced long-term memory without changing the role of working memory in guiding attention compared to baseline. Third, performing stimulation is advantageous since it provides preliminary evidence for the putative origins of the anterior P1 ERP component. Much work is needed to determine whether anterior P1 is a definitive reflection of long-term memory or how its response to familiarity is dependent on the nature of inputs and task requirements. Identifying the sources of anterior P1, as potentially indicated by this study to include medial frontal cortex, might be a good starting point. Fourth, the study suggests that top-down control can be relegated to long-term memory rapidly – within the first two repetitions of repeating the same search behavior. This potentially suggests newer avenues for theories of automaticity as under the right neural conditions (enhanced excitability) the transition to long-term memory can happen rapidly without much need for repetition of instances of search. Finally, using tDCS and observing its effects allow us to determine which of the two memory guidance mechanisms under consideration are best available for manipulation. Under situations where attention is sub-optimal, identification of the prime locus of selection guidance presents itself as a potential target for manipulation to improve any deficits, as will be discussed further in the following section.

5 Identifying and Remediating Memory-Guided Attentional Deficits in Schizophrenia with tDCS

Impairments in cognition are prevalent in various neuropsychiatric disorders [144, 145]. Cognitive impairments often precede and are predictive of later functional disability, such as the mild cognitive impairment observed before the onset of Alzheimer's disease [146] or cognitive symptoms prior to the onset of psychosis in schizophrenia [147]. These impairments remain a defining feature throughout the progression of the disorder. Therefore, identifying the sources of cognitive deficits and finding ways to slow, stop, or reverse these impairments are major goals of psychiatry and translational neuroscience.

Realizing this objective generally requires linking various domains of experimental and theoretical neuroscience. Within cognitive neuroscience, theoretical frameworks and their associated experimental paradigms can be developed, and their validity can first be tested in healthy human samples, observing transient modulations in cognitive functioning with respect to baseline. These human-validated interventions can then be transferred to a patient population, in order to bring the level of cognitive functioning closer to the baseline levels of healthy participants. However, determining a theoretical framework that can be empirically validated in healthy individuals consistently is a difficult feat in itself [148]. Here, we discuss how a framework developed to study memory-guided attention, with supporting theoretical, behavioral, and electrophysiological evidence, can be translated to study cognitive dysfunction and remediation in schizophrenia.

Schizophrenia is a debilitating brain disorder, characterized by complex combinations of positive, negative, and cognitive symptoms that show considerable variability across individuals. Its pathological origins and treatment options are a matter of interest, debate, and profound implications. Despite the variability, cognitive impairments, especially those related to attention, are considered "core" pathophysiological signatures of schizophrenia [148]. In this view, despite the variability of symptoms across patients, deficits in certain basic cognitive processes like attention would be consistently observed throughout the disease progression but might remain uncaptured in conventional neuropsychiatric examinations [148–150]. Since the subtleties of attentional processes are routinely quantified in cognitive neuroscience, we can adapt frameworks from cognitive neuroscience to examine attentional deficits in schizophrenia.

Top-down control of attention during visual search is one framework from cognitive neuroscience found to be impaired in schizophrenia [151]. What is not known for certain, however, is the source of this impairment. The theoretical motivation for how attentional deficits play out in visual search follows from our previous discussion on models of learning and skill acquisition. Cognitive models of schizophrenia suggest that faulty internally generated predictions cause various symptoms due to mismatch between unreliable predictions and the incoming perceptual input [148, 152, 153]. Memory representations of previous experiences are a crucial source of these predictions. If representations are not stored in memory or if they are inaccessible, various cognitive symptoms could arise [148]. At the same time, the top-down control of attention during visual search could be unfavorably affected. According to theoretical models of learning and memory [55, 56], repeated encounters of an object should lead to a greater number of instances being stored in long-term memory. In the absence of any deficits, after searching for the same object repeatedly, attention can be automatically guided by representations of that object in long-term memory. However, if long-term memory representations are impaired, they cannot properly guide attention. In order to perform visual search successfully, schizophrenia patients would have to consistently rely on their working memory resources. Every instance of visual search, in a way, is a new search in itself, having little benefit from previous instances of the same search.

This deficit in the ability of long-term memory to guide attention in schizophrenia was noted in a recent study by Reinhart and colleagues [54]. We now know that memory-guided visual search paradigms when used in conjunction with electrophysiology can be used to distinguish the contributions of working memory and long-term memory in attentional guidance in healthy individuals [28, 45, 47, 52, 53]. In their study with schizophrenia patients, Reinhart and colleagues [54] examined whether the contribution of working memory and long-term memory, as indexed by the CDA and anterior P1, respectively, was different between healthy people and people with schizophrenia. For healthy participants, if the target identity for visual search is fixed for a set of consecutive trials, attention appears to increasingly rely on long-term memory representations and less on working memory representations. This is evidenced by increasing negativity of anterior P1 and reduction in the CDA amplitudes. Reinhart et al. found that schizophrenia patients did not exhibit a reduction in CDA amplitudes despite maintenance of the target identity over a series of consecutive trials. In fact, the patients exhibited even greater CDA amplitudes at baseline relative to healthy controls, and these amplitudes did not change with repetitive search. This demonstrates that schizophrenia patients engage more working memory resources to maintain the same target representations as healthy individuals. Moreover, the anterior P1 amplitude did not change over time unlike the increasing negativity observed in healthy controls. Therefore, prior instances of searching for the same target did not lead to a gradual transfer of attentional control to long-term memory. Instead, every search was like a new search causing the patients to rely on their limited working memory resources. This is perhaps why the CDA amplitudes did not reduce over time, unlike healthy controls. These results support the hypothesis that attentional deficits in schizophrenia could stem from an inability to perform a successful transition from working memory to long-term memory to guide attention, consistent with the well-established structural and functional alterations in long-term memory brain regions such as the hippocampus in schizophrenia [154, 155].

Given research showing how medial-frontal tDCS can be used to preferentially enhance the contribution of long-term memory in attentional guidance in healthy young adults [53], it is possible that a similar stimulation protocol could temporarily reverse the attentional deficits in patients with schizophrenia. Indeed, following 20 min of anodal tDCS over medial frontal cortex, Reinhart and colleagues [54] observed that schizophrenia patients exhibited behavioral and electrophysiological patterns that were indistinguishable from healthy controls at baseline. After stimulation, schizophrenia patients showed faster search times, reduced CDA amplitudes, and increasingly negative anterior P1 amplitudes as a function of target repetitions. Stimulation also improved attentional focus in schizophrenia patients as their N2pc amplitudes went up. This suggests that even downstream attentional selection mechanisms can be improved if deficits in guidance are corrected. Overall, these results imply that 20 min of medial-frontal stimulation was enough to transiently improve attentional performance in people with schizophrenia.

The pattern of results provides a compelling framework of attentional dysfunction in schizophrenia. Attentional abnormalities could arise due to inefficiencies in working memory and an inability to transfer the control between working memory and long-term memory representations in consistent environments. While healthy individuals can successfully transfer control to long-term memory, schizophrenia patients have to continue to rely on the resource-limited working memory throughout. Since patients require greater working memory resources to maintain a given representation, the inefficiency of working memory affects attentional performance. Increased excitability of the medial frontal cortex may correct for the deficits in the long-term memory system that would have otherwise taken over the process of attentional guidance or may modify cognitive control mechanisms that mediate among different memory subsystems in order to execute task demands. More work will be needed to establish the robustness of these effects and validate these hypotheses further.

While other recent studies have looked at the benefits of tDCS on working memory enhancement ([156, 157], but see [158]), no other study has yet examined working memory and long-term memory together, specifically in the context of attentional guidance. Further, even though tDCS has been used in a variety of other neuropsychiatric disorders (see [159, 160] for a review), its use in ameliorating attentional deficits due to impairments in top-down control needs more research in schizophrenia as well as other illnesses. Here, we have discussed in detail the recent evidence that tDCS can transiently remedy such impairments in schizophrenia and can even provide mechanistic understanding to the causes of these impairments. Such line of work, still in its infancy, can have important translational implications going forward for various neuropsychiatric disorders.

6 Conclusions and Future Directions

It is intriguing how we direct our attention to find something of interest and relevant in a complex environment. Over the years, we have gained considerable insight into the various memory systems that can store object representations that serve as a template to guide attention [45, 70]. While theoretical proposals on this matter go back a century [161], we discussed empirical evidence for memory-guided attentional allocation following advancements in cognitive psychology and electrophysiology in the recent decades, and we believe we are now in a position to determine the causal relationships between the phenomena of memory-guided attention and its neural substrates with the advent of noninvasive neuromodulation techniques such as tDCS. With the recent use of tDCS in examining memory-guiding attention, we have gained insight into how long-term memory-related neural dynamics drive improved attention, independent of working memory top-down control. Further, the locus of attentional deficits in schizophrenia was identified in the transition between working memory and long-term memory, and tDCS was found to transiently reverse these deficits. Thus, with tDCS we can test how different memory systems contribute and interact to provide top-down control and guide attention during the analysis of complex visual scenes.

Despite these advances, there is much work to be done. Determining the action mechanisms of tDCS is a major frontier to cross in order to interpret its effects and design better experiments to answer subtle questions. As a noninvasive neuromodulation technique, it is difficult to precisely determine the effects of tDCS on physiology directly. The efforts have primarily centered on in vitro and in vivo stimulation in animal models [115] although a growing body of work uses neuroimaging in humans to observe network-level changes in neuronal functioning and linking them with what we know from animal models [162]. More recently, using human cadavers as a model, scientists have determined that conventionally applied tDCS is quite weak raising concerns about the purported focality and efficacy of stimulation [98]. The fact that we do not completely understand how tDCS affects the underlying tissue in the living human brain, what stimulation parameters are most effective for modulating cognitive processes, and the heterogeneity of cognitive paradigms tested with tDCS might account for some of the observed variability across tDCS studies that have led to contradictory results [163]. Scientists now argue for more integrated experimental designs which leverage information from anatomy and physiology to develop computational models of induced electric field gradients in the brain and of the consequent behavioral effects [98, 164–166] to determine more focal and effective stimulation parameters and protocols. Given that effects of most tDCS designs are limited to the cortex, newer stimulation tools are being developed in animal models which can target deeper, subcortical structures [167], and these designs are currently being adapted for humans. Thus, we can examine the interaction among these memory systems more stringently, as our designs of tDCS and understanding of its mechanisms advance.

Another growing consensus in neuroscience is that large-scale, coordinated activity across populations of neurons underlies cognition [168]. While we discussed the putative ERP measures associated with working memory and long-term memory, a large body of work suggests that patterns of rhythmic, oscillatory activity and their interactions, as observed in EEG and magnetoencephalography (MEG), are also associated with these cognitive phenomena. For example, activity in the alpha frequency band (8–12 Hz) [169–171] and interactions between theta and gamma bands [107, 172–174] have been shown to index contents of working memory. Theta (4–8 Hz) and beta (12–30 Hz) oscillations, among others, have been widely implicated in long-term memory consolidation [175–177]. As oscillations are observed across various spatial scales of neuronal investigation, such phenomena can be more reliably tied to neuronal physiology improving their validity and interpretability compared with ERP components [178]. Thus, scientists are now interested in the effects of tDCS beyond simple excitation and inhibition of a given region, examining how they specifically impact brain oscillations. Examining EEG before and after tDCS can show the stimulation-induced changes in activity across various frequency bands [179]. Recently, mathematical models applied to MEG data have also been used to examine changes in oscillatory activity during tDCS [180]. Using principles similar to tDCS, a growing body of work directly modulates these oscillatory activity patterns, entraining a neuronal population at a certain frequency using transcranial alternating current stimulation (tACS) or modulating the interactions among oscillatory activity patterns across networks [107, 181–186]. Understanding how these transcranial stimulation methods modulate global activity patterns and linking them with changes occurring at a single neuron and synapse level will be crucial in determining their efficacy for scientific and translational objectives. Moving forward, we are hopeful that innovative multimodal approaches will allow us to more definitively test, and potential unify, major theories of attention, learning, and skill acquisition [55–59] and ultimately advance a more robust understanding of human goal-directed cognition and action.

References

1. Corbetta M, Shulman GL (2002) Control of goal-directed and stimulus-driven attention in the brain. Nat Rev Neurosci 3(3):201
2. Wolfe JM, Cave KR, Franzel SL (1989) Guided search: an alternative to the feature integration model for visual search. J Exp Psychol Hum Percept Perform 15(3):419
3. Wolfe JM (1994) Guided search 2.0 a revised model of visual search. Psychon Bull Rev 1 (2):202–238
4. Desimone R, Duncan J (1995) Neural mechanisms of selective visual attention. Annu Rev Neurosci 18(1):193–222
5. Bundesen C (1990) A theory of visual attention. Psychol Rev 97(4):523
6. Bundesen C, Habekost T, Kyllingsbæk S (2005) A neural theory of visual attention: bridging cognition and neurophysiology. Psychol Rev 112(2):291

7. Duncan J, Humphreys GW (1989) Visual search and stimulus similarity. Psychol Rev 96(3):433
8. Theeuwes J (1993) Visual selective attention: a theoretical analysis. Acta Psychol (Amst) 83 (2):93–154
9. Hamker FH (2004) A dynamic model of how feature cues guide spatial attention. Vision Res 44(5):501–521
10. Downing PE (2000) Interactions between visual working memory and selective attention. Psychol Sci 11:467–473
11. Soto D et al (2005) Early, involuntary top-down guidance of attention from working memory. J Exp Psychol Hum Percept Perform 31(2):248
12. Soto D, Humphreys GW, Rotshtein P (2007) Dissociating the neural mechanisms of memory-based guidance of visual selection. Proc Natl Acad Sci U S A 104 (43):17186–17191
13. Soto D et al (2008) Automatic guidance of attention from working memory. Trends Cogn Sci 12(9):342–348
14. Olivers CN (2009) What drives memory-driven attentional capture? The effects of memory type, display type, and search type. J Exp Psychol Hum Percept Perform 35 (5):1275
15. Olivers CN, Meijer F, Theeuwes J (2006) Feature-based memory-driven attentional capture: visual working memory content affects visual attention. J Exp Psychol Hum Percept Perform 32(5):1243
16. Carlisle NB, Woodman GF (2011) When memory is not enough: electrophysiological evidence for goal-dependent use of working memory representations in guiding visual attention. J Cogn Neurosci 23 (10):2650–2664
17. Carlisle NB, Woodman GF (2011) Automatic and strategic effects in the guidance of attention by working memory representations. Acta Psychol (Amst) 137(2):217–225
18. Downing P, Dodds C (2004) Competition in visual working memory for control of search. Vis Cogn 11(6):689–703
19. Woodman GF, Luck SJ (2007) Do the contents of visual working memory automatically influence attentional selection during visual search? J Exp Psychol Hum Percept Perform 33(2):363
20. Han SW, Kim MS (2009) Do the contents of working memory capture attention? Yes, but cognitive control matters. J Exp Psychol Hum Percept Perform 35(5):1292
21. Kiyonaga A, Egner T, Soto D (2012) Cognitive control over working memory biases of selection. Psychon Bull Rev 19(4):639–646
22. Houtkamp R, Roelfsema PR (2006) The effect of items in working memory on the deployment of attention and the eyes during visual search. J Exp Psychol Hum Percept Perform 32(2):423
23. Peters JC, Goebel R, Roelfsema PR (2009) Remembered but unused: the accessory items in working memory that do not guide attention. J Cogn Neurosci 21(6):1081–1091
24. Luria R, Vogel EK (2011) Visual search demands dictate reliance on working memory storage. J Neurosci 31(16):6199–6207
25. Dalvit S, Eimer M (2011) Memory-driven attentional capture is modulated by temporal task demands. Vis Cogn 19(2):145–153
26. Soto D, Humphreys GW (2008) Stressing the mind: the effect of cognitive load and articulatory suppression on attentional guidance from working memory. Percept Psychophys 70(5):924–934
27. Dombrowe I, Olivers CN, Donk M (2010) The time course of working memory effects on visual attention. Vis Cogn 18 (8):1089–1112
28. Reinhart RM, Woodman GF (2014) High stakes trigger the use of multiple memories to enhance the control of attention. Cereb Cortex 24(8):2022–2035
29. Olivers CN et al (2011) Different states in visual working memory: when it guides attention and when it does not. Trends Cogn Sci 15(7):327–334
30. van Moorselaar D, Theeuwes J, Olivers CN (2016) Learning changes the attentional status of prospective memories. Psychon Bull Rev 23(5):1483–1490
31. Logan GD, Gordon RD (2001) Executive control of visual attention in dual-task situations. Psychol Rev 108(2):393
32. Chelazzi L et al (1993) A neural basis for visual search in inferior temporal cortex. Nature 363(6427):345
33. Chelazzi L et al (1998) Responses of neurons in inferior temporal cortex during memory-guided visual search. J Neurophysiol 80 (6):2918–2940
34. Klaver P et al (1999) An event-related brain potential correlate of visual short-term memory. Neuroreport 10(10):2001–2005
35. Vogel EK, Machizawa MG (2004) Neural activity predicts individual differences in visual working memory capacity. Nature 428 (6984):748

36. Vogel EK, McCollough AW, Machizawa MG (2005) Neural measures reveal individual differences in controlling access to working memory. Nature 438(7067):500
37. Ikkai A, McCollough AW, Vogel EK (2010) Contralateral delay activity provides a neural measure of the number of representations in visual working memory. J Neurophysiol 103 (4):1963–1968
38. Machizawa MG, Goh CC, Driver J (2012) Human visual short-term memory precision can be varied at will when the number of retained items is low. Psychol Sci 23 (6):554–559
39. Woodman GF, Vogel EK (2008) Selective storage and maintenance of an object's features in visual working memory. Psychon Bull Rev 15(1):223–229
40. Luria R et al (2010) Visual short-term memory capacity for simple and complex objects. J Cogn Neurosci 22(3):496–512
41. Balaban H, Luria R (in press) Using the contralateral delay activity to study online processing of items still within view. In: Pollmann S (ed) Spatial learning and attention guidance, Neuromethods. Springer Nature, New York
42. Woodman GF, Arita JT (2011) Direct electrophysiological measurement of attentional templates in visual working memory. Psychol Sci 22(2):212–215
43. Carlisle NB et al (2011) Attentional templates in visual working memory. J Neurosci 31 (25):9315–9322
44. Reinhart RM, Carlisle NB, Woodman GF (2014) Visual working memory gives up attentional control early in learning: ruling out interhemispheric cancellation. Psychophysiology 51(8):800–804
45. Woodman GF, Carlisle NB, Reinhart RM (2013) Where do we store the memory representations that guide attention? J Vis 13 (3):1–1
46. Gunseli E, Meeter M, Olivers CN (2014) Is a search template an ordinary working memory? Comparing electrophysiological markers of working memory maintenance for visual search and recognition. Neuropsychologia 60:29–38
47. Gunseli E, Olivers CN, Meeter M (2014) Effects of search difficulty on the selection, maintenance, and learning of attentional templates. J Cogn Neurosci 26(9):2042–2054
48. Eimer M (1996) The N2pc component as an indicator of attentional selectivity. Electroencephalogr Clin Neurophysiol 99(3):225–234
49. Kiss M, Van Velzen J, Eimer M (2008) The N2pc component and its links to attention shifts and spatially selective visual processing. Psychophysiology 45(2):240–249
50. Luck SJ, Kappenman ES (2011) The Oxford handbook of event-related potential components. Oxford University Press, Oxford
51. Luck SJ (2014) An introduction to the event-related potential technique. MIT Press, Cambridge
52. Reinhart RM, McClenahan LJ, Woodman GF (2016) Attention's accelerator. Psychol Sci 27 (6):790–798
53. Reinhart RM, Woodman GF (2015) Enhancing long-term memory with stimulation tunes visual attention in one trial. Proc Natl Acad Sci U S A 112(2):625–630
54. Reinhart RM, Park S, Woodman GF (2018) Localization and elimination of attentional dysfunction in schizophrenia during visual search. Schizophr Bull 45:96–105
55. Logan GD (1988) Toward an instance theory of automatization. Psychol Rev 95(4):492
56. Logan GD (2002) An instance theory of attention and memory. Psychol Rev 109 (2):376
57. Anderson JR (1982) Acquisition of cognitive skill. Psychol Rev 89(4):369
58. Anderson JR (2000) Learning and memory: an integrated approach. Wiley, Hoboken
59. Rickard TC (1997) Bending the power law: a CMPL theory of strategy shifts and the automatization of cognitive skills. J Exp Psychol Gen 126(3):288
60. Schneider W, Shiffrin RM (1977) Controlled and automatic human information processing: I. Detection, search, and attention. Psychol Rev 84(1):1
61. Shiffrin RM, Schneider W (1977) Controlled and automatic human information processing: II. Perceptual learning, automatic attending and a general theory. Psychol Rev 84(2):127
62. Neisser U (1963) Decision-time without reaction-time: experiments in visual scanning. Am J Psychol 76(3):376–385
63. Nickerson RS (1966) Response times with a memory-dependent decision task. J Exp Psychol 72(5):761
64. Woodman GF et al (2007) The role of working memory representations in the control of attention. Cereb Cortex 17(suppl_1): i118–i124
65. Wolfe JM (2012) Saved by a log: how do humans perform hybrid visual and memory search? Psychol Sci 23(7):698–703
66. Moores E et al (2003) Associative knowledge controls deployment of visual selective attention. Nat Neurosci 6(2):182

67. Chun MM (2000) Contextual cueing of visual attention. Trends Cogn Sci 4 (5):170–178
68. Summerfield JJ et al (2006) Orienting attention based on long-term memory experience. Neuron 49(6):905–916
69. Stokes MG et al (2012) Long-term memory prepares neural activity for perception. Proc Natl Acad Sci U S A 109(6):E360–E367
70. Hutchinson JB, Turk-Browne NB (2012) Memory-guided attention: control from multiple memory systems. Trends Cogn Sci 16 (12):576–579
71. Võ MLH, Wolfe JM (2012) When does repeated search in scenes involve memory? Looking at versus looking for objects in scenes. J Exp Psychol Hum Percept Perform 38(1):23
72. Rosen ML et al (2017) Cortical and subcortical contributions to long-term memory-guided visuospatial attention. Cereb Cortex 28(8):2935–2947
73. Rosen ML et al (2015) Influences of long-term memory-guided attention and stimulus-guided attention on visuospatial representations within human intraparietal sulcus. J Neurosci 35(32):11358–11363
74. Rosen ML, Stern CE, Somers DC (2014) Long-term memory guidance of visuospatial attention in a change-detection paradigm. Front Psychol 5:266
75. Brady TF et al (2008) Visual long-term memory has a massive storage capacity for object details. Proc Natl Acad Sci U S A 105 (38):14325–14329
76. Standing L (1973) Learning 10000 pictures. Q J Exp Psychol 25(2):207–222
77. Danker JF et al (2008) Characterizing the ERP old–new effect in a short-term memory task. Psychophysiology 45(5):784–793
78. Paller KA, Lucas HD, Voss JL (2012) Assuming too much from 'familiar' brain potentials. Trends Cogn Sci 16(6):313–315
79. Voss JL, Schendan HE, Paller KA (2010) Finding meaning in novel geometric shapes influences electrophysiological correlates of repetition and dissociates perceptual and conceptual priming. Neuroimage 49 (3):2879–2889
80. Tsivilis D, Otten LJ, Rugg MD (2001) Context effects on the neural correlates of recognition memory: an electrophysiological study. Neuron 31(3):497–505
81. Duarte A et al (2004) Dissociable neural correlates for familiarity and recollection during the encoding and retrieval of pictures. Brain Res Cogn Brain Res 18(3):255–272
82. Friedman D (2007) ERP studies of recognition memory: differential effects of familiarity, recollection, and episodic priming. New Res Cogn Sci:188
83. Reinhart RM et al (2017) Using transcranial direct-current stimulation (tDCS) to understand cognitive processing. Atten Percept Psychophys 79(1):3–23
84. Bikson M et al (2016) Safety of transcranial direct current stimulation: evidence based update 2016. Brain Stimul 9(5):641–661
85. Bindman LJ, Lippold OCJ, Redfearn JWT (1962) Long-lasting changes in the level of the electrical activity of the cerebral cortex produced by polarizing currents. Nature 196 (4854):584
86. Creutzfeldt OD, Fromm GH, Kapp H (1962) Influence of transcortical dc currents on cortical neuronal activity. Exp Neurol 5 (6):436–452
87. Gartside IB (1968) Mechanisms of sustained increases of firing rate of neurones in the rat cerebral cortex after polarization: role of protein synthesis. Nature 220(5165):383
88. Purpura DP, McMurtry JG (1965) Intracellular activities and evoked potential changes during polarization of motor cortex. J Neurophysiol 28(1):166–185
89. Terzuolo CA, Bullock TH (1956) Measurement of imposed voltage gradient adequate to modulate neuronal firing. Proc Natl Acad Sci U S A 42(9):687
90. Radman T et al (2009) Role of cortical cell type and morphology in subthreshold and suprathreshold uniform electric field stimulation in vitro. Brain Stimul 2(4):215–228
91. Reato D et al (2010) Low-intensity electrical stimulation affects network dynamics by modulating population rate and spike timing. J Neurosci 30(45):15067–15079
92. Reinhart RM, Woodman GF (2014) Causal control of medial–frontal cortex governs electrophysiological and behavioral indices of performance monitoring and learning. J Neurosci 34(12):4214–4227
93. Bikson M, Rahman A, Datta A (2012) Computational models of transcranial direct current stimulation. Clin EEG Neurosci 43 (3):176–183
94. Kuo HI et al (2013) Comparing cortical plasticity induced by conventional and high-definition 4 × 1 ring tDCS: a neurophysiological study. Brain Stimul 6(4):644–648
95. Monai H et al (2016) Calcium imaging reveals glial involvement in transcranial direct current stimulation-induced plasticity in mouse brain. Nat Commun 7:11100

96. Ruohonen J, Karhu J (2012) tDCS possibly stimulates glial cells. Clin Neurophysiol 123 (10):2006–2009
97. Gellner AK, Reis J, Fritsch B (2016) Glia: a neglected player in non-invasive direct current brain stimulation. Front Cell Neurosci 10:188
98. Vöröslakos M et al (2018) Direct effects of transcranial electric stimulation on brain circuits in rats and humans. Nat Commun 9 (1):483
99. Datta A et al (2009) Gyri-precise head model of transcranial direct current stimulation: improved spatial focality using a ring electrode versus conventional rectangular pad. Brain Stimul 2(4):201–207
100. Ruffini G et al (2013) Transcranial current brain stimulation (tCS): models and technologies. IEEE Trans Neural Syst Rehabil Eng 21(3):333–345
101. Reinhart RM, Woodman GF (2015) The surprising temporal specificity of direct-current stimulation. Trends Neurosci 38 (8):459–461
102. Poreisz C et al (2007) Safety aspects of transcranial direct current stimulation concerning healthy subjects and patients. Brain Res Bull 72(4–6):208–214
103. Fonteneau C et al (2019) Sham tDCS: a hidden source of variability? Reflections for further blinded, controlled trials. Brain Stimul. https://doi.org/10.1016/j.brs.2018.12.977
104. Boggio PS et al (2008) Prefrontal cortex modulation using transcranial DC stimulation reduces alcohol craving: a double-blind, sham-controlled study. Drug Alcohol Depend 92(1–3):55–60
105. Coffman BA, Trumbo MC, Clark VP (2012) Enhancement of object detection with transcranial direct current stimulation is associated with increased attention. BMC Neurosci 13(1):108
106. Bikson M et al (2018) Rigor and reproducibility in research with transcranial electrical stimulation: an NIMH-sponsored workshop. Brain Stimul 11(3):465–480
107. Reinhart RMG, Nguyen JA (2019) Working memory revived in older adults by synchronizing rhythmic brain circuits. Nat Neurosci. https://doi.org/10.1038/s41593-019-0371-x
108. Richardson JD et al (2014) Toward development of sham protocols for high-definition transcranial direct current stimulation (HD-tDCS). NeuroRegulation 1(1):62–72
109. Liebetanz D et al (2002) Pharmacological approach to the mechanisms of transcranial DC-stimulation-induced after-effects of human motor cortex excitability. Brain 125 (10):2238–2247
110. Nitsche MA et al (2002) Modulation of cortical excitability by transcranial direct current stimulation. Nervenarzt 73(4):332–335
111. Nitsche MA et al (2003) Pharmacological modulation of cortical excitability shifts induced by transcranial direct current stimulation in humans. J Physiol 553(1):293–301
112. Stagg CJ et al (2009) Polarity-sensitive modulation of cortical neurotransmitters by transcranial stimulation. J Neurosci 29 (16):5202–5206
113. Clark VP et al (2011) Transcranial direct current stimulation (tDCS) produces localized and specific alterations in neurochemistry: a 1H magnetic resonance spectroscopy study. Neurosci Lett 500(1):67–71
114. Medeiros LF et al (2012) Neurobiological effects of transcranial direct current stimulation: a review. Front Psych 3:110
115. Pelletier SJ, Cicchetti F (2015) Cellular and molecular mechanisms of action of transcranial direct current stimulation: evidence from in vitro and in vivo models. Int J Neuropsychopharmacol 18(2):pyu047
116. Callan DE et al (2016) Simultaneous tDCS-fMRI identifies resting state networks correlated with visual search enhancement. Front Hum Neurosci 10:72
117. Schestatsky P, Morales-Quezada L, Fregni F (2013) Simultaneous EEG monitoring during transcranial direct current stimulation. J Vis Exp 76:50426
118. Lauro LJR et al (2014) TDCS increases cortical excitability: direct evidence from TMS–EEG. Cortex 58:99–111
119. Roy A, Baxter B, He B (2014) High-definition transcranial direct current stimulation induces both acute and persistent changes in broadband cortical synchronization: a simultaneous tDCS–EEG study. IEEE Trans Biomed Eng 61(7):1967–1978
120. Santarnecchi E et al (2015) Enhancing cognition using transcranial electrical stimulation. Curr Opin Behav Sci 4:171–178
121. Krause MR et al (2017) Transcranial direct current stimulation facilitates associative learning and alters functional connectivity in the primate brain. Curr Biol 27 (20):3086–3096
122. Filmer HL et al (2014) Applications of transcranial direct current stimulation for

understanding brain function. Trends Neurosci 37(12):742–753
123. Kuo MF, Paulus W, Nitsche MA (2014) Therapeutic effects of non-invasive brain stimulation with direct currents (tDCS) in neuropsychiatric diseases. Neuroimage 85:948–960
124. Coffman BA, Clark VP, Parasuraman R (2014) Battery powered thought: enhancement of attention, learning, and memory in healthy adults using transcranial direct current stimulation. Neuroimage 85:895–908
125. Reteig LC et al (2017) Transcranial electrical stimulation as a tool to enhance attention. J Cogn Enhanc 1(1):10–25
126. Moore T, Zirnsak M (2017) Neural mechanisms of selective visual attention. Annu Rev Psychol 68:47–72
127. Nelson JT et al (2014) Enhancing vigilance in operators with prefrontal cortex transcranial direct current stimulation (tDCS). Neuroimage 85:909–917
128. Clarke PJ et al (2014) The causal role of the dorsolateral prefrontal cortex in the modification of attentional bias: evidence from transcranial direct current stimulation. Biol Psychiatry 76(12):946–952
129. Sparing R et al (2009) Bidirectional alterations of interhemispheric parietal balance by non-invasive cortical stimulation. Brain 132 (11):3011–3020
130. Moos K et al (2012) Modulation of top-down control of visual attention by cathodal tDCS over right IPS. J Neurosci 32 (46):16360–16368
131. Soto D, Llewelyn D, Silvanto J (2012) Distinct causal mechanisms of attentional guidance by working memory and repetition priming in early visual cortex. J Neurosci 32 (10):3447–3452
132. Wang M et al (2018) Evaluating the role of the dorsolateral prefrontal cortex and posterior parietal cortex in memory-guided attention with repetitive transcranial magnetic stimulation. Front Hum Neurosci 12:236
133. Sestieri C et al (2014) Domain-general signals in the cingulo-opercular network for visuospatial attention and episodic memory. J Cogn Neurosci 26(3):551–568
134. Pardo JV et al (1990) The anterior cingulate cortex mediates processing selection in the Stroop attentional conflict paradigm. Proc Natl Acad Sci U S A 87(1):256–259
135. Dosenbach NU et al (2006) A core system for the implementation of task sets. Neuron 50 (5):799–812
136. Peterson BS et al (1999) An fMRI study of Stroop word-color interference: evidence for cingulate subregions subserving multiple distributed attentional systems. Biol Psychiatry 45(10):1237–1258
137. Benedict RH et al (2002) Covert auditory attention generates activation in the rostral/ dorsal anterior cingulate cortex. J Cogn Neurosci 14(4):637–645
138. Margulies DS et al (2007) Mapping the functional connectivity of anterior cingulate cortex. Neuroimage 37(2):579–588
139. Johnston K et al (2007) Top-down control-signal dynamics in anterior cingulate and prefrontal cortex neurons following task switching. Neuron 53(3):453–462
140. Rushworth MF et al (2007) Functional organization of the medial frontal cortex. Curr Opin Neurobiol 17(2):220–227
141. Posner MI, Dehaene S (1994) Attentional networks. Trends Neurosci 17(2):75–79
142. Bonini F et al (2014) Action monitoring and medial frontal cortex: leading role of supplementary motor area. Science 343 (6173):888–891
143. Scangos KW et al (2013) Performance monitoring by pre-supplementary and supplementary motor area during an arm movement countermanding task. J Neurophysiol 109:1928
144. Millan MJ et al (2012) Cognitive dysfunction in psychiatric disorders: characteristics, causes and the quest for improved therapy. Nat Rev Drug Discov 11(2):141
145. Trivedi JK (2006) Cognitive deficits in psychiatric disorders: current status. Indian J Psychiatry 48(1):10
146. Bozoki A et al (2001) Mild cognitive impairments predict dementia in nondemented elderly patients with memory loss. Arch Neurol 58(3):411–416
147. Bowie CR, Harvey PD (2006) Cognitive deficits and functional outcome in schizophrenia. Neuropsychiatr Dis Treat 2(4):531
148. Hemsley DR (2005) The development of a cognitive model of schizophrenia: placing it in context. Neurosci Biobehav Rev 29 (6):977–988
149. Barnett W, Mundt C (1992) Are latent thought disorders the core of negative schizophrenia? In: Phenomenology, language & schizophrenia. Springer, New York, pp 240–257
150. Huber G (1986) Psychiatrische Aspekte des Basisstörungskonzeptes. In: Schizophrene Basisstörungen. Springer, Berlin, pp 39–143

151. Gold JM et al (2007) Impaired top–down control of visual search in schizophrenia. Schizophr Res 94(1–3):148–155
152. Weiner I (2003) The "two-headed" latent inhibition model of schizophrenia: modeling positive and negative symptoms and their treatment. Psychopharmacology (Berl) 169 (3–4):257–297
153. Gray JA et al (1991) The neuropsychology of schizophrenia. Behav Brain Sci 14(1):1–20
154. Heckers S et al (1998) Impaired recruitment of the hippocampus during conscious recollection in schizophrenia. Nat Neurosci 1 (4):318
155. Sigurdsson T et al (2010) Impaired hippocampal–prefrontal synchrony in a genetic mouse model of schizophrenia. Nature 464 (7289):763
156. Orlov ND et al (2017) Stimulating cognition in schizophrenia: a controlled pilot study of the effects of prefrontal transcranial direct current stimulation upon memory and learning. Brain Stimul 10(3):560–566
157. Papazova I et al (2018) Improving working memory in schizophrenia: effects of 1 mA and 2 mA transcranial direct current stimulation to the left DLPFC. Schizophr Res 202:203–209
158. Gomes JS et al (2018) Effects of transcranial direct current stimulation on working memory and negative symptoms in schizophrenia: a phase II randomized sham-controlled trial. Schizophr Res Cogn 12:20–28
159. Brunoni AR et al (2012) Clinical research with transcranial direct current stimulation (tDCS): challenges and future directions. Brain Stimul 5(3):175–195
160. Kekic M et al (2016) A systematic review of the clinical efficacy of transcranial direct current stimulation (tDCS) in psychiatric disorders. J Psychiatr Res 74:70–86
161. James W (1890) The principles of psychology. Holt, New York
162. Stagg CJ, Nitsche MA (2011) Physiological basis of transcranial direct current stimulation. Neuroscientist 17(1):37–53
163. Horvath JC, Forte JD, Carter O (2015) Quantitative review finds no evidence of cognitive effects in healthy populations from single-session transcranial direct current stimulation (tDCS). Brain Stimul 8(3):535–550
164. Bestmann S, de Berker AO, Bonaiuto J (2015) Understanding the behavioural consequences of noninvasive brain stimulation. Trends Cogn Sci 19(1):13–20
165. Romei V, Thut G, Silvanto J (2016) Information-based approaches of noninvasive transcranial brain stimulation. Trends Neurosci 39(11):782–795
166. Widge AS (2018) Cross-species neuromodulation from high-intensity transcranial electrical stimulation. Trends Cogn Sci 22 (5):372–374
167. Grossman N et al (2017) Noninvasive deep brain stimulation via temporally interfering electric fields. Cell 169(6):1029–1041
168. Siegel M, Donner TH, Engel AK (2012) Spectral fingerprints of large-scale neuronal interactions. Nat Rev Neurosci 13(2):121
169. Jensen O et al (2002) Oscillations in the alpha band (9–12 Hz) increase with memory load during retention in a short-term memory task. Cereb Cortex 12(8):877–882
170. Foster JJ et al (2015) The topography of alpha-band activity tracks the content of spatial working memory. J Neurophysiol 115 (1):168–177
171. Reinhart RM, Woodman GF (2014) Oscillatory coupling reveals the dynamic reorganization of large-scale neural networks as cognitive demands change. J Cogn Neurosci 26(1):175–188
172. Daume J et al (2017) Phase-amplitude coupling and long-range phase synchronization reveal frontotemporal interactions during visual working memory. J Neurosci 37 (2):313–322
173. Fell J, Axmacher N (2011) The role of phase synchronization in memory processes. Nat Rev Neurosci 12(2):105
174. Sarnthein J et al (1998) Synchronization between prefrontal and posterior association cortex during human working memory. Proc Natl Acad Sci U S A 95(12):7092–7096
175. Backus AR et al (2016) Hippocampal-prefrontal theta oscillations support memory integration. Curr Biol 26(4):450–457
176. Ketz NA, Jensen O, O'Reilly RC (2015) Thalamic pathways underlying prefrontal cortex--medial temporal lobe oscillatory interactions. Trends Neurosci 38(1):3–12
177. Griffiths B et al (2016) Brain oscillations track the formation of episodic memories in the real world. Neuroimage 143:256–266
178. Reinhart RM et al (2012) Homologous mechanisms of visuospatial working memory maintenance in macaque and human: properties and sources. J Neurosci 32(22):7711–7722
179. Polanía R, Nitsche MA, Paulus W (2011) Modulating functional connectivity patterns and topological functional organization of the human brain with transcranial direct current stimulation. Hum Brain Mapp 32 (8):1236–1249

180. Soekadar SR et al (2013) In vivo assessment of human brain oscillations during application of transcranial electric currents. Nat Commun 4:2032
181. Herrmann CS et al (2013) Transcranial alternating current stimulation: a review of the underlying mechanisms and modulation of cognitive processes. Front Hum Neurosci 7:279
182. Antal A, Herrmann CS (2016) Transcranial alternating current and random noise stimulation: possible mechanisms. Neural Plast 2016:1
183. Reato D et al (2013) Effects of weak transcranial alternating current stimulation on brain activity—a review of known mechanisms from animal studies. Front Hum Neurosci 7:687
184. Reinhart RM (2017) Disruption and rescue of interareal theta phase coupling and adaptive behavior. Proc Natl Acad Sci U S A 114 (43):11542–11547
185. Helfrich RF et al (2014) Selective modulation of interhemispheric functional connectivity by HD-tACS shapes perception. PLoS Biol 12 (12):e1002031
186. Polanía R et al (2012) The importance of timing in segregated theta phase-coupling for cognitive performance. Curr Biol 22 (14):1314–1318

Part III

Functional Imaging Methods

Neuromethods (2020) 151: 209–238
DOI 10.1007/7657_2019_18
© Springer Science+Business Media, LLC 2019
Published online: 13 April 2019

Lesion-Behavior Mapping in Cognitive Neuroscience: A Practical Guide to Univariate and Multivariate Approaches

Hans-Otto Karnath, Christoph Sperber, Daniel Wiesen, and Bianca de Haan

Abstract

Lesion-behavior mapping is an influential and popular approach to anatomically localize cognitive brain functions in the human brain. The present chapter provides a practical guideline for each step of the typical lesion-behavior mapping study pipeline, ranging from patient and imaging data, lesion delineation, spatial normalization, and statistical testing to the anatomical interpretation of results. An important aspect of this guideline at the statistical level will be to address the procedures related to univariate as well as multivariate voxelwise lesion analysis approaches.

Keywords Lesion analysis, Univariate voxel-based lesion-symptom mapping, Multivariate voxel-based lesion-symptom mapping, Multivariate pattern analysis, VLSM, VLBM, MLBM, MVPA, Brain behavior inference, Stroke, Human

1 Introduction

Within the spectrum of neuroscience methods, the analysis of pathological behavior following brain injury is an important source of knowledge regarding the function and anatomy of the healthy brain. Lesion analysis belongs to the brain interference methods, i.e.—in contrast to correlative methods as, e.g., fMRI, EEG, or MEG—to those methods where inferences about the functional brain architecture are drawn from changes in behavior following focal brain lesions. The prominent advantage of these methods is that they allow us to determine whether a brain region is required for a specific task and function [1]. If a stroke causes a behavioral deficit, we know that the brain territory affected is essential for the normal functioning of this behavior.

In principle, the lesion-behavior approach can be applied to various research questions. One of its main purposes is to investigate the functional architecture of the healthy human brain. In the following, we will concentrate on its use for this fundamental neuroscientific research question. We will provide a practical guideline for each step of the lesion-behavior mapping pipeline for this purpose. Multiple considerations, ranging from patient and imaging data, lesion delineation, spatial normalization, and statistical testing to the anatomical interpretation of results, are necessary to arrive at meaningful conclusions in lesion-behavior mapping studies. For an overview of the required preprocessing steps, see Fig. 1. At the statistical level, recent developments allow us to choose from univariate approaches (voxel-based lesion-behavior mapping [VLBM]) as well as multivariate methods (multivariate lesion-behavior mapping [MLBM]) for lesion-behavior analyses. We will focus on both types of statistical approaches and provide arguments for their use and implementation.

2 Patient Selection

Ideally, researchers should *a priori decide on a reasonable patient recruitment time period and unselectively include all suitable patients* that present during that time period in the study. For example, if the researcher is interested in a function known to be located in the human left hemisphere from previous work (e.g., language functions), all patients with a left-sided brain lesion are included over a certain time period (e.g., 2 years). Exclusion criteria are, e.g., dementia, psychiatric disorders, or additional diffuse brain lesions since it would be difficult to determine which aspect(s) of the behavioral deficit can be attributed to the newly occurred brain lesion or might instead be due to these additional disorders. In contrast, absence of the behavioral deficit of interest is not to be used as an exclusion criterion. The inclusion of patients that do not have the behavioral deficit of interest is essential, as this allows us to differentiate between areas of the brain where damage is associated with the deficit of interest and areas of the brain where damage merely reflects increased vulnerability to injury [1].

The appropriate etiology of brain damage in studies aiming at the understanding of the functional architecture of the healthy human brain is stroke. Other etiologies, e.g., traumatic brain injury and tumors, have also been used in this context, but a considerable body of work suggests that such etiologies are less suitable or might even derive misleading results [2, 3]. There is no doubt that for clinical research questions as, e.g., the evaluation of brain tumor treatment [4], lesion-behavior mapping can also be a useful tool but—as mentioned in the introduction section above—such clinical questions are beyond the scope of the present chapter.

Stroke lesions are well suited for lesion-behavior mapping analyses aiming at the healthy brain's functional architecture because they go along with (1) structural imaging data which (in most cases) clearly reflect the area of infarcted and irretrievably lost neural tissue as well as (2) clear behavioral consequences that can, for the large part, be directly linked to the original function of the impaired

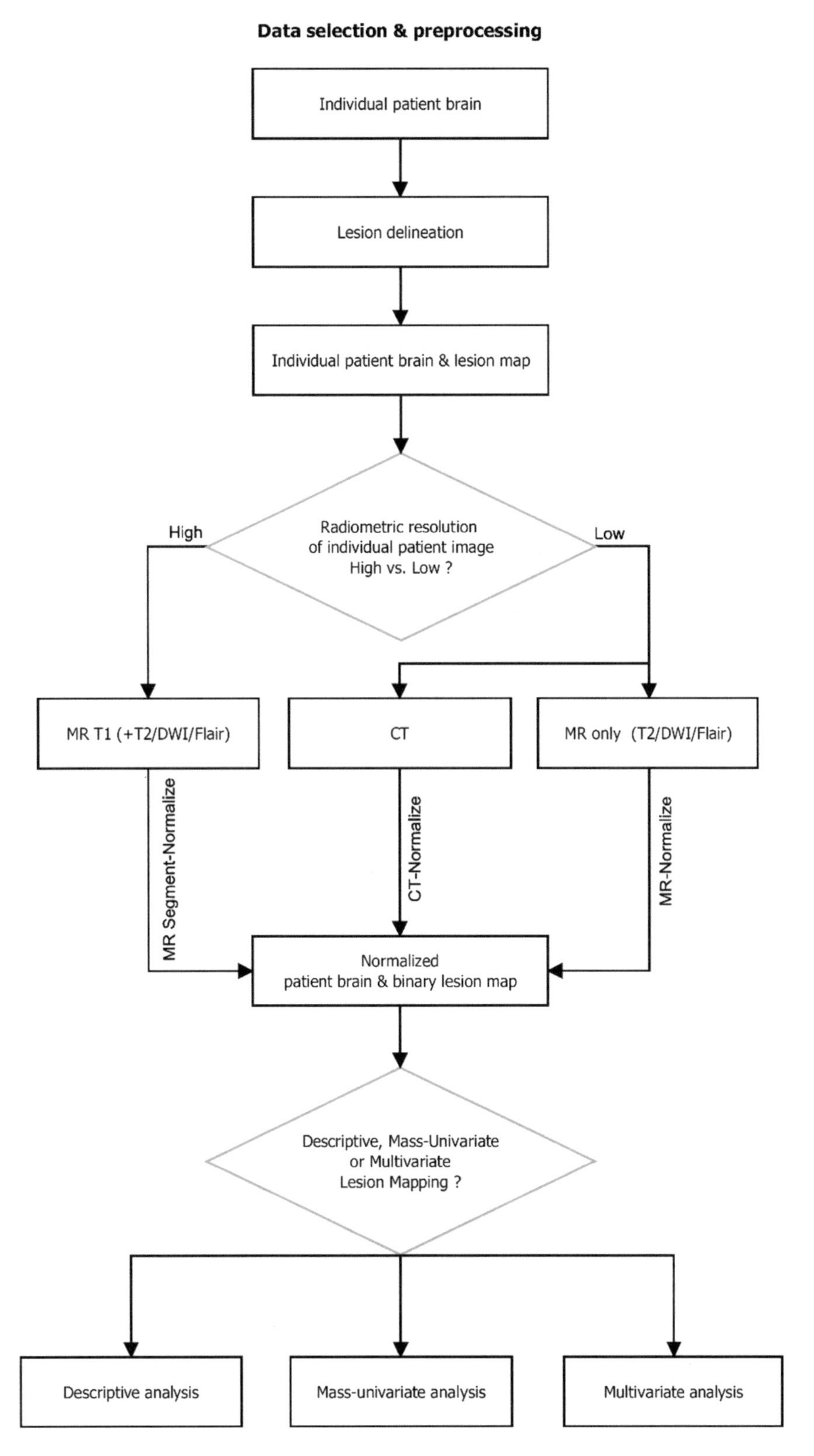

Fig. 1 Pipeline for data selection and preprocessing. When performing a lesion-behavior mapping analysis, different preprocessing steps need to be followed and decisions need to be made, respecting the type of the data

part of the brain. This is particularly true for strokes in the acute phase in which the brain has not yet had time to functionally reorganize. *Lesion-behavior mapping thus should prefer acute stroke patients*, while chronic stroke patients should remain to investigate the neural correlate of chronic dysfunction. Importantly, *combining both acute and chronic stroke patients in the same lesion-behavior mapping study is to be avoided*. Lesion-behavior mapping analyses assume that each patient is assessed at the same point in time following stroke onset and that thus the contribution of a certain brain area to a certain cognitive function is directly reflected in the behavioral scores in each patient. Combining both acute and chronic stroke patients in the same lesion-behavior mapping study violates this assumption.

3 Imaging Data

Imaging data should be obtained in the acute stroke phase, regardless of whether the aim is to study the functional architecture of the brain or to study the neural correlates of chronic cognitive deficits. For the study of the functional architecture of the brain, it has been shown that only the combination of acute behavioral scores and acute structural imaging precisely identifies the targeted brain areas. In contrast, lesion-behavior mapping analyses based on chronic behavior, in combination with either chronic or acute imaging, hardly detected any of the targeted substrates [5]. On the other hand, if combined with acute structural imaging, the behavioral data from chronic stroke patients allows us to make long-term clinical predictions based on the location of the acute brain damage [6–8].

In acute stroke patients, *the lesion can be visualized using either CT or MRI*. The development of CT templates for spatial normalization of individual patient images [9] has removed the main reason to disregard CT for lesion-behavior mapping studies. Moreover, modern spiral CT scanners provide high-resolution images, and in many clinical institutions, CT remains the dominant imaging modality of choice at admission. Importantly, the choice between administering CT and MRI to patients at admission is not random but follows specific clinical criteria. As a consequence, the systematic exclusion of patients with CT images only implements a selection bias, typically influencing important factors such as lesion size, general clinical status, severity of cognitive deficits, etc. (for a detailed discussion [10]). As such, lesion-behavior mapping studies aiming at the healthy brain's functional architecture should unselectively recruit all suitable patients, irrespective of whether they have received CT or MR imaging at admission.

We suggest the following practical guideline: In *patients with CT imaging only*, use noncontrast CT images to visualize the brain lesion. In *patients with MR imaging only*, use diffusion-weighted images (DWI) to visualize the lesion if imaging is performed less than 48 h following stroke onset, and use T2FLAIR images if imaging is performed more than 48 h following stroke onset. In *patients with both CT and MR images*, the researcher is in the privileged situation to choose the best from both modalities, i.e., to use those images where the lesion is most conspicuous.

4 Lesion Delineation

Following CT or MR data acquisition, the lesion needs to be delineated on each slice of the patient's brain image. Manual lesion delineation, which can be done using programs like MRIcroN (https://www.nitrc.org/projects/mricron; [11]) or ITK-SNAP (http://www.itksnap.org; [12]), is time-consuming and potentially observer-dependent. To address these disadvantages, both fully automated and semiautomated lesion delineation methods have been developed. Fully automated methods do not require any user interaction but also may be more susceptible to imaging artefacts (for discussion [3]). Given this downside, *semiautomated lesion delineation methods* that combine fully automated steps with mandatory user interaction *provide an optimal compromise*. While several semiautomated lesion delineation approaches exist (e.g., [13]), the semiautomated approach *Clusterize* (https://www.medizin.uni-tuebingen.de/kinder/en/research/neuroimaging/software/; [14]) has been shown to be capable of significantly speeding up lesion delineation, without loss of either lesion delineation precision or lesion delineation reproducibility in acute stroke patients scanned in both CT and a range of common MRI modalities [15]. The principle of *Clusterize* is simple: On the basis of local intensity maxima and iterative region growing, the whole CT or MR brain image is fully automatically clusterized, including the lesioned area. The user subsequently manually selects those clusters that correspond to the lesion.

5 Spatial Normalization of Patient Brain and Lesion Map

We now have a 3D binary lesion map reflecting the voxels where brain function is impaired for each patient (see Fig. 1). However, all brains differ in orientation, size, and shape. As such, before we can perform statistical comparisons, we need to spatially normalize the patient brains and lesion maps into standard stereotaxic space. Spatial normalization can be performed with programs such as BrainVoyager (http://www.brainvoyager.com/; [16]), SPM (http://www.fil.ion.ucl.ac.uk/spm/), FSL (https://fsl.fmrib.ox.ac.uk/fsl/fslwiki; [17]), AFNI (https://afni.nimh.nih.gov/; [18, 19]), or ANTs (http://stnava.github.io/ANTs/; [20]). The analysis package SPM is widely used due to its platform independence, free obtainability, and availability of many add-ons. As such, we here focus on the spatial normalization routines of SPM, as implemented in the *Clinical Toolbox* (https://www.nitrc.org/projects/clinicaltbx/; [9]). This toolbox provides specialized templates that *allow spatial normalization of both CT and MR brain images of elderly, stroke-aged populations.* As such, the Clinical Toolbox is ideally suited to be used in lesion-behavior mapping studies where the patients included are typically older, and where different modalities, i.e., CT as well as MR images, are present in different patients. Moreover, the *Clinical Toolbox* provides both the traditional normalization procedure and the unified segmentation and normalization approach [21] which are required to process the different types of imaging data typically present in clinical patient samples.

Normalization of DWI Data: DWI data are usually collected with different b-values (typically b0, b500, and b1000). While the b1000 image best suited to visualize the lesion in acute stroke patients has a low radiometric resolution, the additionally collected b0 image often has a relatively high radiometric resolution. The typical approach thus is to first coregister the image best suited to visualize the lesion but with a low radiometric resolution (e.g., the b1000 DWI) to the image with a high radiometric resolution (e.g., the b0 DWI or—if available—a T1 image). Subsequently, the image with the high radiometric resolution is normalized using the unified segmentation and normalization approach [21]. This approach has been shown to be superior to the traditional normalization approach [22, 23], but requires an image with a high radiometric resolution. The unified segmentation and normalization approach combines tissue classification (i.e., segmentation), bias correction, and image registration (i.e., spatial normalization) in a single model. Once the necessary image transformations have been estimated, they are applied to the patient's brain image(s) and the lesion map, bringing all images in standard stereotaxic space. Since the average age in clinical stroke samples investigated for lesion-behavior mapping studies is typically above 60 years, the *Clinical Toolbox* provides template-based tissue probability maps for use with the unified segmentation and normalization approach [9] derived from elderly adults (mean age: 61.3 years).

Normalization of CT and T2FLAIR Data: If only CT or only T2 FLAIR data are available, we use the traditional normalization procedure in which linear (affine) and nonlinear (nonaffine) image transformations are applied to match the orientation, size, and shape of each patient brain to the orientation, size, and shape of a template brain in standard stereotaxic space. In these cases, the template image to be selected in the *Clinical Toolbox* [9] should ideally have the same image modality as the patient image that is spatially normalized, as the accuracy of this approach depends on how similar the voxel intensities of a given brain area are between the patient and the template image. Moreover, the template image should roughly match the population of the lesion mapping study. That is, if the lesion-behavior mapping study is performed in elderly stroke patients, the template or template-based tissue probability maps used should have been derived from an elderly population. The *Clinical Toolbox* provides a template for CT imaging data [9] derived from elderly adults (mean age: 61.3 years), as well as a template for T2FLAIR imaging data (https://brainder.org/download/flair/; [24]) derived from a wide range of adults (mean age 35.4, range 18–69 years) for use with the traditional normalization procedure.

In patients with MR acquisition at admission, sometimes T1 images, i.e., data providing high radiometric resolution, are available in addition to the T2FLAIR data. In these cases, we can apply the unified segmentation and normalization approach instead of the traditional normalization procedure to optimize the normalization result [22]. The image with a low radiometric resolution, i.e., the T2FLAIR data, is coregistered to the image with a high radiometric resolution (the T1 image). The image with a high radiometric resolution then is normalized using the unified segmentation and normalization approach. The necessary image transformations are estimated and applied to the patient's brain image(s) and the lesion map, bringing all images in standard stereotaxic space.

5.1 Correcting for the Lesion During Spatial Normalization

In both the traditional normalization procedure and the unified segmentation and normalization approach, the area of brain damage creates a large mismatch between the patient brain and the template brain. The two dominant solutions to this problem are cost function masking [25] and enantiomorphic normalization [26]. The *Clinical Toolbox* provides both options to address this problem. During cost function masking, lesioned voxels are excluded during spatial normalization. As such, the image transformations necessary to bring the patient's brain image(s) and the lesion map in standard stereotaxic space are derived from intact areas of the brain only. During enantiomorphic normalization, on the other hand, the lesion is "corrected" by replacing it with brain tissue from the lesion homologue in the intact hemisphere of the brain. As such, the image transformations necessary to bring the patient's brain image(s) and the lesion map in standard stereotaxic space are effectively derived from a brain image without a lesion.

Logically, one can expect enantiomorphic normalization to perform better than cost function masking when lesions are large and unilateral, as spatial normalization with cost function masking becomes less accurate as lesion size increases (as the area from which the image transformations can be derived decreases with increasing lesion size), while enantiomorphic normalization does not. Cost function masking can, however, be expected to perform better than enantiomorphic normalization when lesions are bilateral and affect similar areas in both hemispheres, as enantiomorphic normalization would in this case replace the lesion with the likewise lesioned homologue. Moreover, as enantiomorphic normalization assumes that the brain is essentially symmetric, enantiomorphic normalization might be suboptimal in areas known to be considerably asymmetric (e.g., the planum temporale). The decision to use *either cost function masking or enantiomorphic normalization for normalization* thus should be made on an individual, patient-to-patient basis. A comparison of both methods, however, highlighted a generally better precision of enantiomorphic normalization [26]. Thus, if no clear preference exists, enantiomorphic normalization should be chosen.

6 Voxelwise Lesion Analysis

Following lesion delineation and spatial normalization, we have a spatially normalized binary lesion map for each patient. Moreover, we have a behavioral measurement for each patient. With these two sources of information, we are ready to perform a voxelwise lesion-behavior mapping analysis to relate lesion location and patient behavior (see Fig. 1). Over the last decades, various techniques for lesion analyses evolved, spanning from subtraction plots over mass-univariate mapping to recently developed multivariate methods. The choice of one or the other of these techniques is largely depending on the fulfilment of specific prerequisites and own research interests. The simplest type of voxelwise analysis is a *lesion subtraction analysis*. In fact, this method is not a statistical but *a descriptive method that should be reserved for studies aiming at rare disorders*, i.e., for studies that a priori are expected to include only a small number of patients. If more patient data is available, it is possible to conduct either a mass-univariate or a multivariate lesion analysis. If we have good reasons to assume that the *function of interest is represented in a widely distributed network in the human brain, multivariate approaches are to be preferred*, as in these cases the ability of univariate techniques to detect all of the network modules might be limited [27, 28]. In the following, we introduce all three types of possible voxelwise analysis techniques.

6.1 Voxelwise Comparison by a Lesion Subtraction Analysis

In a lesion subtraction analysis, the lesion overlap map of patients without the cognitive deficit of interest is subtracted from the lesion overlap map of patients showing the cognitive deficit of interest. The typical pipeline is shown in Fig. 2. To account for potential

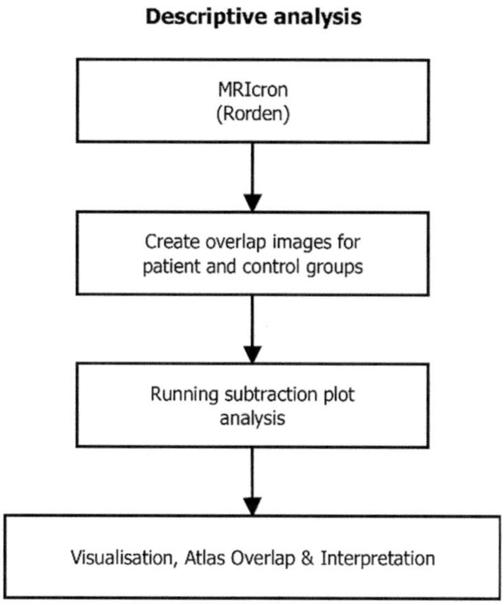

Descriptive analysis

MRIcron (Rorden) → Create overlap images for patient and control groups → Running subtraction plot analysis → Visualisation, Atlas Overlap & Interpretation

Fig. 2 Pipeline for voxelwise comparison by a lesion subtraction analysis. Flowchart showing steps to follow in order to conduct a descriptive subtraction lesion analysis

sample size differences between the two patient groups, these subtraction analyses need to use proportional values. That is, for each voxel the percentage of patients without the cognitive deficit of interest that have a lesion at the voxel is subtracted from the percentage of patients with the cognitive deficit of interest that have a lesion at the voxel. The result of the subtraction analysis is a map with the percentage relative frequency difference between these two groups for each voxel (Fig. 3). This map highlights those areas of the brain where lesions are more frequent in patients with than in patients without the cognitive deficit of interest and so distinguishes between regions that are merely often damaged in strokes and regions that are specifically associated with the deficit of interest [1]. To control for neuropsychological comorbidity, the two patient groups contrasted in a lesion subtraction analysis can be made comparable with respect to additional neurological impairments (of no interest), e.g., paresis, visual field defects, etc., or lesion size.

6.1.1 Tools for a Lesion Subtraction Analysis

Lesion subtraction analyses can be performed with programs such as MRIcron (https://www.nitrc.org/projects/mricron; [29]). MRIcron works best on a Windows OS, but is also available for Linux and Macintosh OSX computers. After downloading and unzipping the archive, double-click the MRIcron application to launch the program. To perform a lesion subtraction analysis, follow the routine and tips below:

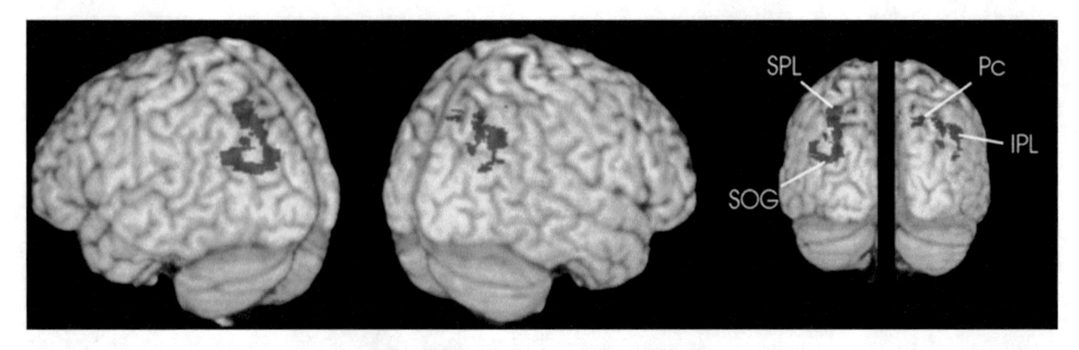

Fig. 3 Example of statistical topography resulting from lesion subtraction analysis. Medial and lateral surface views of a lesion subtraction plot, generated with MRIcron (http:/people.cas.sc.edu/rorden/mricron/index.html). The subtraction plot demonstrates the percentage of lesion overlap of two experimental groups (ten patients with unilateral left brain lesions, six patients with unilateral right hemisphere lesions) after the subtraction of the overlap of two respective control groups. It thus highlights those areas of the brain where lesions are more frequent in patients with than in patients without the cognitive deficit of interest and so distinguishes between regions that are merely often damaged in strokes and regions that are specifically associated with the deficit of interest (Reproduced from Karnath and Perenin [61] with permission from Oxford University Press)

(a) First the binary normalized lesion images need to be copied into a single directory.

(b) It is recommended to convert these .nii files into .voi files. This can be done by selecting "Convert" ("NII->VOI") from the "Draw" menu (if necessary, select "Show drawing tools" from the "Draw" menu to make all options in the "Draw" menu visible).

(c) Create a lesion overlap image for each patient subgroup (usually one patient subgroup with the deficit of interest and a control patient subgroup without the deficit of interest) by selecting "Create overlap images" from the "Statistics" submenu of the "Draw" menu.

(d) Once we have a lesion overlap image for each patient subgroup, we can perform a subtraction analysis by selecting "Subtraction Plots" from the "Statistics" submenu of the "Draw" menu. Under "Select POSITIVE overlap image," select the overlap image of the patient subgroup with the deficit of interest, and under "Select NEGATIVE overlap image," select the overlap image of the control patient subgroup without the deficit of interest. On non-Windows computers, saving the subtraction plot sometimes needs to be repeated several times (just keep reselecting the originally chosen filename and ignore the "overwrite" warning).

(e) The resulting subtraction plot can be loaded as an overlay (over a suitable template image) by selecting "Add" from the "Overlay" menu. The range of values to be shown can be adjusted by modifying the lower bound and upper bound intensity value in the boxes to the right of the "Autocontrast" button (). Values between 1 and 100 show voxels more often damaged in patients with the deficit of interest than in patients without the deficit of interest, whereas values between −1 and −100 show the reverse. These values reflect relative frequency of damage, e.g., a value of 50 reflects that that voxel is damaged 50% more frequently in patients with the deficit of interest than in patients without the deficit of interest. Assuming $n = 10$ in both groups, this could mean that the voxel is damaged in 9 of 10 patients (i.e., 90%) with the deficit of interest and in 4 of 10 patients (i.e., 40%) without the deficit of interest (or 7 of 10 vs. 2 of 10, etc.).

6.2 Univariate Voxelwise Statistical Comparisons

The spatially normalized binary lesion map of each patient together with the behavioral measurement for each patient that we derive from lesion delineation and spatial normalization also is the basis for a univariate voxelwise statistical lesion-behavior mapping analysis. The typical pipeline of such an analysis is shown in Fig. 4. In contrast to a descriptive subtraction procedure (see above), we here perform a statistical test at each voxel to relate voxel status (lesioned/non-lesioned) and patient behavior and generate a voxelwise map of statistical significance (Fig. 5). There are no clear guidelines about the minimum required sample size for performing univariate lesion-behavior mapping. Such analyses are *commonly performed with sample sizes ranging between 30 and 100 patients.* Yet, a recent study investigated the impact of sample size on the reproducibility of voxel-based lesion-behavior mapping and showed that low-powered studies (due to small sample sizes) produce heterogeneous results and might thus over- or underestimate the true effect size in the population [30]. However, they also demonstrated that higher sample sizes increase the probability of small or even trivial effect sizes becoming significant. Therefore, the authors argue to choose a sample size as high as possible to address the first issue and to report effect sizes to address the second.

Continuous Behavioral Data: When the behavioral data is continuous, the behavioral data of the group of patients in whom a given voxel is damaged is statistically compared to the behavioral data of the group of patients in whom that same voxel is intact. This is traditionally done with a *two-sample t-test*, which assumes that the behavioral data is normally distributed and measured on an interval scale. Unfortunately, however, behavioral data from patient populations is often not normally distributed. During behavioral assessment, patients without the deficit of interest will typically all demonstrate close to maximum performance, whereas performance in patients with the deficit of interest will typically be poorer and more variable over patients. As a consequence, the distribution of

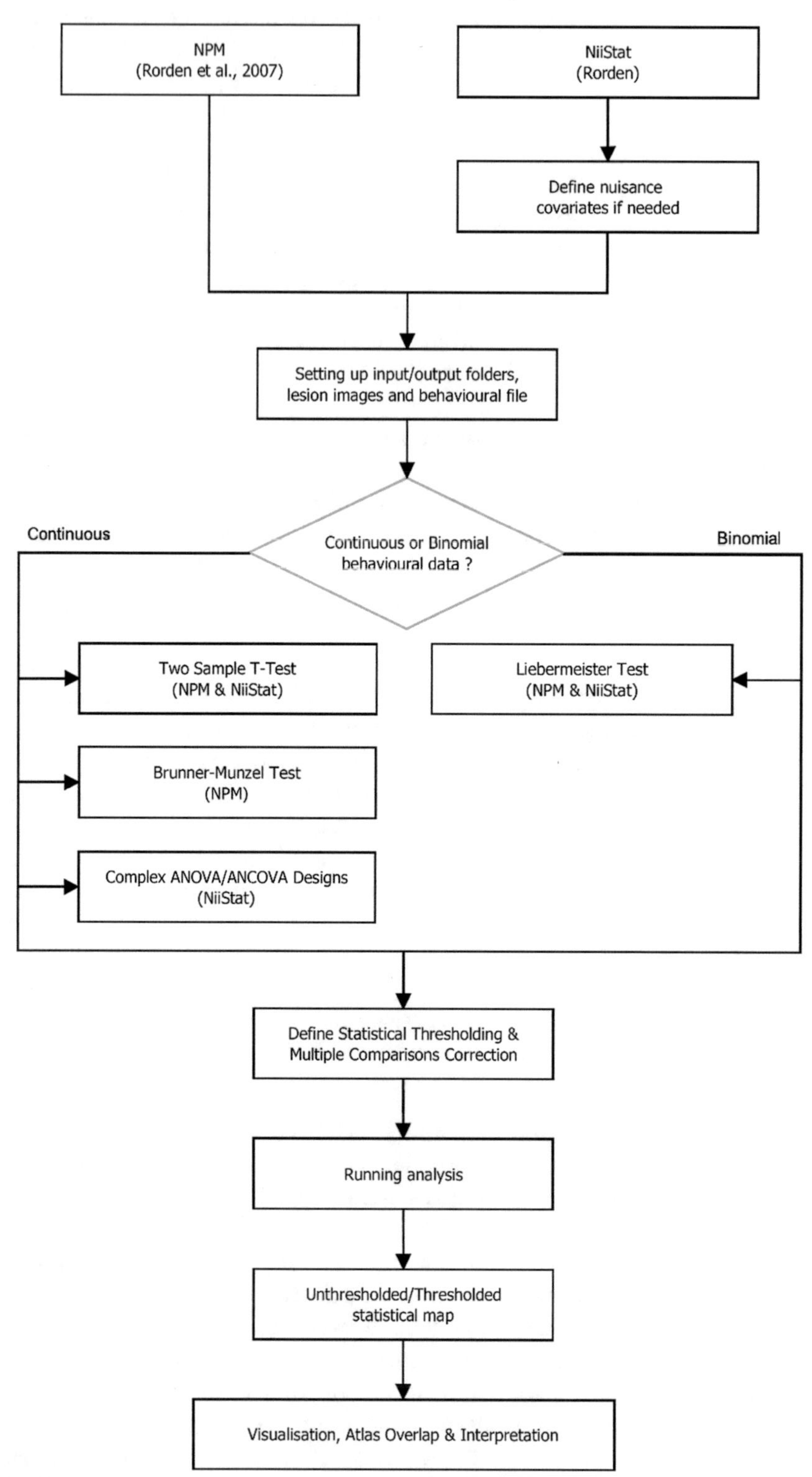

Fig. 4 Pipeline for univariate voxelwise lesion analysis. Flowchart showing steps to follow and decisions to be made in order to conduct a univariate voxelwise lesion analysis

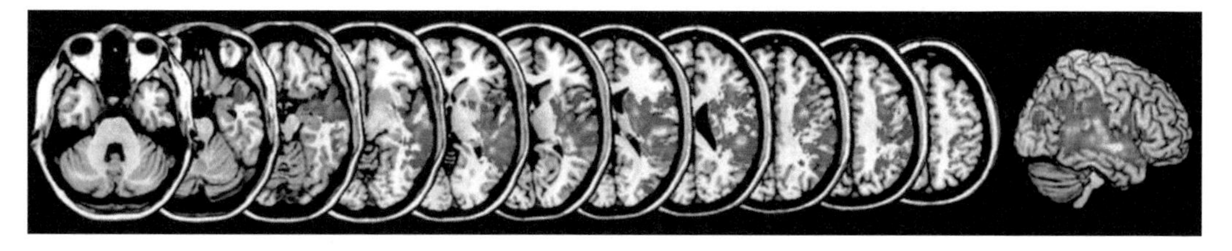

Fig. 5 Example of statistical topography resulting from univariate voxelwise lesion analysis. Statistical voxelwise lesion-behavior mapping result using the *t*-test statistic. The analysis included 54 patients with right hemisphere damage and was based on a continuous behavioral variable. Shown are axial views and a three-dimensional rendering of a sagittal view, generated with MRIcron (http:/people.cas.sc.edu/rorden/mricron/index.html) (Reproduced from Karnath et al. [6] with permission from Oxford University Press)

the behavioral data from patient populations is often negatively skewed. Moreover, behavioral data from patient populations is often not measured on an interval but on an ordinal scale. The use of the assumption-free rank order test proposed by Brunner and Munzel [31] is more appropriate in these situations. In fact, this so-called Brunner-Munzel test has been shown to have higher statistical power than the *t*-test while offering similar protection against false positives in situations where the distribution of the behavioral data is skewed [29]. The Brunner-Munzel test is provided in the NPM software for lesion-behavior analyses (https://www.nitrc.org/projects/mricron).

Binomial Behavioral Data: When the behavioral data is binomial (i.e., when the deficit is either present or absent, as in, e.g., hemianopia), we statistically assess, for each voxel, whether the variables "voxel status" (voxel lesioned vs. voxel intact) and "behavioral status" (deficit present vs. absent) are associated or independent. The statistical test typically used in these situations is the Pearson's chi-square test. In many lesion-behavior mapping analyses, however, expected cell frequencies are lower than five to ten in at least some voxels, resulting in inflated false-positive rates when using Pearson's chi-square test. A statistical test more appropriate in these situations is the quasi-exact test proposed by Liebermeister [32]. In a lesion-behavior mapping analysis with simulated data, observed false-positive rates for this so-called Liebermeister test closely approximated the set false-positive threshold [29]. The Liebermeister test is provided in the NPM software for lesion-behavior analyses (https://www.nitrc.org/projects/mricron).

6.2.1 Tools for Univariate Voxelwise Statistical Comparisons

Univariate voxelwise statistical analyses can be performed with programs or toolboxes such as VLSM (http://aphasialab.org/vlsm; [33]), VoxBo (https://www.nitrc.org/projects/voxbo/; [34]), NPM (https://www.nitrc.org/projects/mricron; [29]), and/or NiiStat (https://www.nitrc.org/projects/niistat/). Of these, NPM and NiiStat are the most up-to-date and will thus be the focus of this section.

NPM [29]

The program NPM (nonparametric mapping) is included in the MRIcron download (https://www.nitrc.org/projects/mricron). A brief manual is given at http://people.cas.sc.edu/rorden/mricron/stats.html. NPM is capable of performing lesion analyses using either continuous behavioral data (two-sample *t*-test and Brunner-Munzel test) or binomial behavioral data (Liebermeister test). As such, NPM is an easy-to-use program in situations where there is only a single dependent variable of interest per statistical test. In situations where there are multiple dependent variables and you want to use a regression approach (e.g., additional nuisance covariates; see Sect. 7.1), the program NiiStat should be used. To correct for multiple comparisons, NPM supports Bonferroni, FDR, and permutation-based thresholding (see Sect. 7.2). Like MRIcron, NPM works best on a Windows OS, but is also available for Linux and Macintosh OSX computers. After downloading and unzipping the archive, double-click the NPM application to launch the program. To perform a voxelwise statistical lesion analysis, follow the routine and tips below:

(a) First the binary normalized lesion images need to be copied into a single directory.

(b) It is recommended to convert these .nii files into .voi files. This can be done in MRIcron by selecting "Convert" ("NII-> VOI") from the "Draw" menu (if necessary, select "Show drawing tools" from the "Draw" menu to make all options in the "Draw" menu visible).

(c) In NPM, we first need to specify the design. To do this, select "Design..." from the "VLSM" menu, and click the "Design IVs" button. The amount of predictors (i.e., the behavioral variables) and the predictor names can be adapted if necessary. It is important to realize that entering multiple predictors here simply means that the statistical test will be performed separately on each predictor (i.e., the program will not implement a regression approach). To select the lesion images, click the button "Select Images." To ensure sufficient minimum lesion overlap, voxels damaged in a very low percentage of the patient sample should be excluded. Typically, voxels damaged in less than 5–10% of the patient sample are excluded. Finally, after clicking the "OK" button, the behavioral scores for each patient can be added. In NPM, higher behavioral scores are assumed to reflect better performance. In other words, e.g., binomial behavioral data, patients with the deficit of interest should be coded as "0," and patients without the deficit of interest should be coded as "1." To save this specified design as an MRIcron .val file, select "Save" from the "File" menu.

(d) Under the "Options" menu, the amount of permutations used to control the family-wise error rate using permutation thresholding can be selected. It is recommended to set this to the maximum value (8000). Moreover, in the case of continuous behavioral data, the "Tests" submenu allows you to select whether you want to perform a two-sample *t*-test, a Brunner-Munzel test, or both.

(e) To run the analysis, select "Binary images, binary groups (lesions)" from the "VLSM" menu when the behavioral data is binomial. When the behavioral data is continuous, select "Binary images, continuous groups (vlsm)" from the "VLSM" menu. Select the saved specified design (MRIcron .val file). When it asks for "Base Statistical Map," enter a filename. The results from the analysis will then be saved as <filename><StatisticalTestUsed><Predictorname>. If desired, an explicit mask can be selected to restrict the analysis to a region of interest in the brain. If you do not want to restrict your analysis to a region of interest, simply press "Cancel."

(f) For each behavioral predictor and statistical test used, the program creates an unthresholded z-map (<filename><StatisticalTestUsed><Predictorname>.nii.gz). Additionally, for each behavioral predictor, the program creates a text file (<filename>Notes<Predictorname>.txt) containing the range of unthresholded *z*-values obtained for each statistical test, as well as the critical *z*-values for several commonly used *p*-values following different methods of correcting for multiple comparisons (using Bonferroni, FDR, or permutation-based thresholding). The unthresholded z-map can be loaded as an overlay (over a suitable template image) in visualization software packages such as MRIcron (see Sect. 6.1) by selecting "Add" from the "Overlay" menu. To visualize the statistically significant voxels in MRIcron, the range of values to be shown can be adjusted by modifying the lower bound and upper bound intensity value in the boxes to the right of the "Autocontrast" button ( ). The lower bound intensity value should be set to the critical *z*-value.

NiiStat (C. Rorden; https:/ www.nitrc.org/projects/ niistat/)

The Niistat toolbox can be downloaded from https://www.nitrc.org/projects/niistat/. This toolbox requires MATLAB (The MathWorks, Inc., Natick, MA; version 9.0 or higher for graphical user interface) and SPM (Wellcome Institute of Imaging Neuroscience, London, UK; version SPM12 or later). A detailed description of the toolbox and installation instructions is given at https://www.nitrc.org/plugins/mwiki/index.php/niistat:MainPage. This page also includes a well-documented manual and tutorial. The NiiStat toolbox allows the use of a wide range of neuroimaging techniques, including univariate lesion-behavior mapping.

(a) After successful installation, run the following command in MATLAB to launch the toolbox:

```
NiiStatGUI; %for more sophisticated graphical user interface
NiiStat; %for a rather simple user interface
```

(b) Like NPM, NiiStat is capable of performing lesion analyses using either continuous behavioral data or binomial behavioral data. Unlike NPM, however, NiiStat comes with general linear model functionality, allowing the inclusion of nuisance covariates, as well as performing ANOVAs.

(c) In situations where there is only a single dependent variable of interest per statistical test, NiiStat computes the Liebermeister test when the behavioral data is binomial and computes statistics using the general linear model when the behavioral data is continuous (in which case results will be identical to the results of a two-sample *t*-test). Thus, barring minor rounding error differences, when there is only a single dependent variable of interest, results will be identical to those generated using NPM.

(d) To correct for multiple comparisons, NPM supports Bonferroni, FDR, and permutation-based thresholding (see Sect. 7.2.).

6.3 Multivariate Voxelwise Statistical Comparisons

Due to the limited ability of univariate lesion-behavior mapping methods *to identify brain modules that are complex—like brain networks*—multivariate lesion-behavior mapping (MLBM) approaches have been proposed [35–39]. The typical pipeline of such analyses is shown in Fig. 6. In a recent simulation study [40], it has been demonstrated that a *patient sample size of 100 to 120 patients provides a satisfying trade-off* between model quality and feasibility, using an approach termed "support vector regression based multivariate lesion-symptom mapping" (SVR-LSM; [36]). In another simulation study, Pustina and colleagues [39] demonstrated that the performance of another multivariate approach, namely, the multivariate sparse canonical correlations technique (SCCAN), was generally superior to mass-univariate lesion analysis already at small sample sizes. However, the authors did not determine a minimum sample size needed to generate a reliable anatomical mapping. Hence, more studies with different patient groups, sample sizes, and different MLBM methods are needed to conclude on a specific gold standard concerning sample size. In the following, we will introduce MLBM methods that were implemented in publicly available software and that underwent thorough validation. We will mainly focus on the SVR-LSM approach [36] which is based on support vector regression (SVR)

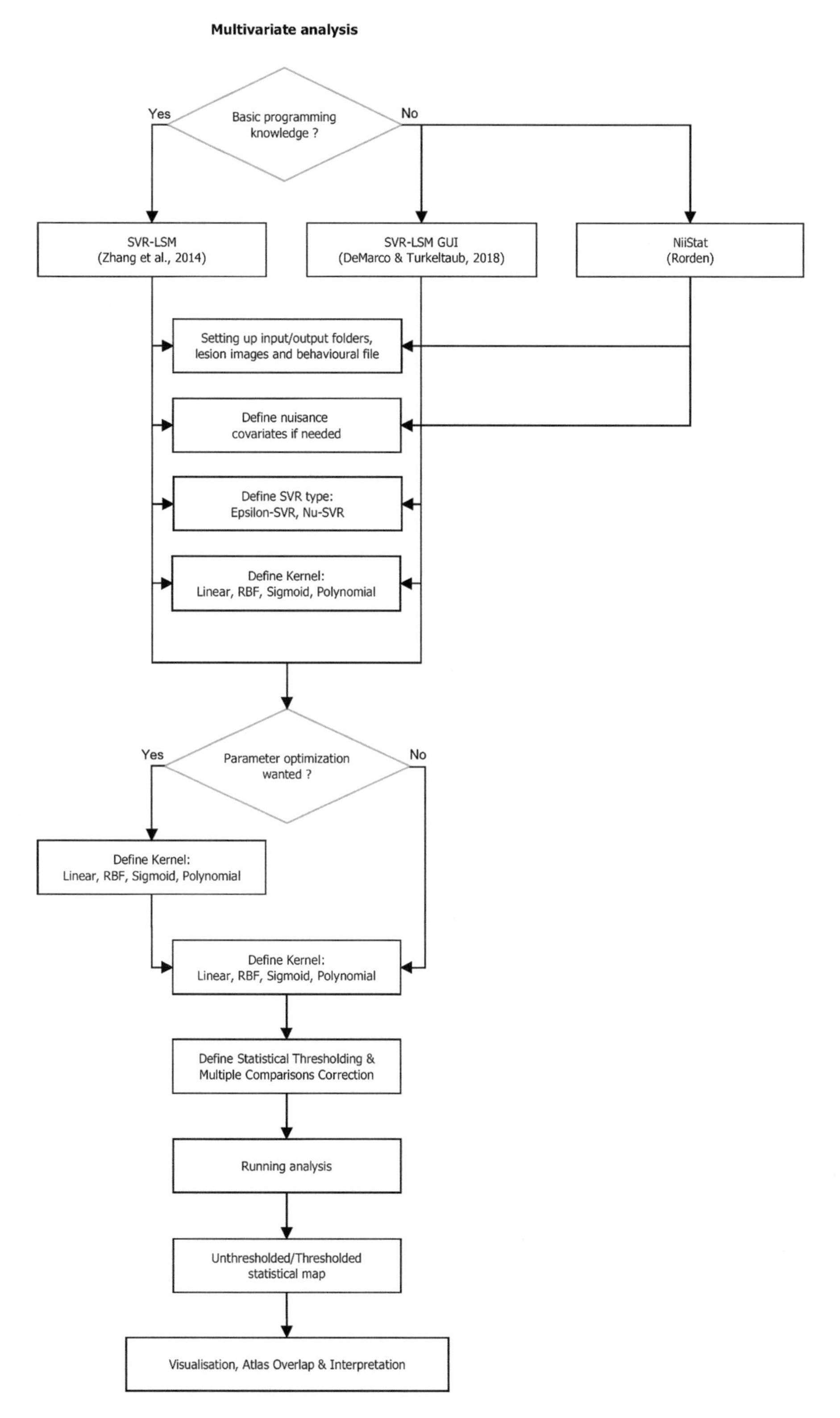

Fig. 6 Pipeline for multivariate voxelwise lesion analysis. Flowchart showing steps to follow and decisions to be made in order to conduct a multivariate voxelwise lesion analysis

and has already been used in several studies. Further, we will introduce a tool to perform the strategically entirely different SCCAN approach based on sparse canonical correlations [39].

6.3.1 Support Vector Regression-Based Multivariate Lesion-Symptom Mapping (SVR-LSM)

The basic idea behind SVR-LSM is to compute a support vector regression (SVR) in order to model a behavioral variable based on the lesion status of each voxel. If a voxel significantly contributes to this model, it is then assessed via permutation testing. In more detail, first, the continuous behavioral variable is modelled based on the lesion status of all voxels included in the analysis. The model assigns to each included voxel a beta-parameter, a so-called feature weight. These beta-values, however, cannot be interpreted directly. To evaluate the significance of single beta-weights, SVRs are performed for a large amount of random permutations of the behavioral data. These latter analyses reveal what feature weights can be expected with random data. Like commonly performed in permutation testing, statistical thresholds can be inferred from these analyses to assess statistical significance of feature weights in the analysis of real data. The resulting topography is a voxelwise map of statistical significance (Fig. 7).

Several parameters are relevant in such analyses and need to be adjusted. First, machine learning algorithms often use parameters that define aspects of how models are generated. These parameters, so-called hyperparameters (e.g., C and γ in SVR with an RBF kernel), have to be chosen a priori. Generally, these hyperparameters can be optimized through grid search before the actual analysis. Optimization in SVR-LSM, however, has proven to be tricky, because a model with high fit does not necessarily provide generalizable beta-parameters. Therefore, unambiguous maximization/minimization strategies are not available (for more details see Zhang et al. [36] and Rasmussen et al. [41]). Another parameter

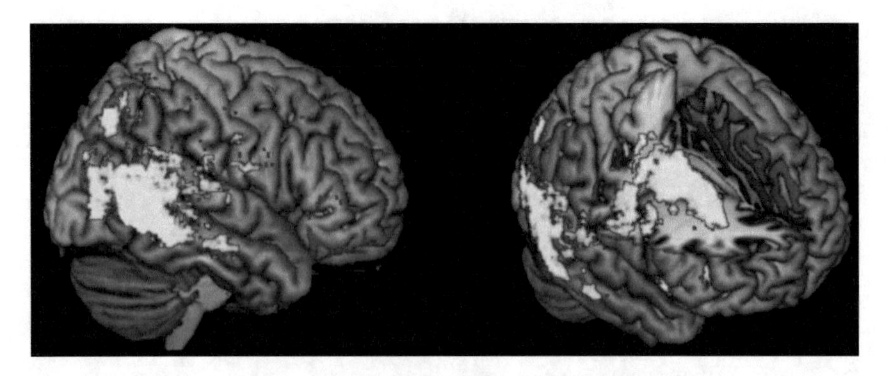

Fig. 7 Example of statistical topography resulting from multivariate voxelwise lesion analysis. Three-dimensional renderings of a support vector regression-based multivariate lesion-symptom mapping result using the 3D-interpolation algorithm provided by MRIcron for visualization (http:/people.cas.sc.edu/rorden/mricron/index.html; 8 mm search depth) with sagittal view (left) and inside view (right). The analysis included 203 patients with right hemisphere damage and was based on a continuous behavioral variable (From [60])

to be adjusted is the number of permutations. Large numbers require computational power, but they should be chosen as high as possible. Permutation numbers of at least several thousands are required; 10,000 permutations should be an ideal number. Next, we need to control for lesion size. While the optimal approach for this control is still a matter of debate [42], it is clear that at least any kind of such control is required, such as nuisance regression or direct total lesion volume control [36]. Further, it was shown that a correction for multiple comparisons is required [40]. The reason is that statistical significance in the current implementation of SVR-LSM is assessed for each voxel individually, i.e., many statistical tests are performed at once. There is no consensus yet on what correction suits best; available software programs offer several approaches such as false discovery rate.

Tools for SVR-LSM Analyses

Various tools and strategies allowing to perform SVR-LSM analyses are available, differing in functionality and the analysis algorithms they are based on. Among those the MATLAB toolbox provided by Zhang and colleagues [36] is interesting because it contains a set of MATLAB scripts which can be easily adapted (by advanced MATLAB users) to one's own needs, allowing a simple MLBM analysis but also more advanced research questions. The following section will also address two other toolboxes, SVR-LSM GUI [42] and NiiStat (www.nitrc.org/projects/niistat), which extended the initial toolbox from Zhang and colleagues [36], adding a graphical user interface, which will make the analysis technique more accessible to clinicians and non-experienced MATLAB users. As these two tools are well documented and are rather straightforward, we will not go into details.

SVR-LSM MATLAB Toolbox [36]: The SVR-LSM toolbox by Zhang and colleagues [36] can be downloaded from https://github.com/yongsheng-zhang/SVR-LSM or https://github.com/dmirman/SVR-LSM. The latter contains additionally a five-fold cross-validation routine for finding the optimal set of the SVR parameters and was added by Daniel Mirman. Requirements for the use of this toolbox include MATLAB (The MathWorks, Inc., Natick, MA; version 7.12 or higher), SPM (Wellcome Institute of Imaging Neuroscience, London, UK; version SPM8 or SPM12), and libSVM (http://www.csie.ntu.edu.tw/~cjlin/libsvm/; version 3.18 or higher). Note that libSVM comes already with the toolbox, but needs to be compiled before being able to be used. A common error during the compilation process involves the miss of a suitable compiler in MATLAB. In most cases the installation of a further compiler following the instructions in the MATLAB Add-On Explorer should solve the problem. As a detailed description of the toolbox and installation instructions can be found in the manual of the toolbox, we will only summarize the essential information and give some hints.

After successful installation of the toolbox, you can follow the routine and tips below to conduct a simple SVR-LSM analysis. Note that this toolbox does not provide any graphical user interface and requires command line interaction. This will guarantee a maximum of flexibility for using parts of the scripts for own implementations and specific research questions. If you are not familiar with this type of use, we recommend using the SVR-LSM GUI toolbox (https://github.com/atdemarco/svrlsmgui; [42]) or NiiStat (www.nitrc.org/projects/niistat).

(a) First the binary normalized lesion images (.nii files) need to be copied to the "\lesion_imgs" folder and the behavioral variable files (.csv file) need to be set up and copied into the "score" folder.

(b) Open the main script called "SVR-LSM-toolbox.m" in the MATLAB editor. Here, all the required parameters for the analysis can be adapted to your needs. Follow the instructions in the manual to find out which parameters you need to specify.

(c) Out of the box, the analysis is running an epsilon-SVR using an RBF (radial basis function) kernel. This type of regression is nonlinear and has proven its validity in the initial investigations. The "degree" of nonlinearity is largely dependent on several user-defined parameters, namely, C (for cost), ε (for epsilon-SVR), and γ (for the RBF kernel). These parameters define among others how the features (voxels) are treated and thus will contribute to the overall model quality.

(d) The toolbox sets automatically the C, ε, and γ values to 20, 0.01, and 2 respectively. As the optimal parameters are related to the data in question, we can recommend to perform a grid search with k-fold cross-validation to find the optimal parameters for a specific dataset. The SVR-LSM toolbox from Daniel Mirman includes a function called "optimize_parameters.m" which is able to do that for C and γ. To do so, the "SVR_LSM_toolbox.m" script needs to be run until line 88, while omiting line 61. After completion, the necessary data is loaded in the MATLAB workspace and can be send to the optimize_parameters.m function by entering the following commands into the MATLAB command line:

```
svr_cost = 1:50;
svr_gamma = 0.5:10;
[cost_best, gamma_best, acc] = optimize_parameters(variables, svr_cost, svr_gamma);
```

This will run a fivefold cross-validation grid search with a parameter range of 1–50 for C and 0.5–10 for γ in steps of 1. If another parameter range is needed, svr_cost and svr_gamma should be modified before running the "optimize_parameters.m" function. cost_best and gamma_best will return the best C and ɣ for the dataset. Additionally, the variable acc is returned which provides information about the prediction accuracy. After that, the final analysis can be run by first changing the general parameter variables for C and ɣ, running the following commands:

```
parameters.cost = cost_best;
parameters.gamma = gamma_best;
```

Now the rest of the "SVR_LSM_toolbox.m" script can be run from line 90 on.

(e) If a different type of analysis is needed (e.g., a linear SVR) or a different kernel is required, the "run_svr_lsm_beta_map.m" function can be opened up in a text editor, and the values in line 10 can be changed. For a detailed description on what value corresponds to which setting, we refer to explanations given in the libSVM toolbox (https://www.csie.ntu.edu.tw/~cjlin/libsvm/). In this case, please do not forget to change also the values at line 21 in the "optimize_parameters.m" function accordingly.

(f) As explained in Zhang et al. [36], it is important to control for lesion size. Without changing anything, the toolbox is running automatically direct total lesion volume control (dTLVC). Nevertheless, it is still possible to include lesion size as a covariate without using dTLVC and regressing out lesion size out of the behavioral score. If needed, the dTLVC correction can be turned off at line 23 setting the flag to 0 in the "SVR_LSM_toolbox.m" script while adding the name of the lesion size variable at line 19. Note that the lesion size variable should then be included as an additional column in the behavioral score (.csv) file.

(g) Before running the analysis, we recommend to document the chosen parameters or to save a copy of the "SVR_LSM_toolbox.m" and the "run_svr_lsm_beta_map.m" scripts in case you modified something; the toolbox does not output any report of the chosen parameters.

(h) The duration of the final analysis is largely dependent on the used hardware and the chosen permutation number and hence can take up to several hours.

(i) Finally, the toolbox creates an unthresholded probability map which can be opened up in visualization software packages as MRIcron. If some voxels are surviving the FDR thresholding, the critical p is saved to the generated report. For visualizing the inverse probability map in, for example, MRIcron, you need to set the lower bound intensity to 1—critical p—and the upper bound to 1.

SVR-LSM GUI MATLAB Toolbox [42]: The setup of the SVR-LSM GUI toolbox is straightforward as it provides a graphical user interface and the user is able to click through the different settings and parameters. The toolbox can be downloaded from https://github.com/atdemarco/svrlsmgui. Requirements for the use of this toolbox include MATLAB (The MathWorks, Inc., Natick, MA; recommended version 2017a), SPM (Wellcome Institute of Imaging Neuroscience, London, UK; version SPM12), MATLAB Parallel Computing Toolbox (for parallelization functionality) and MATLAB Statistics and Machine Learning Toolbox (for MATLAB's SVR functionality), or libSVM (for libSVM SVR functionality, as in Zhang et al. [36]; http://www.csie.ntu.edu.tw/~cjlin/libsvm/; version 3.18 or higher). Note that libSVM comes already with the toolbox in a compiled variant for Linux (64-bit) and Mac (64-bit) Systems and hence doesn't need any further consideration unless you are using a different OS. Then you need to build the libSVM as explained in the libsvm-3.XX Readme. The benefit of this toolbox is additionally that it does not rely necessarily on the libSVM toolbox, as the authors implemented also the SVR functionalities out of the MATLAB Statistics and Machine Learning Toolbox. This is especially interesting in case of persisting errors for compiling the libSVM toolbox. Moreover, the authors recently showed that there are no major differences in using the libSVM or MATLAB Machine Learning Toolbox-based SVR algorithms [42].

(a) After successful installation, run the following command in the MATLAB command line interface to open up the GUI, allowing you to specify all necessary parameters by clicking through it:

```
svrlsmgui;
```

(b) In comparison to the SVR-LSM toolbox by Zhang et al. [36], the SVR-LSM GUI toolbox is currently only providing an epsilon-SVR with a nonlinear RBF kernel but might be extended with a linear kernel in future releases.

(c) Another benefit of this tool is that it is able to use the MATLAB Parallel Toolbox and perform the analysis using a parallel pool, which provides a remarkable speed benefit. For additional information about timings, we refer to the original work by DeMarco and Turkeltaub [42].

(d) In difference to the SVR-LSM toolbox by Zhang et al. [36], the SVR-LSM GUI is very flexible in what kind of correction for lesion size is required. Hence, the authors implemented all four techniques, dTLVC and nuisance regression on behavior, on lesion data, or on both.

(e) Moreover, the toolbox does not support FDR thresholding but rather implements by default a permutation-based approach which first thresholds the output on a voxelwise level and additionally applies an FWE-corrected cluster threshold. This procedure has been adapted from functional neuroimaging analysis [43] and controls the rate of a single false-positive cluster, instead of a single positive voxel. This technique is relatively new for MLBM. For more details, see DeMarco and Turkeltaub [42] or Mirman et al. [44]. Nevertheless, in a recent investigation from Mirman et al. [44], the authors showed—at least for univariate testing—that this correction technique can produce clusters that extend well beyond the correct region.

(f) As FDR correction is not based on the individual permutations used for the transformation of the SVR beta-parameters to a probability map, there is always the option to use freely available scripts to apply FDR-based multiple comparison correction on the unthresholded probability maps after termination of the SVR-LSM GUI analysis. Note that FDR correction is kind of a trade-off between allowing some false positives and minimizing false negatives. Moreover, for the univariate setting, Mirman and colleagues [44] demonstrated that FDR tends to produce anti-conservative results for smaller samples and thus might be rather interesting for studies with larger samples.

(g) Continuous family-wise error rate (FWER) is an additional implementation which has been developed and tested in the univariate field [44], but which has also been integrated into the SVR-LSM GUI toolbox for MLBM. This procedure uses a user-defined most extreme voxelwise test statistic (v) for setting the upper limit of how many false-positive voxels are tolerated. Inferentially, this procedure is similar to FDR correction, as it allows a certain number of false positives and hence a better trade-off between very conservative and very lenient thresholding. Nevertheless, this procedure is again based on a user-defined criterion, and as there is no gold standard on the choice of v, the selection might be modified until the desired outcome shows up. The selection of the exact v value should thus be carried out with caution to prevent "p-hacking."

NiiStat (www.nitrc.org/projects/niistat): The NiiStat toolbox can be downloaded from https://www.nitrc.org/projects/niistat/. Requirements for the use of this toolbox include MATLAB (The MathWorks, Inc., Natick, MA; version 9.0 or higher for Graphical User Interface), SPM (Wellcome Institute of Imaging Neuroscience, London, UK; version SPM12 or later), and libSVM (http://www.csie.ntu.edu.tw/~cjlin/libsvm/; version 3.18 or higher). The toolbox already includes libSVM. A detailed description of the toolbox and installation instructions is given at https://www.nitrc.org/plugins/mwiki/index.php/niistat:MainPage. The NiiStat toolbox allows the use of a wide range of neuroimaging techniques, including univariate lesion-behavior mapping. Only recently a first implementation of support vector regression has been published. So far the SVR-LSM analysis in NiiStat is rather rudimentary, compared to the other toolboxes, and offers only a reduced customizability through the GUI. As NiiStat is one of the most extensive tools for neuroimaging research and lesion-behavior mapping, we expect large improvement in the next releases.

(a) After successful installation, one can run the following command in the MATLAB command line interface to open up the GUI, allowing you to specify various parameters by clicking through it:

```
NiiStatGUI; %for a more sophisticated graphical user interface
NiiStat; %for a rather simple user interface
```

(b) This toolbox currently runs an epsilon-SVR with a linear kernel as default and does not provide an optimization routine to find the optimal SVR-LSM hyperparameter cost (C). Hence, it relies on a default setting of 1 for C and 0.01 for ε.

(c) NiiStat allows FDR or Bonferroni for multiple comparisons correction and performs dTLVC as default for lesion volume correction.

6.3.2 Sparse Canonical Correlation-Based Lesion-Symptom Mapping (SCCAN)

Besides the SVR-based MLBM, another well-validated technique exists based on sparse canonical correlations [39]. SCCAN is a sparse solution for multidimensional canonical correlation analysis, which allows one to symmetrically relate sets of variables to each other, hence to find sets of variables that covary. In the context of lesion-behavior mapping, SCCAN is based on an optimization procedure that gradually selects a multivariate model of sets of voxels which maximize correlations with behavioral scores. The outcome of SCCAN largely depends on preselected parameters as, for example, sparseness, which should be selected carefully. Currently the only implementation of SCCAN in a MLBM tool (LESYMAP) is based on such optimization algorithm that finds the best set by evaluation of different parameter combinations via cross-validation. The final model is then calculated with the optimal parameter set and evaluated by assessing the model's predictive performance which is then statistically tested for significance.

Tools for Performing SCCAN

SCCAN is implemented in the LESYMAP package (https://dorianps.github.io/LESYMAP/) and based on R programming language. The installation of the LESYMAP package can be burdensome, as it relies on a relatively high number of R packages. Requirements for the use of this toolbox include R (version 3.0 or above), OS (Linux, Mac, or Windows Linux Subsystem), and several R packages including their dependencies (ITKR, ANTsRCore, ANTsR). Most errors encountered during installation will probably be based on missing dependencies of different R sub-packages. Hence, in case of any installation errors, the first step would be to check the error output and to install the missing dependencies individually. In some cases, some of the packages cannot be built as you need to install a newer version of CMake (https://cmake.org/) through your OS. After successful installation the use of the toolbox is straightforward, as it comes with a well-documented manual and an example dataset. Note that this toolbox is also able to perform univariate lesion-behavior mapping and will be complemented by SVR-LSM in a future release.

7 Statistical Thresholding and Correction Factors in Lesion Analyses

7.1 Inclusion of Nuisance Covariates and Ensuring Sufficient Minimum Lesion Overlap

One variable known to correlate strongly with the severity of behavioral deficit in stroke populations is lesion volume: the larger the lesion, the more likely it is that a patient will show a behavioral deficit. Thus, to avoid identifying brain areas where damage is related simply to lesion volume instead of patient behavior, lesion volume should be controlled, e.g., as a nuisance covariate by using regression approaches. These or other approaches are implemented in most software packages both for univariate and multivariate lesion mapping. Moreover, to ensure sufficient statistical power, voxels damaged in a very low percentage of the patient sample should be excluded. *Correcting for lesion volume as well as ensuring sufficient minimum lesion overlap* has been shown to improve the anatomical validity in univariate voxelwise statistical analyses [45]. Finally, similarly as done for lesion volume, other nuisance covariates can additionally be included, such as the severity of frequently co-occurring deficits that may correlate with the cognitive function of interest or fiber tract disconnection likelihood [46].

7.2 Correcting for Multiple Comparisons

During a voxelwise statistical lesion-behavior mapping analysis, the same statistical test is performed at many individual voxels. This is the case for all univariate approaches and likewise for SVR-LSM. A *correction for multiple comparisons* thus *is needed*. The traditional method to correct for multiple comparisons is the Bonferroni correction. While this method offers excellent control of the family-wise error rate, it is also considered as very conservative for neuroimaging data where individual voxels are not truly independent; it severely reduces the statistical power to detect an effect. As such, considerable efforts have been made to develop alternative, less conservative, ways to correct for multiple comparisons.

Such an attempt to control the family-wise error rate (FWER), but without sacrificing statistical power, is *permutation thresholding* based on the maximum statistic [43]. The underlying logic is that if an observed test statistic is truly due to the difference in voxel status, similar or more extreme test statistics would be unlikely to arise in situations where the pairing of behavioral data and voxel status is scrambled. Modern computers allow to use large numbers of permutations, and several thousands (e.g., 10,000) should be done. Permutation thresholding offers the same control of the family-wise error rate as the Bonferroni correction. Importantly, however, in situations where the individual voxels are not truly independent, permutation thresholding offers better statistical power than the Bonferroni correction. So far, various ways of implementing permutation thresholding exist. For example, there are techniques based on cluster-wise or voxelwise correction. Further, a more recent approach called "continuous FWER" is available. It is also based on permutation thresholding and was mentioned in one of the toolboxes introduced above (see Sect. 6.3.1.1). For a detailed discussion about these different techniques, we refer to recent work by Mirman and colleagues [44].

Another less conservative approach to correcting for multiple comparisons is offered by *false discovery rate (FDR) thresholding* [47–49]. Here, the goal is not to control the family-wise error rate but to control the proportion of false positives among observed positives. As a consequence, a false discovery rate threshold of 5% means that up to 5% (e.g., 1 out of 20) of the observed positives might be false positives. In situations where no positives are observed, false discovery rate thresholding will provide the same control of the family-wise error rate as the Bonferroni correction. However, in situations where positives are observed, false discovery rate thresholding will result in more positives surviving the correction for multiple comparisons than either the Bonferroni correction or permutation thresholding. In fact, as the amount of observed positives increases, the false discovery rate threshold decreases. This adaptiveness of false discovery rate thresholding, however, comes at the price of reduced control of the family-wise error rate (as up to 5% of the positives surviving the correction for multiple comparisons could be false positives). In situations where control of the family-wise error rate is paramount, permutation thresholding should thus be preferred to correct for multiple comparisons.

8 Anatomical Interpretation of Lesion Analysis Results

Following a univariate or a multivariate voxelwise statistical lesion-behavior mapping analysis, we obtain a statistical map highlighting the voxels where voxel status (lesioned vs. non-lesioned) and patient behavior are significantly related. In the case of a lesion subtraction analysis, on the other hand, we obtain a map highlighting areas of the brain where lesions are descriptively more frequent in patients with than in patients without the cognitive deficit of interest (often thresholded to isolate those percentage relative frequency difference values thought to be meaningful [with typical threshold values of 20–50%]). Anatomical interpretation then consists of describing the location of these significant or meaningful voxels, typically with the help of a brain atlas. For practicability, coordinates of peak voxels or a coordinate range can be provided to characterize the findings resulting from the voxelwise lesion-behavior mapping analysis or subtraction analysis results. However, it should be noted that all voxels identified as "statistically significant" in statistical lesion-behavior mapping analyses or as "meaningful" in lesion subtraction analyses have the same impact and thus should be given the same weighting when interpreting the results.

Nowadays, there are many different cortical atlases to choose from. Whereas atlases derived from single-subject data remain popular (the Brodmann atlas or the AAL atlas by Tzourio-Mazoyer et al. [50]), probabilistic atlases derived from multi-subject data should be preferred, as these are able to quantify the intersubject variability in location and extent of each anatomical area. Within these multi-subject atlases, a second division can be made based on the brain characteristics used to parcellate distinct areas in different atlases. Whereas some probabilistic multi-subject atlases are based on macroscopical landmarks such as gyri and sulci [51, 52], others are based on histology [53] and on functional connectivity patterns [54] or are created via surface-based registrations of multimodal MRI acquisitions [55]. Finally, in addition to these cortical atlases, multi-subject atlases exist for white matter fiber tracts, based on either DTI fiber tracking [56, 57] or on histology [58]. Which atlas to choose for the anatomical interpretation of the results of a lesion-behavior mapping study is not a trivial issue. It is important to realize that different atlases might result in very different anatomical interpretations of the same lesion-behavior mapping results [59]. A pragmatic solution to this problem may be to interpret the data resulting from a lesion-behavior mapping analysis with respect to more than only one atlas. For example, due to the marked variance between DTI- and histology-based white matter atlases [59], Wiesen et al. [60] recently decided to interpret a multivariate voxelwise statistical lesion-behavior mapping analysis according to a probabilistic cytoarchitectonic atlas [58] and a DTI-based probabilistic fiber atlas [57] simultaneously, using a common threshold of $p \geq 0.3$ before overlaying the data on the statistical topographies.

Acknowledgments

This work was supported by the Deutsche Forschungsgemeinschaft (KA 1258/23-1). Daniel Wiesen was supported by the Luxembourg National Research Fund (FNR/11601161).

References

1. Rorden C, Karnath H-O (2004) Using human brain lesions to infer function: a relic from a past era in the fMRI age? Nat Rev Neurosci 5:813–819
2. Karnath H-O, Steinbach JP (2011) Do brain tumours allow valid conclusions on the localisation of human brain functions?—Objections. Cortex 47:1004–1006
3. de Haan B, Karnath H-O (2018) A hitchhiker's guide to lesion-behaviour mapping. Neuropsychologia 115:5–16
4. Wick W, Stupp R, Beule A-C, Bromberg J, Wick A, Ernemann U, Platten M, Marosi C, Mason WP, van den Bent M, Weller M, Rorden C, Karnath H-O, The European Organisation for Research and Treatment of Cancer and the National Cancer Institute of Canada Clinical Trails Group (2008) A novel tool to analyse MRI recurrence patterns in glioblastoma. Neuro Oncol 10:1019–1024
5. Karnath H-O, Rennig J (2017) Investigating structure and function in the healthy human brain: validity of acute versus chronic lesion-symptom mapping. Brain Struct Funct 222:2059–2070
6. Karnath H-O, Rennig J, Johannsen L, Rorden C (2011) The anatomy underlying acute versus chronic spatial neglect: a longitudinal study. Brain 134:903–912
7. Abela E, Missimer J, Wiest R, Federspiel A, Hess C, Sturzenegger M, Weder B (2012) Lesions to primary sensory and posterior parietal cortices impair recovery from hand paresis after stroke. PloS One 7:e31275
8. Wu O, Cloonan L, Mocking SJT, Bouts MJRJ, Copen WA, Cougo-Pinto PT, Fitzpatrick K, Kanakis A, Schaefer PW, Rosand J, Furie KL, Rost NS (2015) Role of acute lesion topography in initial ischemic stroke severity and long-term functional outcomes. Stroke 46:2438–2444
9. Rorden C, Bonilha L, Fridriksson J, Bender B, Karnath H-O (2012) Age-specific CT and MRI templates for spatial normalization. Neuroimage 61:957–965
10. Sperber C, Karnath H-O (2018) On the validity of lesion-behaviour mapping methods. Neuropsychologia 115:17–24
11. Rorden C, Brett M (2000) Stereotaxic display of brain lesions. Behav Neurol 12:191–200
12. Yushkevich PA, Piven J, Hazlett HC, Smith RG, Ho S, Gee JC, Gerig G (2006) User-guided 3D active contour segmentation of anatomical structures: significantly improved efficiency and reliability. Neuroimage 31:1116–1128
13. Wilke M, de Haan B, Juenger H, Karnath HO (2011) Manual, semi-automated, and automated delineation of chronic brain lesions: a comparison of methods. Neuroimage 56:2038–2046
14. Clas P, Groeschel S, Wilke M (2012) A semi-automatic algorithm for determining the demyelination load in metachromatic leukodystrophy. Acad Radiol 19:26–34
15. de Haan B, Clas P, Juenger H, Wilke M, Karnath H-O (2015) Fast semi-automated lesion demarcation in stroke. Neuroimage Clin 9:69–74
16. Goebel R (2012) BrainVoyager—past, present, future. Neuroimage 62:748–756
17. Jenkinson M, Beckmann CF, Behrens TE, Woolrich MW, Smith SM (2012) FSL. Neuroimage 62:782–790
18. Cox RW (1996) AFNI: software for analysis and visualization of functional magnetic resonance neuroimages. Comput Biomed Res 29:162–173
19. Cox RW (2012) AFNI: what a long strange trip it's been. Neuroimage 62:743–747
20. Avants BB, Tustison NJ, Song G, Cook PA, Klein A, Gee JC (2011) A reproducible evaluation of ANTs similarity metric performance in brain image registration. Neuroimage 54:2033–2044
21. Ashburner J, Friston KJ (2005) Unified segmentation. Neuroimage 26:839–851
22. Crinion J, Ashburner J, Leff A, Brett M, Price C, Friston K (2007) Spatial normalization of lesioned brains: performance evaluation

and impact on fMRI analyses. Neuroimage 37:866–875
23. Klein A, Andersson J, Ardekani BA, Ashburner J, Avants B, Chiang MC, Christensen GE, Collins DL, Gee J, Hellier P, Song JH, Jenkinson M, Lepage C, Rueckert D, Thompson P, Vercauteren T, Woods RP, Mann JJ, Parsey RV (2009) Evaluation of 14 nonlinear deformation algorithms applied to human brain MRI registration. Neuroimage 46:786–802
24. Winkler AM, Kochunov P, Glahn DC. FLAIR templates. http://glahngroup.org
25. Brett M, Leff AP, Rorden C, Ashburner J (2001) Spatial normalization of brain images with focal lesions using cost function masking. Neuroimage 14:486–500
26. Nachev P, Coulthard E, Jäger HR, Kennard C, Husain M (2008) Enantiomorphic normalization of focally lesioned brains. Neuroimage 39:1215–1226
27. Karnath H-O, Sperber C, Rorden C (2018) Mapping human brain lesions and their functional consequences. Neuroimage 165:180–189
28. Xu T, Jha A, Nachev P (2018) The dimensionalities of lesion-deficit mapping. Neuropsychologia 115:134–141
29. Rorden C, Karnath H-O, Bonilha L (2007) Improving lesion-symptom mapping. J Cogn Neurosci 19:1081–1088
30. Lorca-Puls DL, Gajardo-Vidal A, White J, Seghier ML, Leff AP, Green DW, Crinion JT, Ludersdorfer P, Hope TMH, Bowman H, Price CJ (2018) The impact of sample size on the reproducibility of voxel-based lesion-deficit mappings. Neuropsychologia 115:101–111
31. Brunner E, Munzel U (2000) The nonparametric Behrens-Fisher problem: asymptotic theory and a small-sample approximation. Biom J 42:17–25
32. Liebermeister C (1877) Über Wahrscheinlichkeitsrechnung in Anwendung auf therapeutische Statistik. Samml Klin Vorträge (Innere Medizin No. 31-64) 110:935–962
33. Bates E, Wilson SM, Saygin AP, Dick F, Sereno MI, Knight RT, Dronkers NF (2003) Voxel-based lesion-symptom mapping. Nat Neurosci 6:448–450
34. Kimberg DY, Coslett HB, Schwartz MF (2007) Power in voxel-based lesion-symptom mapping. J Cogn Neurosci 19:1067–1080
35. Smith DV, Clithero J, Rorden C, Karnath H-O (2013) Decoding the anatomical network of spatial attention. Proc Natl Acad Sci U S A 110:1518–1523
36. Zhang Y, Kimberg DY, Coslett HB, Schwartz MF, Wang Z (2014) Multivariate lesion-symptom mapping using support vector regression. Hum Brain Mapp 35:5861–5876
37. Mah Y-H, Husain M, Rees G, Nachev P (2014) Human brain lesion-deficit inference remapped. Brain 137:2522–2531
38. Toba MN, Zavaglia M, Rastelli F, Valabrégue R, Pradat-Diehl P, Valero-Cabré A, Hilgetag CC (2017) Game theoretical mapping of causal interactions underlying visuo-spatial attention in the human brain based on stroke lesions. Hum Brain Mapp 3471:3454–3471
39. Pustina D, Avants B, Faseyitan OK, Medaglia JD, Coslett HB (2018) Improved accuracy of lesion to symptom mapping with multivariate sparse canonical correlations. Neuropsychologia 115:154–166
40. Sperber C, Wiesen D, Karnath H-O (2019) An empirical evaluation of multivariate lesion behaviour mapping using support vector regression. Hum Brain Mapp 40:1381–1390
41. Rasmussen PM, Hansen LK, Madsen KH, Churchill NW, Strother SC (2012) Model sparsity and brain pattern interpretation of classification models in neuroimaging. Pattern Recognit 45:2085–2100
42. DeMarco AT, Turkeltaub PE (2018) A multivariate lesion symptom mapping toolbox and examination of lesion-volume biases and correction methods in lesion-symptom mapping. Hum Brain Mapp 21:2461–2467
43. Nichols TE, Holmes AP (2002) Nonparametric permutation tests for functional neuroimaging: a primer with examples. Hum Brain Mapp 15:1–25
44. Mirman D, Landrigan J-F, Kokolis S, Verillo S, Ferrara C, Pustina D (2018) Corrections for multiple comparisons in voxel-based lesion-symptom mapping. Neuropsychologia 115:112–123
45. Sperber C, Karnath H-O (2017) Impact of correction factors in human brain lesion-behavior inference. Hum Brain Mapp 38:1692–1701
46. Rudrauf D, Mehta S, Grabowski TJ (2008) Disconnection's renaissance takes shape: formal incorporation in group-level lesion studies. Cortex 44:1084–1096
47. Benjamini Y, Hochberg Y (1995) Controlling the false discovery rate: a practical and powerful approach to multiple testing. J R Stat Soc Ser B Methodol 57:289–300
48. Benjamini Y, Yekutieli D (2001) The control of the false discovery rate in multiple testing under dependency. Ann Statistics 29:1165–1188

49. Genovese CR, Lazar NA, Nichols T (2002) Thresholding of statistical maps in functional neuroimaging using the false discovery rate. Neuroimage 15:870–878
50. Tzourio-Mazoyer N, Landeau B, Papathanassiou D, Crivello F, Etard O, Delcroix N, Mazoyer B, Joliot M (2002) Automated anatomical labeling of activations in SPM using a macroscopic anatomical parcellation of the MNI MRI single-subject brain. Neuroimage 15:273–289
51. Hammers A, Allom R, Koepp MJ, Free SL, Myers R, Lemieux L, Mitchell TN, Brooks DJ, Duncan JS (2003) Three-dimensional maximum probability atlas of the human brain, with particular reference to the temporal lobe. Hum Brain Mapp 19:224–247
52. Shattuck DW, Mirza M, Adisetiyo V, Hojatkashani C, Salamon G, Narr KL, Poldrack RA, Bilder RM, Toga AW (2008) Construction of a 3D probabilistic atlas of human cortical structures. Neuroimage 39:1064–1080
53. Zilles K, Schleicher A, Langemann C, Amunts K, Morosan P, Palomero-Gallagher N, Schormann T, Mohlberg H, Bürgel U, Steinmetz H, Schlaug G, Roland PE (1997) Quantitative analysis of sulci in the human cerebral cortex: development, regional heterogeneity, gender difference, asymmetry, intersubject variability and cortical architecture. Hum Brain Mapp 5:218–221
54. Joliot M, Jobard G, Naveau M, Delcroix N, Petit L, Zago L, Crivello F, Mellet E, Mazoyer B, Tzourio-Mazoyer N (2015) AICHA: an atlas of intrinsic connectivity of homotopic areas. J Neurosci Methods 254:46–59
55. Glasser MF, Coalson TS, Robinson EC, Hacker CD, Harwell J, Yacoub E, Ugurbil K, Andersson J, Beckmann CF, Jenkinson M, Smith SM, Van Essen DC (2016) A multimodal parcellation of human cerebral cortex. Nature 536:171–178
56. Zhang Y, Zhang J, Oishi K, Faria AV, Jiang H, Li X, Akhter K, Rosa-Neto P, Pike GB, Evans A, Toga AW, Woods R, Mazziotta JC, Miller MI, van Zijl PCM, Mori S (2010) Atlas-guided tract reconstruction for automated and comprehensive examination of the white matter anatomy. Neuroimage 52:1289–1301
57. Thiebaut de Schotten M, Ffytche DH, Bizzi A, Dell'Acqua F, Allin M, Walshe M, Murray R, Williams SC, Murphy DGM, Catani M (2011) Atlasing location, asymmetry and inter-subject variability of white matter tracts in the human brain with MR diffusion tractography. Neuroimage 54:49–59
58. Bürgel U, Amunts K, Battelli L, Mohlberg H, Gilsbach JM, Zilles K (2006) White matter fiber tracts of the human brain: three-dimensional mapping at microscopic resolution, topography and intersubject variability. Neuroimage 29:1092–1105
59. de Haan B, Karnath H-O (2017) 'Whose atlas I use, his song I sing?'—The impact of anatomical atlases on fiber tract contributions to cognitive deficits. Neuroimage 163:301–309
60. Wiesen D, Sperber C, Yourganov G, Rorden C, Karnath H-O (2019) Using machine learning-based lesion behavior mapping to identify anatomical networks of cognitive dysfunction: spatial neglect and attention bioRxiv 556753. https://doi.org/10.1101/556753
61. Karnath H-O, Perenin M-T (2005) Cortical control of visually guided reaching: evidence from patients with optic ataxia. Cereb Cortex 15:1561–1569

Neuromethods (2020) 151: 239–255
DOI 10.1007/7657_2019_23
© Springer Science+Business Media, LLC 2019
Published online: 11 May 2019

Topographic Mapping of Parietal Cortex

Summer Sheremata

Abstract

The parietal cortex supports a vast number of cognitive functions including visual attention, short-term memory, and decision-making. In particular, the intraparietal sulcus (IPS) is central to many of these tasks. Functional magnetic resonance imaging (fMRI) can identify at least six areas in IPS within individuals based upon topographic representations of the visual field. Recent studies have utilized novel mapping techniques to increase the feasibility of defining these topographic areas in individual participants with a high degree of reliability. This chapter introduces a method for demonstrating topographic maps that has been used across several different magnets and head coils to quickly and efficiently demonstrate occipital and parietal topographic maps. By increasing the efficiency with which laboratories can functionally map the parietal cortex, a greater number of studies can utilize objective identification of within-individual regions of interest. Increasing the accessibility of defining these topographic maps will further the goal of understanding the cognitive processes subserved by the parietal cortex.

Keywords MRI, fMRI, Parietal cortex, Intraparietal sulcus, Retinotopic, Topographic, Regions of interest (ROIs)

1 Introduction

Topographic mapping studies were among the earliest demonstrations of the feasibility of fMRI for measuring brain activity in humans. Topography in the visual system was suggested by an early twentieth-century report that lesions to early visual cortex led to deficits in perceiving contralateral and inverted visual field locations [1], leading to a clear, testable model of the visual system. For example, damage to the dorsal occipital cortex in the left hemisphere led to the inability to perceive stimuli in the lower-right visual quadrant. This model was supported by electrophysiological studies suggesting similar relations between coordinates in cortical space and visual field locations in nonhuman subjects [2, 3]. Electrophysiological studies, however, were limited by the vast number of neurons required to sample (400–600) due to the variability in the shape of cortical areas [2]. In contrast, fMRI allows

The original version of this chapter was revised. The correction to this chapter is available at https://doi.org/10.1007/7657_2019_29

for the sampling of a large extent of cortical space in a short amount of time. Furthermore, due to the minimally invasive nature of fMRI, these studies could be conducted in human participants, thereby reducing the reliance on homology across species.

Early studies of retinotopic mapping delineated the expected visual field to cortical location mapping in early visual cortex, visual areas 1–3 (V1–3, [4–6]). These retinotopically defined areas quickly became important in studies that investigated the properties not only of visual processing [7] but also higher-order cognitive functions such as visual attention [8, 9]. In the past 20 years, the number of retinotopically defined areas has increased exponentially, revealing topographic maps throughout occipital, temporal, parietal, and frontal cortex as well as in the lateral geniculate and pulvinar nuclei of the thalamus [10, 11]. This has allowed for the objective definition of maps across multiple cortical areas and cognitive domains.

Within the parietal cortex, the ability to determine a mapping structure has aided in the understanding of many cognitive functions such as visual attention, short-term memory [12, 13], numerosity [13], long-term memory [14], and decision-making [15]. In particular, there has been a focus of the role of the parietal cortex in memory mechanisms [16, 17]. The objective definition of these topographically defined areas has been fundamental to understanding brain mechanisms involved in visual short-term and working memory [12, 18, 19], as well as long-term memory [14]. Advances in methodologies have resulted in reliable techniques that can be used to demonstrate these maps across individuals across multiple imaging sites. This chapter describes a specific technique that has demonstrated parietal maps across several 3-Tesla magnets using a variety of head coils and has been adapted and utilized for retinotopic mapping on a 7-Tesla magnet.

2 Materials

The materials listed below are those needed for using AFNI [20] to preprocess and fit a population receptive field (pRF) [21] model as implemented in AFNI by Silson and colleagues [22]. This method is combined with a stimulus previously used to demonstrate parietal maps [23] with the following adaptations. First, because an earlier study [24] demonstrated that maps of parietal cortex are more reliable when the mapping stimulus is covertly attended (participants attend the stimulus while maintaining fixation), the analysis here uses only the attend stimulus condition. Covert attention is counterintuitive for many participants and will thus likely require at least one training session for individual participants outside the scanner. There may be specific paradigms in which using a relatively easy attend-fixation task may render better mapping structure, for instance, when using naïve participants unaccustomed to deploying covert attention. Second, four runs of this condition were run, in contrast to the six runs previously used. The number of runs is reduced here as analyses within the laboratory demonstrated that gains in signal-to-noise decreased after completing four runs for most participants. If time allows, however, additional runs may increase the signal-to-noise ratio, especially within naïve participants.

There are numerous alternative methods for demonstrating retinotopic maps in the parietal cortex, and the choice of paradigm and stimulus is described below in the Notes section. The use of a covert attention task with a traveling bar stimulus and pRF mapping analysis was chosen to maximize the number of occipital and parietal areas mapped in a relatively short period of time (~20 min). However, delayed saccade techniques may be preferable for localizing parietal and frontal topographic map structures [25–27], and different stimuli may render optimal maps for early and ventral cortex. However, for studies that require mapping of visual and parietal cortex quickly, the method described below is ideally suited.

The data demonstrated here are from a typical participant with average retinotopic maps to demonstrate expected results of retinotopic mapping. The data were collected at University MRI in Boca Raton, FL, on a 3T General Electric Signa scanner with an eight-channel receive-only coil. Table 1 demonstrates the command line codes used to analyze the data in native space. Table 2 includes command line codes for displaying the topographic maps on the cortical surface. Finally Fig. 1 demonstrates the resultant parietal maps. Note that using this code requires that FreeSurfer and AFNI are downloaded and installed properly with variables set for locating required files. Also, displaying the data on the cortical surface requires that the participant's cortical surface has been reconstructed and is in a directory with appropriate pointers to the structural directory.

1. 3T MRI scanner with at least 60 min time slot for acquiring high-resolution T1-weighted image and four 4-min, 12-s population receptive field (pRF) mapping runs.
2. High-capacity storage device or data transfer system for images.
3. Linux or Mac computer for image processing and analyses.
4. DICOM to NIFTI converter software (dcm2niix; https://github.com/rordenlab/dcm2niix/releases).
5. FreeSurfer software [28] for reconstruction of the cortical surface.

Table 1
Command lines for preprocessing and running population receptive field mapping using AFNI software

```
# Create anatomy file and copy to output directory

mri_convert ${subj_anat_dir}/brain.mgz ${subj_anat_dir}/mri/brain.nii

3dcopy ${subj_anat_dir}/mri/brain.nii ${subj}_anat_stripped+orig
```

```
# Motion correct

touch out.pre_ss_warn.txt

foreach run($runs)

   3dvolreg -verbose -zpad 1 -base pb00.${subj}.r01.tcat+orig.'[2]' \
      -1Dfile dfile.r${run}.1D -prefix pb01.${subj}.r${run}.volreg \
      -cubic pb00.$subj.r$run.tcat+orig

end
```

```
# Detrend

foreach run($runs)

   3dTstat -prefix tempMean pb01.${subj}.run${run}.volreg

   3dDetrend -prefix tempDetrend -vector dfile.r${run}.1D \
      -polorrt 2 pb01.${subj}.r${run}.volreg+orig

   3dcalc -a tempDetrend+orig -b tempMean+orig -expr "(a+b)" -float \
      -prefix pb02.${subj}.r${run}.detrended

   rm temp*

end
```

```
# Smoothing

foreach run($runs)

   3dmerge -1blur_fwhm 4.0 -doall -prefix pb03.${subj}.r${run}.smoothing \
      pb02.${subj}.r${run}.detrend+orig

end
```

(continued)

Table 1
(continued)

```
# Average

3dcalc -a pb03.retina2.r01.smoothing+orig.HEAD \
       -b pb03.retina2.r02.smoothing+orig.HEAD \
       -c pb03.retina2.r03.smoothing+orig.HEAD \
       -d pb03.retina2.r04.smoothing+orig.HEAD \
       -expr '(a+b+c+d)/4' -prefix average.ret
```

```
# Automask and remove mean

3dcopy average.ret  AFNI_pRF+orig.BRIK

3dTstat -mean -prefix AFNI_pRFmean AFNI_pRF+orig.BRIK

3dAutomask -prefix automask AFNI_pRF+orig.BRIK

3dcalc -a AFNI_pRF+orig. -b AFNI_pRFmean+orig. -c automask+orig. \
       -expr '100*c*(a-b)/b'  -prefix e.scale.demean
```

```
# Run model

3dNLfim -input e.scale.demean+orig -mask automask+orig -noise Zero \
        -signal Conv_PRF
        -sconstr 0 -10.0 10.0  \
        -sconstr 1 -1.0 1.0    \
        -sconstr 2 -1.0 1.0    \
        -sconstr 3 0.0 1.0     \
        -BOTH -nrand 10000 -nbest 5 -bucket 0 Buckslow.PRF -snfit snfitslow.PRF

3dcalc -a Buckslow.PRF+orig'[1]' -b Buckslow.PRF+orig'[2]' -expr 'sqrt(a^2+b^2)' \
        -prefix polarslow.m

3dcalc -a Buckslow.PRF+orig.'[1]' -b Buckslow.PRF+orig.'[2]' -expr 'atan2(a,-b)' \
        -prefix polarslow.ph

3dbucket -glueto Buckslow.PRF+orig.  polarslow.m+orig.  polarslow.ph+orig.
```

Table 2
Command lines for projecting the data onto the cortical surface using AFNI and SUMA software

```
# Deoblique and align to epi

3dWarp -prefix Buckslow.PRF_warped+orig -deoblique Buckslow.PRF+orig

3dWarp -prefix PRF_mc_warped+orig -deoblique pb01.${subj}.r01.volreg+orig

align_epi_anat.py -anat ${subj}_anat_stripped+orig -anat_has_skull no \
     -epi PRF_mc_warped+orig -epi_strip 3dAutomask -cost lpc+ZZ \
      -multi_cost nmi lpa -suffix _al2epi -epi_base 48 -prep_off
```

```
# Align to SUMA

@SUMA_Make_Spec_FS -NIFTI -fspath ${subj_anat_dir}/ -sid ${anat_subj}

@SUMA_AlignToExperiment -wd -exp_anat anat_stripped_al2epi+orig.BRIK     \
 -surf_anat ${subj_anat_dir}/SUMA/${anat_subj}_SurfVol+orig.     \
 -align_centers -strip_skull surf_anat \
 -prefix ${anat_subj}_SurfVol_ns_centre_AInd_Exp


foreach hemi (lh rh)

3dVol2Surf -spec ${subj_anat_dir}/SUMA/std.141.${anat_subj}_${hemi}.spec   \
          -surf_A smoothwm -surf_B pial  \
          -sv ${anat_subj}_SurfVol_ns_centre_AInd_Exp+orig     \
          -grid_parent $output_dir/Buckslow.PRF_warped+orig     \
          -oob_value 0-map_func ave -f_steps 10 -f_index nodes.   \
          -out_niml std_${hemi}_Buckslow.PRF_warped+orig.BRIK.niml.dset

end
```

6. AFNI [20] and SUMA (http://afni.nimh.nih.gov/afni/suma/) software for analysis and display.
7. Software (i.e., Python or MATLAB) for creating stimulus files and storing them in four-dimensional files to be read by AFNI.

3 Methods

3.1 Data Acquisition

The following protocol has been demonstrated on Siemens 3-T Tim Trio [23] using the posterior 8 channels of a 16-channel head coil as well as a General Electric 3-T Signa (Fig. 1) scanner with an 8-channel receive-only head coil. The Notes section and Fig. 2 demonstrate topographic maps on the Siemens 3-T Tim Trio and Siemens Magnetom 7T magnet using 24-channel radiofrequency head coil (Fig. 2). Depending on the magnet and visual presentation, differences in scanning platforms and protocols may require some adjustments to the protocol. For current imaging, the acquisition protocol is as follows:

1. Localizer
2. 1 mm isotropic T1-weighted sequence
3. A minimum of four 3-mm isotropic T2*-weighted runs (125 TRs, 2.0s/TR)

3.2 fMRI Processing and Analysis

3.2.1 Transformation of Data into 4d Files

Once the images from the scanner are transferred (via storage devices or transfer protocols), the functional data are stored in four-dimensional files. Standard MRI software will store the data in three-dimensional files, either a single file for each slice at each time point or all slices for a single time point. Proper transfer of files from three-dimensional DICOM files to four-dimensional files (typically NIFTI or BRIK files) is essential to confirm. Each data file should be inspected using FreeSurfer's mri_info command or by reading the header file for each BRIK file. The first two dimensions of the volume size reflects the field of view of each slice

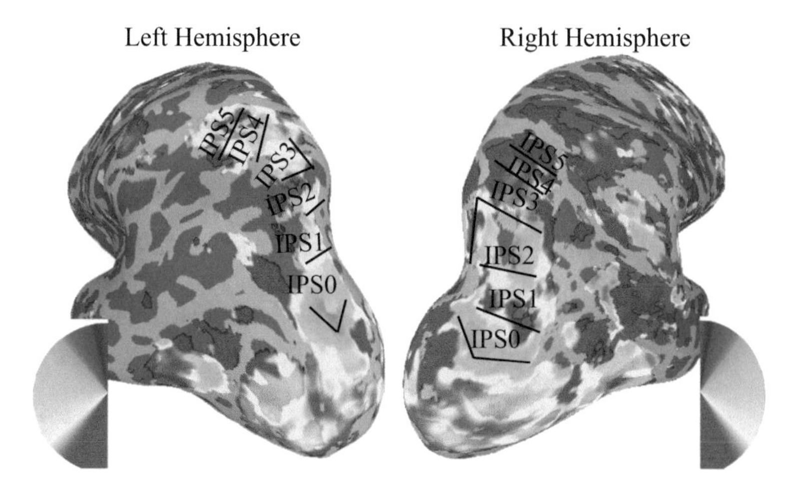

Fig. 1 Topographic maps of a typical subject using the population receptive field mapping technique and a covert attention paradigm

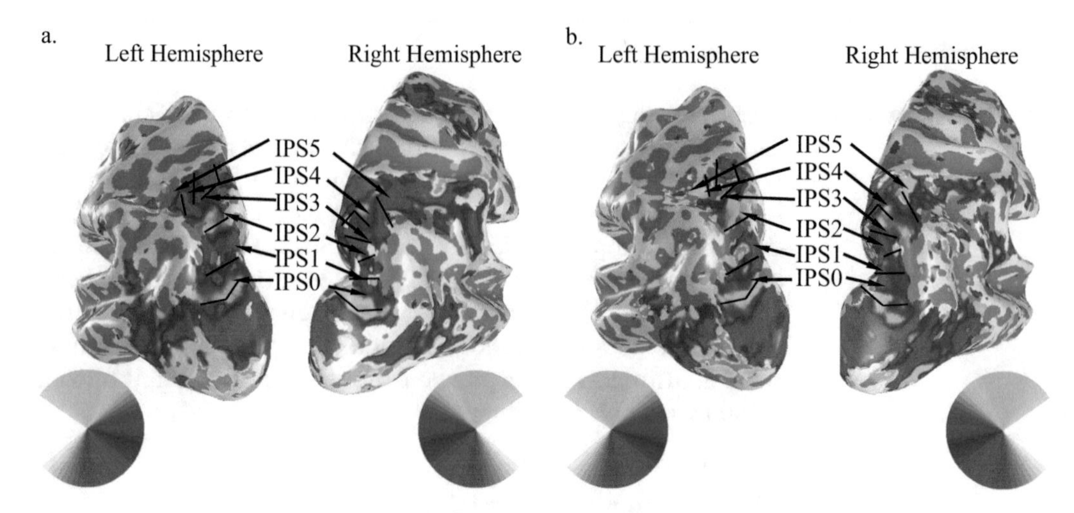

Fig. 2 Topographic map structure of a participant across imaging sessions in a 3T and 7T magnet

divided by the in-plane resolution (i.e., 240 mm/3 mm = 80), the second dimension reflects the number of slices, and the fourth dimension reflects the number of time points – here 125 TRs.

3.2.2 Cortical Surface Reconstruction

FreeSurfer is generally considered the software of choice for reconstructing the cortical surfaces [28]. This process determines an intensity value for separating white and gray matter, identifying voxels of each type, creating a cortical ribbon, and then using that ribbon to reconstruct the cortical surface. A more detailed description of this process is available on the FreeSurfer website (https://surfer.nmr.mgh.harvard.edu). This is the most time-intensive step for each session, requiring several hours depending upon the quality of the T1 image and the speed of the computer. The automated reconstruction is often sufficient in young, healthy adults but must occasionally be amended due to topological defects in the cortical surface.

3.2.3 Preprocessing of the Functional Data

A population receptive field (pRF) is the purported average of the receptive fields from the neurons underlying a single voxel's response. Due to the noise intrinsic in the BOLD signal and due to the possibility of movement across scans, it is important to remove any predictable noise. Several analysis packages offer software for this purpose, but here I will describe the analysis using the AFNI processing stream.

Table 1 includes code for running this analysis from the point at which the reconstruction in FreeSurfer has been completed and the data are saved in the directory where the analysis will be completed. The code is written to run in a C shell in a Macintosh terminal. The following variables need to be set using setenv in the terminal:

```
setenv subj subject_code
setenv output_dir path_to_analysis_directory
setenv anat_subj subject_anatomical_code
setenv subj_anat_dir $SUBJECTS_DIR/${anat_subj}/
```

The first step in preprocessing of the data is motion correction, 3dvolreg in the AFNI software package. In this step a participant's head position is registered to a base volume. A single frame from a single run is taken as the base, or standard, to which head position is compared for each run, each time point. When deviations from the standard are detected at a given time point a rigid body matrix transformation is used to correct that time point to the standard.

Detrending the data is a second step that models and removes noise. Noise signals such as scanner drift are most commonly modeled as a linear or exponential signal. In this analysis, three programs are used to complete this step. 3dTstat computes the mean, 3dDetrend models the noise signal, and 3dcalc computes the detrended signal.

A third optional preprocessing step includes spatial blurring (also referred to as smoothing) of the data, 3dmerge in AFNI. The use of spatial blurring in retinotopic mapping is debated. One side of the debate argues that each voxel represents a distinct population of neurons and that averaging the data across voxels is not recommended. However, reasons for spatial blurring do exist. The primary argument is that because contiguous voxels show an orderly progression of spatial preference, the response of each voxel is necessarily related to the response of neighboring voxels. In this case, a small degree of smoothing is unlikely to adversely affect the model fit and can indeed increase the signal to noise of the output. In this analysis, a small 4-mm spatial blurring kernel is used to increase signal strength. It is the experience of the author that a small amount of smoothing has little effect on retinotopic mapping in the parietal cortex. The smoothed runs are then averaged together (3dcalc) to create a single file.

The next step in preprocessing is demeaning the data. This step is done by three separate programs. 3dTstat is used to create a mean signal, 3dautomask creates a mask of the voxels within the brain, and finally 3dcalc computes the regression of the mean values. Masking the data to include only voxels within the brain can reduce processing time in a pRF analysis. This analysis uses the least squares estimate to fit the data to models based upon the size and preferred location for each voxel. Each voxel for which the data must be modeled increases the processing demands substantially. Therefore, using a mask decreases the time required by limiting the computational requirements inherent in the data analysis to those voxels that correspond to areas within the brain.

Once the data have been preprocessed, a nonlinear model is fit to each voxel, 3dNLfim in AFNI. This step requires the following variables:

setenv AFNI_CONVMODEL_REF *gamma_model (created by 3dDeconvolve)*

setenv AFNI_MODEL_PRF_STIM_DSET *stimulus_file (shows stimulus locations over TRs)*

setenv AFNI_MODEL_PRF_ON_GRID *YES (limits possible solutions and processing time)*

setenv AFNI_MODEL_DEBUG *2 (verbosity level 0-3)*

The inputs to this model, in addition to the preprocessed data, are the stimulus locations and the gamma function that models the hemodynamic response to a stimulus. The gamma function can be measured within individual participants or a conventional gamma model can be used. The example uses a conventional gamma fit created by 3dDeconvolve.

The output 3dNLfim model is a file called Buckslow, from AFNI's use of the term bucket for storing fMRI data. There are several output variables. AMP is the amplitude of the signal at each voxel. *X*, *Y* are the Cartesian coordinates for the preferred location in the visual field, on a grid between −1 and 1. Sigma is the width of the Gaussian. R^2 and *F*-stat *R* are statistical variables that offer a measure of the goodness of fit for the model at each voxel. Finally, the analysis computes the preferred location in terms of eccentricity and polar coordinates to map the visual field using 3dcalc and then add these parameters to the bucket using 3dbucket.

Stimulus files can be created using software such as MATLAB or Python that can work with large data sets and aid in visualization of the data. The stimulus file demonstrates the location of the stimulus across the visual field at each time point (TR). Time points in which the stimulus is not presented are also included in the stimulus file as blank (usually all-zero) matrices.

3.3 Visualization of the Data

At this point in the analysis, the model has been fit to the data in native space, or slices that were originally acquired in the scanner. The next steps are to deoblique the data (3dWarp), align the anatomical file to the data (align_epi_anat.py), create the SUMA surfaces (@SUMA_Make_Spec_FS), then align the surface to the data (@SUMA_AlignToExperiment), and then create the surface file (3dVol2Surf) (Table 2).

3.4 Defining Boundaries Between IPS Regions

The first step to defining retinotopic areas is to choose the statistical map of the preferred locations in polar coordinates (the last BRIK) and project it onto an inflated or flattened cortical surface. Early visual cortex maps are often demonstrated on a flattened surface. These maps were defined by cutting along the calcarine sulcus, then along a line on the lateral and medial parietal cortex. Parietal maps, however, are easily visualized on the inflated cortical surface, and most researchers are accustomed to viewing the data on an inflated surface for demonstrating the map structure.

Choosing an appropriate statistical threshold is important and will depend upon a number of factors to visualize the statistical map. pRF models are thresholded based upon the r^2 value, or the variance explained by the model. The data in Fig. 1 are thresholded to display the model fit for any voxel demonstrating at least 10% of the variance explained by the model. The phase of polar representations ranges from $-\pi$ to π. The equation for converting Cartesian coordinates in AFNI or MATLAB to polar coordinates is:

$$a \tan 2(y, x)$$

where x and y represent the distance from the fovea in horizontal and vertical dimensions. However, this equation results in the division of the visual field into the upper and lower visual hemifields. Therefore, for mapping purposes, the equation is modified as follows:

$$a \tan 2(x, -y)$$

so that negative values occur in the left visual field and positive values occur in the right visual field. The choice of a color map should be determined by selecting a distribution of colors sufficient to differentiate between phases in a map. Figure 1 shows the parietal map locations using the color wheel in SUMA restricted to the contralateral visual hemifield. The color progressions will be reversed for the two hemispheres. In the right hemisphere, 0° represents the lower vertical meridian, and π represents the upper vertical meridian. Phases therefore become more positive upon approaching the upper vertical meridian. However, in the right hemisphere, values become more negative as the phases approach the upper vertical meridian ($-\pi$). Here I have inverted the left hemisphere so that the colors in both hemispheres represent the same locations vertically.

Areas V1–V3 in early visual cortex demonstrate quarter visual field representations. For instance, the border between V1 dorsal (V1d) and V1 ventral (V1v) occurs at the horizontal meridian. The border between areas is selected as the middle of the representation of the horizontal meridian. Spatial representations in V1d progress in terms of polar angle from the horizontal to the lower vertical meridian representation. Then, at the boundary with area V2 dorsal (V2d), there is a phase reversal and the representations progress from the lower vertical to the horizontal meridian. At the boundary of V2d and V3 dorsal (V3d), there is another phase reversal at the horizontal meridian.

Beyond early visual cortex, retinotopic maps are defined as hemifield representations, rather than quarter-field representations. For most maps, boundaries between areas still occur at polar phase reversals, however these reversals are invariably at the upper or lower vertical meridian. However, as is apparent in Fig. 1, often times the hemifield representation will not show a phase reversal directly at the vertical meridian, but at a polar angle in the upper or lower visual field biased toward the horizontal meridian. One reason is an under-representation of the vertical meridian in parietal cortex [29, 30] and to a lesser extent in visual cortex [29]. In dorsal parietal cortex, there is also a lower visual field bias resulting in less pronounced upper visual field representations in the parietal cortex [23]. Therefore, particularly in parietal cortex, it is common to define the boundary at the polar angle representation that most closely approaches the vertical meridian.

Borders of topographic maps are defined by visual inspection by a person familiar with the map structure of the parietal cortex. When uncertainty regarding the boundary exists, four different strategies are helpful in resolving the uncertainty. First, a researcher may choose different color maps and map thresholds (described below) to define the regions of interest. When the difficulty arises from distinguishing phases, a color map that better demarcates the boundary in question can be helpful. When a map appears to continue in two directions, increasing the statistical threshold may demonstrate that one direction is more reliable and therefore more likely a reported topographic area. Alternatively, if another researcher is present, both researchers may define the boundaries and then discuss how each map structure matches the prototypical map structure of that area. Finally, certain software packages, such as mrVista (https://github.com/vistalab/vistasoft), contain tools for defining phase progression across an area.

Dorsal to area V3d, a map cluster V3A/B occurs, with both areas V3A and V3B showing a similar polar hemifield map. The ventral border of V3A/B is a phase reversal with representations starting at the lower vertical meridian and progressing through visual field locations up until the upper vertical meridian. Areas V3A and V3B demonstrate similar phase progressions in terms of polar phase. However, the two areas share a foveal representation (known as a foveal confluence) with the two areas diverging with more peripheral eccentricities.

Area IPS0 is the first of the IPS maps and begins at the phase reversal that defines the dorsal border of V3A/B, along the transverse occipital sulcus. This phase reversal has a characteristic "V" shape, although in certain individuals it appears more rounded. Consistent with the macaque literature, this area was originally described as V7 due to its functional similarity with its purported homologue in nonhuman primates [31]. However, due to the fact that it is located on the medial bank of the intraparietal sulcus, dorsal to the transverse occipital sulcus, and the fact that it shares a foveal confluence with IPS1, many consider this area to be more appropriately named IPS0.

A dorsal and anterior progression of maps along the medial bank of the intraparietal sulcus occurs from IPS0 to IPS2 in most participants, with the topographic representations on the medial bank of the intraparietal sulcus. The sulcus then demonstrates a lateral turn, and retinotopic maps follow the turn of the sulcus. This turn usually occurs near the boundary of IPS2 and IPS3, while superior parietal area (SPL1) extends medially.

Defining boundaries between areas can sometimes prove difficult with only one image. For instance, the boundaries between areas IPS3-5 on the right hemisphere of Fig. 1 are not readily apparent with the color scale and threshold that best demonstrates the other maps across both of the hemispheres. Expanding the range across all phases ($-\pi$ to π) to create a whole field color map, increasing or decreasing the threshold of the statistical map, or alternating between color maps may be necessary for ambiguous phase reversals.

Spatial representations in visual and parietal cortex often demonstrate biases also evident in behavioral outcomes. One such bias is an overrepresentation of the lower as compared to upper visual field. This bias, evident in Fig. 1, was reported previously in retinotopic IPS as well as lateral occipital cortex [23]. Therefore boundaries that occur at the upper visual meridian are sometimes more difficult to define than the lower visual field boundaries and may require visualizing the maps using multiple color scales and thresholds.

Finally, while the maps of polar angle are evident across participants, the eccentricity map structure is more difficult to demonstrate in the parietal cortex. An early, seminal paper described a progression from foveal to peripheral from the fundus of the sulcus toward the medial bank [29], and this has recently been replicated using the pRF method [32]. However, previous findings suggest that eccentricity preferences for individual voxels vary based upon attention demands [23]. This raises the question whether eccentricity map structure itself is consistent across tasks.

4 Notes

The mapping procedure delineated here has been demonstrated across several imaging platforms and allows for consistent mapping across both hemispheres in all participants. Figure 2 shows retinotopic mapping of the parietal sulcus by the same participant in a 3T Siemens Tim Trio magnet with the posterior portion of a 16-channel head coil (Fig. 2a) and a 7T Siemens Magnetom magnet with a 24-channel radiofrequency head coil (Fig. 2b). Despite

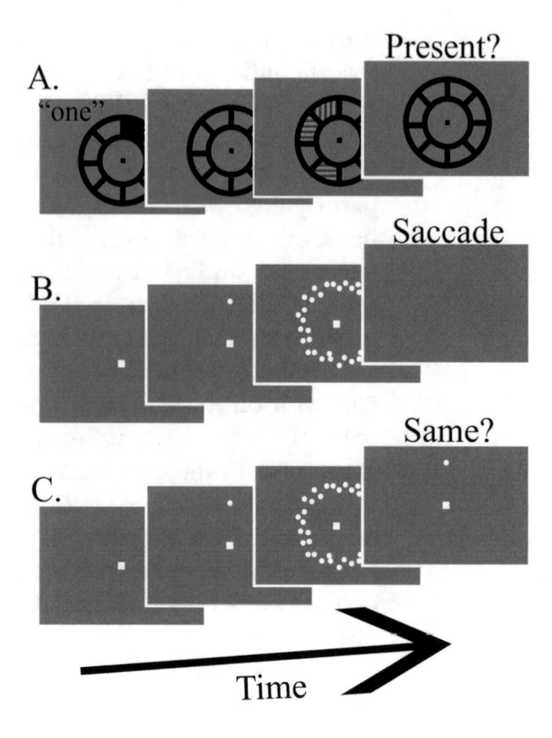

Fig. 3 Trial structure for paradigms that have been used to demonstrate maps in parietal cortex. In each of the tasks, participants attend to or remember stimuli at sequentially clockwise and/or counterclockwise locations. (**a**) Covert attention paradigm adapted from Silver et al. [30]. Participants are presented with a ring divided into eight sections and instructed to direct attention to one of the eight segments (here filled in black) and determine whether a target was present in the attended segment. (**b**, **c**) Memory-guided saccade (**b**) and spatial working memory (**c**) tasks, adapted from Kastner et al. [27]. In both tasks, participants are directed to maintain fixation, while a stimulus is presented in a peripheral location followed by a mask. In the delayed saccade task, when the fixation point disappears, participants make an eye movement to the remembered location. In the spatial working memory task, the participant has to determine whether a memory probe appears in the same or a different position than the remembered stimulus

modifications to the retinotopic mapping procedure and stimulus itself, as well as differences in head coil and magnet, parietal maps demonstrate remarkably similar structure across sessions.

While the pRF method outlined here is very effective for demonstrating topographic maps in the parietal cortex, the choice of mapping paradigm should depend upon the areas that are needed for the analysis (Fig. 3). Delayed saccades have been used from the seminal mapping paper in parietal cortex and are effective in demonstrating maps in the frontal eye fields [25]. Visual stimuli may be titrated for demonstrating maps in early and ventral visual cortex (but see [23, 32] for visual cortical maps using random motion stimuli). Finally, working memory paradigms have demonstrated maps in prefrontal and parietal cortex [26]. The choice of stimulus and analysis procedure should be determined by the needs of the study.

The pRF method has been used to delineate maps across many cortical areas [33]. One of the strengths of pRF mapping is that it models baseline activity level. Outside of early visual cortex, visually responsive neurons have relatively large receptive fields and, within parietal and frontal cortex, may show substantial activity in the absence of visual stimulation. These two properties can both result in what appears to be a uniform distribution of activity in traditional phase-encoded designs. However, the use of baseline periods allows pRF mapping to differentiate between visual stimulation and baseline activation. This baseline activation, independent of visual stimulation, would otherwise conflate the size of the visual field that induces activity. Indeed, the use of baseline periods has been attributed to the success of this method in demonstrating retinotopic maps [34].

Recent studies have extended pRF mapping to estimating noncircular Gaussian models [35]. Representations of space have been shown to be "pulled" toward attended items in nonhuman primates [36]. Studies have demonstrated that attention affects the size and preferred location of visual representations within human participants but were unable to show directionality of attention mechanisms [23, 37]. These advanced methods will likely extend our understanding of how visual attention and short-term memory affect representations of space and objects.

In conclusion, the definition of cortical areas within individual participants has been a boon to visual and cognitive neuroscience. The motion pRF stimulus described here is effective in delineating cortical areas that can be used to objectively define occipital and parietal maps within a relatively short fMRI session. Topographic mapping has elucidated properties of attention and short-term memory such as hemispheric asymmetries in the representation of space [12, 18, 23] that remained elusive in fMRI of healthy individuals. Topographic mapping alongside relatively new imaging techniques such as forward encoding models and functional connectivity serves as a powerful method for understanding the representations underlying visual attention and short-term memory.

References

1. Inouye T (1909) Die Sehstörungen bei Schußverletzungen der kortikalen Sehsphäre: Nach Beobachtungen an Verwundeten der letzten japanischen Kriege. W. Engelmann Verlag, Leipzig
2. Sereno MI, McDonald CT, Allman JM (1994) Analysis of retinotopic maps in extrastriate cortex. Cereb Cortex 4:601–620
3. Tootell RB, Silverman MS, Valois DR (1981) Spatial frequency columns in primary visual cortex. Science 214:813–815
4. Sereno MI, Dale A, Reppas JB et al (1995) Borders of multiple visual areas in humans revealed by functional magnetic resonance imaging. Science 268(5212):889–893
5. DeYoe EA, Carman GJ, Bandettini P et al (1996) Mapping striate and extrastriate visual areas in human cerebral cortex. Proc Natl Acad Sci U S A 93:2382–2386
6. Engel SA, Rumelhart DE, Wandell BA et al (1994) fMRI of human visual cortex. Nature 369:525
7. Tootell BR, Reppas JB, Kwong KK et al (1995) Functional analysis of human MT and related visual cortical areas using magnetic resonance imaging. J Neurosci 15:3215–3230
8. Kastner S, De Weerd P, Desimone R et al (1998) Mechanisms of directed attention in the human extrastriate cortex as revealed by functional MRI. Science 282:108–111
9. Somers DC, Dale AM, Seiffert AE et al (1999) Functional MRI reveals spatially specific attentional modulation in human primary visual cortex. Proc Natl Acad Sci U S A 96:1663–1668
10. Wandell BA, Winawer J (2011) Imaging retinotopic maps in the human brain. Vision Res 51:718–737
11. Benson NC, Jamison KW, Arcaro MJ et al (2018) The HCP 7T retinotopy dataset: description and pRF analysis. Biorxiv. https://doi.org/10.1101/308247
12. Sheremata SL, Bettencourt KC, Somers DC (2010) Hemispheric asymmetry in visuotopic posterior parietal cortex emerges with visual short-term memory load. J Neurosci 30:12581–12588
13. Harvey BM, Klein BP, Petridou N et al (2013) Topographic representation of numerosity in the human parietal cortex. Science 341:1123–1126
14. Hutchinson BJ, Uncapher MR, Weiner KS et al (2012) Functional heterogeneity in posterior parietal cortex across attention and episodic memory retrieval. Cereb Cortex 24:bhs278
15. Liu T, Pleskac TJ (2011) Neural correlates of evidence accumulation in a perceptual decision task. J Neurophysiol 106:2383–2398
16. Todd JJ, Marois R (2004) Capacity limit of visual short-term memory in human posterior parietal cortex. Nature 428:751–754
17. Xu Y, Chun MM (2006) Dissociable neural mechanisms supporting visual short-term memory for objects. Nature 440:91–95
18. Sheremata SL, Somers DC, Shomstein S (2018) Visual short-term memory activity in parietal lobe reflects cognitive processes beyond attentional selection. J Neurosci 38:1511–1519
19. Serences JT, Ester EF, Vogel EK et al (2009) Stimulus-specific delay activity in human primary visual cortex. Psychol Sci 20:207–214
20. Cox RW (1996) AFNI: software for analysis and visualization of functional magnetic resonance neuroimages. Comput Biomed Res 29:162–173
21. Dumoulin SO, Wandell BA (2008) Population receptive field estimates in human visual cortex. Neuroimage 39:647–660
22. Silson EH, Chan AWY, Reynolds RC et al (2015) A retinotopic basis for the division of high-level scene processing between lateral and ventral human occipitotemporal cortex. J Neurosci 35:11921–11935
23. Sheremata SL, Silver MA (2015) Hemisphere-dependent attentional modulation of human parietal visual field representations. J Neurosci 35:508–517
24. Bressler DW, Fortenbaugh FC, Robertson LC et al (2013) Visual spatial attention enhances the amplitude of positive and negative fMRI responses to visual stimulation in an eccentricity-dependent manner. Vision Res 85:104–112
25. Sereno MI, Pitzalis S, Martinez A (2001) Mapping of contralateral space in retinotopic coordinates by a parietal cortical area in humans. Science 294:1350–1354
26. Hagler DJ, Sereno MI (2006) Spatial maps in frontal and prefrontal cortex. Neuroimage 29:567–577
27. Kastner S, DeSimone K, Konen CS et al (2007) Topographic maps in human frontal cortex revealed in memory-guided saccade and spatial working-memory tasks. J Neurophysiol 97:3494–3507
28. Dale AM, Fischl B, Sereno MI (1999) Cortical surface-based analysis I. Segmentation and surface reconstruction. Neuroimage 9:179–194

29. Swisher JD, Halko MA, Merabet LB et al (2007) Visual topography of human intraparietal sulcus. J Neurosci 27:5326–5337
30. Silver MA, Ress D, Heeger DJ (2005) Topographic maps of visual spatial attention in human parietal cortex. J Neurophysiol 94:1358–1371
31. Tootell RBH, Hadjikhani N, Hall EK et al (1998) The retinotopy of visual spatial attention. Neuron 21:1409–1422
32. Mackey WE, Winawer J, Curtis CE (2017) Visual field map clusters in human frontoparietal cortex. Elife 6:e22974
33. Wandell BA, Winawer J (2015) Computational neuroimaging and population receptive fields. Trends Cogn Sci 19:349–357
34. Amano K, Wandell BA, Dumoulin SO (2009) Visual field maps, population receptive field sizes, and visual field coverage in the human MT+ complex. J Neurophysiol 102:2704–2718
35. Silson EH, Reynolds RC, Kravitz DJ et al (2018) Differential sampling of visual space in ventral and dorsal early visual cortex. J Neurosci 38:2294–2303
36. Womelsdorf T, Anton-Erxleben K, Pieper F et al (2006) Dynamic shifts of visual receptive fields in cortical area MT by spatial attention. Nat Neurosci 9:1156–1160
37. de HB, Schwarzkopf DS, Anderson EJ et al (2014) Perceptual load affects spatial tuning of neuronal populations in human early visual cortex. Curr Biol 24:R66–R67

Neuromethods (2020) 151: 257–279
DOI 10.1007/7657_2019_27
© Springer Science+Business Media, LLC 2019
Published online: 5 September 2019

Population-Level Analysis of Human Grid Cell Activation

Matthias Stangl, Thomas Wolbers, and Jonathan P. Shine

Abstract

The groundbreaking discovery of grid cells in the rodent medial entorhinal cortex has led to unparalleled understanding of the neural underpinnings of spatial navigation. These cells show remarkably regular firing patterns, with neighboring firing fields arranged in 60° intervals, and it has been suggested that they provide the spatial metric underlying different cognitive functions, from path integration to the organization of conceptual knowledge. The physiological properties of grid cells mean that the putative signature of these neurons can be observed at the macroscopic level in the blood oxygenation level-dependent response using functional magnetic resonance imaging and local field potentials from intracranial implant patients. In this chapter, we provide a step-by-step guide as to the methods used to assess the activity from populations of grid cells (i.e., grid-cell-like representations) in humans, from the preprocessing of data to the calculation of grid cell metrics. Furthermore, we outline different variants of this analysis as well as reviewing the extant data relating to grid cell function in humans. Finally, we provide an overview as to the future directions of grid cell research in humans.

Keywords Electrophysiology, fMRI, General linear model, Grid cells, Grid-cell-like representations, Human

1 Introduction

Navigating the world is a vitally important behavior for both humans and animals. Understanding how the brain supports this complex cognitive function is, therefore, a key question for neuroscience research. The discovery of a number of functionally distinct neurons in the mammalian brain, with different spatially modulated firing properties, has helped identify the brain's spatial representation system and elucidate the neural mechanisms underpinning spatial navigation [1]. Moreover, recent work suggests that the same neural mechanisms may provide the spatial framework to support cognition in humans more generally [2].

In a seminal study, the first demonstration of spatially modulated neurons was the discovery of place cells in the rodent hippocampus [3]. These neurons were found to code for the animal's location in space, with different place cells coding for specific locations in the environment. The firing of a different cell, so-called grid cells, is also modulated by an animal's location within the environment. Unlike place cells, however, grid cells code for multiple locations across the environment via repetitions of distinct firing fields. Originally discovered in the rat medial entorhinal cortex [4], the firing of grid cells demonstrates remarkably regular organization, forming tessellating equilateral triangles that effectively "tile" the world's navigable surface in a hexagonal lattice (Fig. 1). The equally spaced and repetitive firing means that, for each firing field of a grid cell, the six adjacent fields are arranged in 60° intervals, creating a sixfold symmetry. Later studies provided evidence of grid cells in other brain regions such as the pre- and parasubiculum [5] and in different mammalian species, including bats [6], nonhuman primates [7], and humans [8].

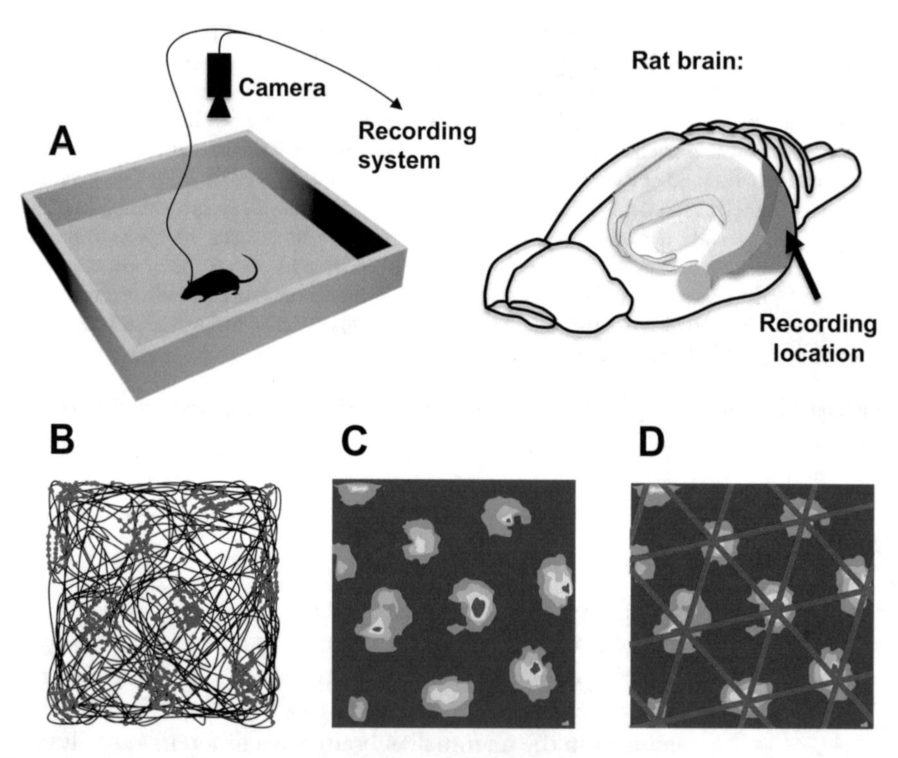

Fig. 1 When an animal is navigating an environment, the firing fields of a grid cell show a remarkably regular organization, forming tessellating equilateral triangles that effectively "tile" the world's navigable surface in a hexagonal lattice. (**a**) Schematic representation of a square environment (left panel) in which a rat is moving during electrophysiological recordings from the medial portion of the entorhinal cortex where grid cells were first recorded and are most populous. Color codes for the schematic representation of a rat brain (right panel): entorhinal cortex, red; hippocampus, yellow; perirhinal cortex, green; postrhinal cortex, blue; amygdala, gray. (**b**) The square environment shown from top-down perspective. Firing locations (red dots) of a typical grid cell are superimposed on the animal's trajectory (black lines) in the environment. (**c**) The firing pattern of a grid cell demonstrated as a "rate map"; warmer colors indicate increased cell firing (red indicates maximum firing; dark blue indicates zero). (**d**) Connecting the centers of the grid cell's firing fields in the rate map, it is possible to observe their repeating, grid-like pattern. Figure adapted from Kesner [38] and Grieves and Jeffery [39]

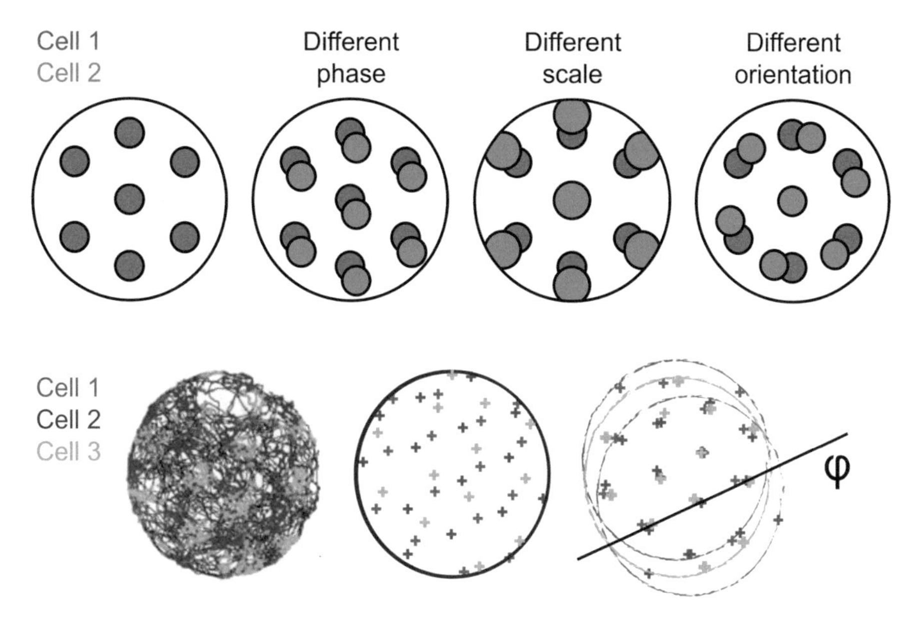

Fig. 2 Properties of grid cell firing patterns. Schematic representations of grid cell firing patterns in a circular environment from top-down perspective. Top panel: Schematic representation of two grid cell firing patterns (Cell 1, green, versus Cell 2, orange). The firing patterns of different grid cells can differ in several ways, such as their phase (displacement in Cartesian coordinates relative to an external reference point in the environment), scale (distance between two neighboring firing fields), and orientation (angular tilt relative to an external reference axis in the environment). Bottom panel: There is evidence that the firing pattern of neighboring grid cells (e.g., Cell 1, red; Cell 2, blue; Cell 3, green) is similar in terms of their orientation (φ), even though their phase and spatial scale may differ [4, 9]. Colored crosses indicate centers of grid cell firing fields. Figure (bottom panel) adapted from Hafting et al. [4]

Despite their regular firing fields, the firing patterns of grid cells can differ in several ways (Fig. 2), such as their phase (displacement in Cartesian coordinates relative to an external reference point in the environment), spatial scale (distance between two neighboring firing fields), and orientation (angular tilt relative to an external reference axis in the environment). While the phase of grid cells appears to randomly vary across cells, there is evidence that neighboring grid cells are more similar in terms of their orientation and spatial scale in comparison to grid cells located further apart [4]. There appears also to be a gradient in spatial scale, such that representations become less fine-grained as one moves from dorsomedial-to-ventrolateral medial entorhinal cortex in the rat [4]. It is still possible, however, to observe coherence in the orientation of distal grid cells even though they may code at different spatial scales [9]. In addition, it has been shown that grid cells are often modulated by the animal's heading direction [10] and velocity [11].

The vast majority of our knowledge about grid cells, such as their precise location in the brain and how they support navigation, stems from electrophysiological work in nonhuman animals, where researchers are able to record directly from individual neurons via implanted electrodes. Understanding how grid cells support higher-order cognitive functions such as spatial navigation in humans requires the acquisition of similar measures from the human brain, which allows us to characterize both cross-species similarities and critical differences between humans and nonhuman animals. Given the invasive nature of electrophysiology, however, investigating grid cells in humans can prove difficult. For example, there are relatively few studies in which grid cells are studied via intracranial recordings from the entorhinal cortex of patients with drug-resistant epilepsy. Consequently, a limited number of research institutions worldwide have the opportunity to apply this method and studies often comprise small sample sizes. In one such study, Jacobs et al. [8] acquired intracranial recordings from epilepsy patients while they completed a spatial navigation task in a virtual environment shown on a computer screen. Consistent with the rat electrophysiology, there was evidence of cells with a sixfold symmetry in their firing rate, demonstrating that grid cells appear preserved across different mammalian species, including humans.

In contrast to electrophysiology, functional magnetic resonance imaging (fMRI) is accessible to a wide range of researchers, and it is possible to investigate the neural correlates of higher-order cognitive processes in large samples of healthy subjects. Consequently, fMRI has been used to study the putative neural correlates of grid cells in healthy human subjects [12]. Although fMRI does not provide single-cell resolution, and the data reflects only macroscopic changes in the blood oxygenation level-dependent (BOLD) response at the voxel level, properties of grid cell firing suggested it would be possible to detect the putative firing signature from a population of grid cells (henceforth referred to as grid-cell-like representations) in the fMRI signal. Specifically, even though grid cells are arranged topographically, the grid orientation of distal cells can be coherent [9]. Moreover, a subpopulation of grid cells, known as “conjunctive grid cells,” show modulation of their firing rate according to the animal’s movement direction in the environment. Similar to head direction cells, conjunctive grid cells show increased firing when the animal travels in the cell’s “preferred” direction relative to other travel directions [10], and this is aligned with the main axes of the grid [12]. As a result of these cell firing dynamics, it was argued that it may be possible to observe the sixfold sinusoidal, or hexadirectional, pattern in the BOLD response when participants performed translations either aligned or misaligned with the grids’ axes (Fig. 3).

This hexadirectional pattern of data was first observed in fMRI using an object-place memory task in a virtual environment [12]. In line with the results of electrophysiological recordings in rodents, the BOLD signal in several brain regions, including the entorhinal

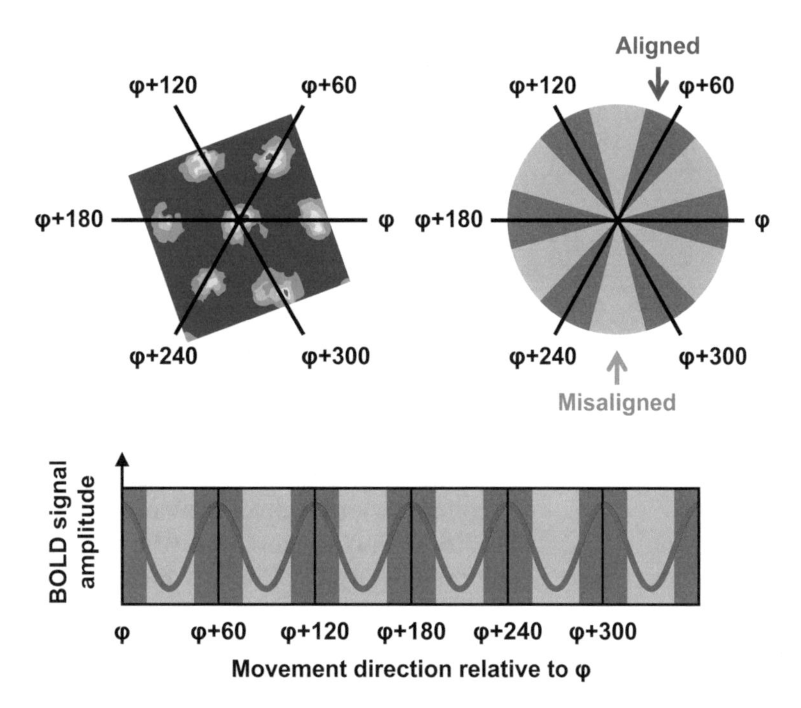

Fig. 3 The analysis logic for measuring grid-cell-like representations in human fMRI signals. Top left: Rate map of a typical grid cell firing pattern in a square environment (shown from top-down perspective). The firing pattern shows a specific orientation (φ) relative to an external reference axis in the environment. Top right: Movement directions within the environment can be categorized either as aligned (green) or misaligned (gray) with the grid cell's firing orientation φ. Bottom: The red curve shows the expected pattern of the BOLD signal amplitude modulated by movement direction relative to φ. High signal peaks are expected for movements aligned with φ or a 60° multiple of φ (green sectors). Figure adapted from Doeller et al. [12]

cortex, showed a sixfold symmetry, with greater activity associated with translations in which the travel direction was aligned with the mean grid orientation, compared to when the travel path was misaligned with a grid axis. This study was critical in demonstrating that fMRI could provide a viable way to study the activation of grid cells at the population level in humans.

Despite growing interest in the role of grid cells in human cognition, the research field is in its infancy, and only recently have open access tools been developed to complement traditional fMRI analysis packages, allowing researchers study grid-cell-like representations [13].

In this chapter, we aim to provide a comprehensive methodological review of grid cell research in humans, which stems from both functional magnetic resonance imaging and recordings from intracranial implant patients. For researchers new to the field of human grid cell research, we will first explain the analysis method used to study the putative grid cell signal using fMRI, detail the different ways in which the fMRI data can be modeled, and describe the metrics derived from the analysis of grid-cell-like representations. We will then detail how this analysis has been extended for use in electrophysiology studies before describing the current state of research in both fMRI and intracranial methods, as well as providing an overview as to the currently outstanding questions for grid cell research in humans.

2 General Analysis Method

We start here with an overview of the general analysis method used for the examination of grid-cell-like representations in fMRI, before individual steps are discussed in more detail in the following sections. Furthermore, we will outline different variants of the method, as well as alternative analysis approaches that have been described in the literature.

The general analysis method follows the approach that was first introduced by Doeller et al. [12]. This involves first defining events of interest (henceforth referred to as "grid events") in the timeline of the experimental task and imaging time series. The imaging data are then partitioned into estimation and test datasets, and a general linear model (GLM) is fit to the estimation dataset in order to estimate the putative grid orientation separately for each voxel (GLM1). These voxel-wise grid orientations are then averaged over voxels in a region of interest (ROI), in order to calculate a mean grid orientation. Once the mean grid orientation has been computed in the estimation dataset, this information can then be used to generate a second GLM in which grid events of the test dataset are modeled with respect to their alignment with the mean grid orientation (GLM2). Finally, metrics that describe the grid-cell-like representations are computed, such as their magnitude in a given region as well as measures of their stability and coherence. An overview of the general analysis method pipeline is shown in Fig. 4.

As noted above, initial evidence of grid-cell-like representations in human fMRI was found using a spatial navigation paradigm. Consequently, in the next sections, we will describe the analysis pipeline assuming that an experiment has been run also using navigation in a virtual environment. Throughout this chapter, however, it will become clear that grid-cell-like representations may not be limited to spatial navigation, and therefore the reader should be aware that this analysis method can be extended to more abstract cognitive paradigms.

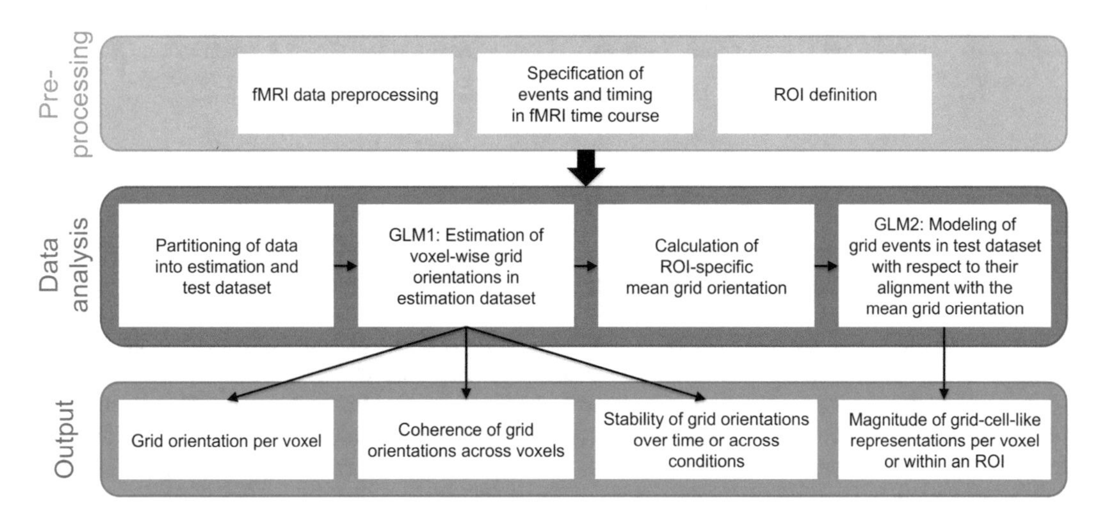

Fig. 4 Analysis pipeline for the general analysis method to investigate grid-cell-like representations. The researcher takes fMRI data (that may have undergone some preprocessing such as normalization or smoothing) together with information regarding events of interest during the fMRI time course and corresponding timing information. The data are then partitioned into estimation and test datasets, before the voxel-wise grid orientations are calculated in a first general linear model (GLM1) using the estimation data. Voxel-wise grid orientations are stored and metrics of grid-cell-like representations assessed, such as coherence of grid orientations across voxels and stability of grid orientations over time or across conditions. In a second general linear model (GLM2), the events in the test dataset are then modeled with respect to their alignment with the ROI-specific mean grid orientation, in order to quantify the magnitude of grid-cell-like representations for individual brain voxels or averaged over voxels within an ROI

2.1 Functional Image Preprocessing for the Analysis of Grid-Cell-like Representations

It is possible to analyze grid-cell-like representations using either the participant's normalized and smoothed functional images (cf. [12, 14, 15]) or functional images in the individual subject's native functional space (cf. [16, 17]). Normalizing to standard space prior to analysis makes it easier to examine group-level, cluster statistics, whereas spatial distortions or interpolation errors resulting from normalization to a standard template can be avoided by performing the analysis in the participant's native space. Apart from the decision as to whether to analyze the data in standard or native space, the preprocessing steps for the analysis of grid-cell-like representations are consistent with normal fMRI, which can include, for example, the realignment of functional images, as well as the removal of physiological noise components. Covariates of these procedures (e.g., the realignment parameters) can then be added to GLM1 and GLM2 as nuisance regressors.

2.2 Specifying Grid Events

To detect grid-cell-like representations, it is necessary first to specify grid events within the fMRI time course. Grid events could comprise, for example, periods of translational movement within a virtual environment (e.g., [12, 16, 17]), imagined movement (e.g., [15]), or two-dimensional movement in abstract space

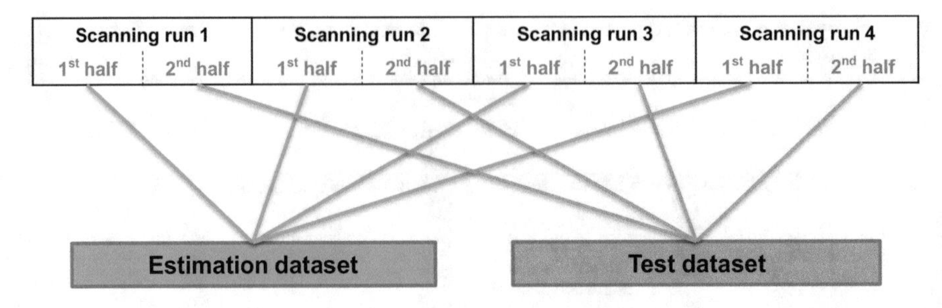

Fig. 5 Example fMRI data partitioning scheme for the analysis of grid-cell-like representations. In the study of Stangl et al. [17], fMRI data of four consecutive scanning runs were partitioned in two data halves: an estimation and a test dataset. The estimation dataset comprised the data from the first half of each scanning run and was used for estimating voxel-wise grid orientations in GLM1. The test dataset comprised the data from the second half of each scanning run and was used for testing the estimated grid orientations in GLM2 and for calculating the magnitude of grid-cell-like representations

(e.g., [14]). For each grid event, an angle (henceforth referred to as "grid event angle") relative to a nominal 0° reference point (e.g., a fixed landmark in the virtual environment) is then defined.

2.3 Partitioning the fMRI Data into Estimation and Test Datasets

Partitioning the data into estimation and test datasets can be performed in a number of different ways. One method is to split the data from multiple scanning runs into odd and even runs [12, 16], performing the estimation in the odd runs and testing in the even ones (or vice versa). Another possibility, within each run, is to split the data into a number of temporal bins and perform the analysis on these odd/even bins [15]. As an example, Stangl et al. [17] split the data from their four scanning runs into estimation and test dataset by using the first half of each scanning run as the estimation dataset and the second half as test dataset (Fig. 5).

Similarly, one could split trials in each run into odd and even events (e.g., translations in a virtual environment) and carry out the estimation in one half before testing on the other half of the events. It should be noted, however, that the BOLD signal for subsequent events might not be independent, especially if there is only a short time interval between them. This may result in an increased correlation between estimation and test datasets, which may in turn affect beta estimates both for GLM1 and for GLM2. Specifically, depending on how movement directions are sampled across trials, grid orientations might be incorrectly estimated in GLM1, and parameter estimates in GLM2 could be inflated or suppressed. Therefore, if researchers want to perform a data split based on odd versus even events, care must be taken in the design of trial presentation.

2.4 Estimating the Putative Grid Orientation per Voxel

Grid events in the estimation dataset are modeled in GLM1 in order to estimate the putative grid orientation separately for each voxel. This involves fitting two parametric modulation regressors that model each grid event with respect to its grid event angle (i.e., the direction associated with this event, such as the heading orientation during a period of movement in an environment). Mathematically, one parametric modulation regressor models grid events within the estimation dataset with

$$\sin(a_t * 6)$$

(i.e., the sine regressor) and another parametric modulation regressor models the same grid events with

$$\cos(a_t * 6)$$

(i.e., the cosine regressor), where a_t represents the grid event angle. By including the multiplication term (*6) in these functions, all grid event angles are transformed into 60° space, which accounts for the sixfold symmetry characteristic of grid cell firing patterns. Consequently, the modulation of the fMRI signal for each voxel at a given time point (t) is given in GLM1 by

$$s_t = \beta_{\sin} * \sin(a_t * 6) + \beta_{\cos} * \cos(a_t * 6)$$

where s_t is the measured fMRI (BOLD) signal and $\beta_{\sin}$ and $\beta_{\cos}$ are the parameter estimates (i.e., the beta weights estimated for a regressor in the GLM; higher parameter estimates indicate a better model fit) for the sine and cosine regressors.

Using the parameter estimates derived from GLM1, the putative grid orientation for each voxel can then be computed using the inverse tangent function

$$\arctan(\beta_{\sin} / \beta_{\cos}) / 6$$

where $\beta_{\sin}$ and $\beta_{\cos}$ are the voxel's parameter estimates of the sine and cosine parametric modulation regressors.

2.5 Analyzing the Stability and Coherence of Grid Orientations

The voxel-wise estimates of the putative grid orientations can be used to derive metrics of the stability and coherence of grid-cell-like representations. For example, estimating grid orientations separately for different time bins (e.g., for the first and second half of an experiment) and comparing these grid orientations between time bins can serve as a measure of the stability of grid-cell-like representations over time (Fig. 6, left panel). Similarly, one could estimate voxel-wise grid orientations separately for different experimental conditions (e.g., navigation in two different environments) and compare grid orientations per voxel across conditions. Previous studies (e.g., [16, 17]) classified grid orientations for single voxels as either "stable" or "unstable," depending on whether the orientations of two different time bins or conditions were within ±15° of one another ("stable") or whether the difference in orientation between the time bins or conditions was greater than ±15°

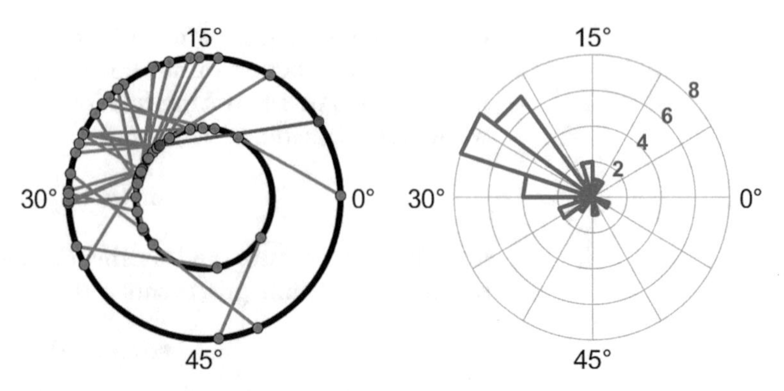

Fig. 6 Examples of grid orientation stability and coherence of 50 voxels within an ROI. Left: Two black rings represent two different temporal time bins (e.g., scanning sessions 1 and 2) or two different experimental conditions, for which voxel-wise grid orientations have been estimated separately. Each voxel's grid orientation is indicated with a circular marker, and a line connects a voxel's orientation for each of the two time bins or conditions. It is evident that the majority of voxels in this example show "stable" orientations (green lines), while only one voxel shows a change of more than $\pm 15°$ and is therefore classified as "unstable" (red line). Right: The coherence of grid orientations within an ROI is typically illustrated with a polar histogram plot, in which the length of each bar indicates the number of voxels sharing a similar grid orientation, with the green numbers indicating the number of voxels represented by each ring of the polar plot. It is evident in this example that the majority of the 50 grid orientations in the ROI show clustering around 25°. Statistical significance of this clustering can be tested using Rayleigh's test for nonuniformity of circular data

("unstable"). The stability of grid-cell-like representations across multiple voxels within a specific brain region was then indicated by the proportion of stable voxels within the region. An alternative measure of stability is to calculate the absolute change in grid orientation between data bins or conditions for each voxel separately and then calculate the average change in orientation over voxels in a brain region (as demonstrated in [17]).

A different characterization of grid-cell-like representation stability is to calculate the coherence of grid orientations across multiple voxels within a brain region (Fig. 6, right panel). For example, it has been reported that voxels in the entorhinal cortex show similar grid orientations [12]. In order to test whether the estimated grid orientations of different voxels cluster around one direction, rather than reflecting a uniform distribution of orientations, these values can be submitted to Rayleigh's test for nonuniformity of circular data. The coherence (or so-called spatial stability) of grid orientations across different voxels within a region is then expressed statistically by the resulting Rayleigh's z-value (with higher z-values indicating higher coherence). It is likely, however, that the spatial smoothing during fMRI data preprocessing gives rise to artificially inflated levels of coherence in grid orientations for neighboring voxels. It is sensible, therefore, to estimate grid orientations and calculate their coherence across voxels also in a control analysis using unsmoothed fMRI data.

2.6 Calculating the Mean Grid Orientation within an ROI

Having calculated the putative grid orientation for a single voxel, the mean grid orientation can also be estimated across multiple voxels within an ROI. For n number of voxels, this is implemented by taking the parameter estimates associated with the sine and cosine regressors and computing

$$\arctan(\text{mean}(\beta_{\sin\_1:n})/\text{mean}(\beta_{\cos\_1:n}))/6$$

where $\beta_{\text{sin\_1}:n}$ is a vector of n parameter estimates for the sine regressor and $\beta_{\text{cos\_1}:n}$ is a vector of n parameter estimates for the cosine regressor, with each vector of parameter estimates being averaged before the inverse tangent function is computed. It is then possible to model individual grid events with respect to their alignment with the ROI's mean grid orientation, as carried out in GLM2 in order to determine the magnitude of grid-cell-like representations within a particular brain region.

2.7 ROI Selection

As described in the previous section, the mean grid orientation can be calculated in a given ROI. Informed by the rodent literature, the majority of previous grid cell studies in humans have used the entorhinal cortex (Fig. 7) as their main ROI [12, 15–17]. Theoretically, however, other regions could be used. For example, human fMRI studies found evidence for grid cell activation outside of the entorhinal cortex, in regions such as the ventromedial prefrontal cortex, and the posterior cingulate cortex [14]. The selection of ROIs will depend primarily upon the researcher's experimental question.

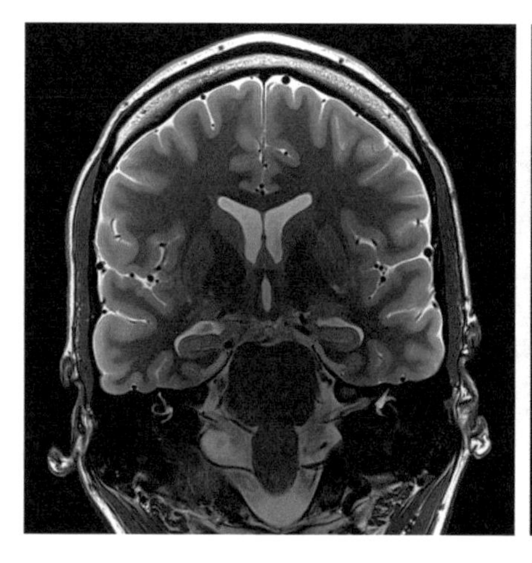

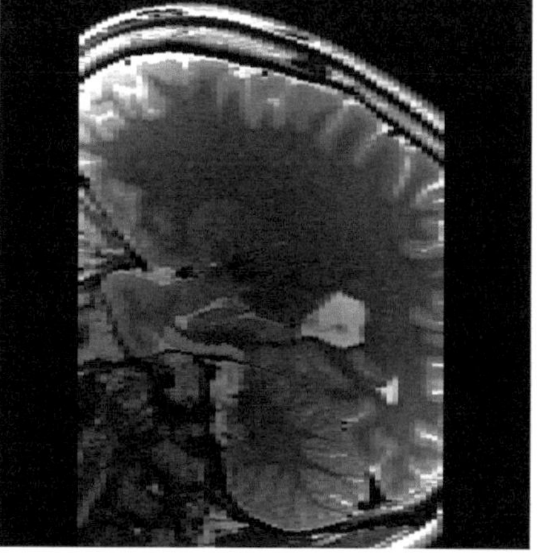

Fig. 7 Entorhinal cortex ROI mask. Example ROI mask for the bilateral entorhinal cortex (red) of one participant in Stangl et al. [17], shown on one subject's individual T2-weighted structural image in coronal (left) and sagittal (right) view

When selecting ROIs based upon anatomy, it is common to create anatomical ROI masks by manually tracing the ROI boundaries on a participant's structural image (e.g., a T2-weighted image). The manual delineation of ROI boundaries is typically performed following published segmentation protocols (e.g., [18]) and using openly available software tools, such as ITK-SNAP [19]. Alternatively, automated tools are available that apply machine learning algorithms in order to enable automatic segmentation of specific brain regions (e.g., [20]).

Instead of using anatomically defined masks, however, one might also wish to use a functionally defined mask from an orthogonal contrast (e.g., [14]) or a separate localizer dataset.

Importantly, in order to allow analysis of grid-cell-like representations in ROI voxels, ROI mask images need to be co-registered to the participant's functional imaging data.

2.8 Quantifying the Magnitude of Grid-Cell-like Representations

A primary measure used to assess grid-cell-like representations is the magnitude of response within an ROI. The magnitude is calculated in GLM2 where grid events in the test dataset are modeled with respect to their alignment with the ROI-specific mean grid orientation. As noted earlier, in a brain region where grid-cell-like representations are present, the BOLD signal amplitude is expected to follow a sinusoidal model, in which the signal amplitude is directly proportional to the alignment of the movement direction with one of the grid's main axes (i.e., the mean grid orientation or a 60° multiple of it). It can be tested to what extent the BOLD signal amplitude within a region follows this sinusoidal model by fitting a parametric modulation regressor to the grid events in the test dataset (Fig. 8).

This parametric regressor is calculated by taking each grid event angle (a_t) and determining its difference from the mean grid orientation (φ) by calculating

$$\cos\left(6*(a_t - \varphi)\right)$$

Following this approach, the resulting parametric modulation values reflect the difference between the individual grid event angle and the mean grid orientation. These values range between "1" for grid event angles that are perfectly aligned with the mean grid orientation (or a 60° multiple of it) and "−1" for completely misaligned events (i.e., mean grid orientation +30°, plus any 60° multiple of this value). The fMRI signal modulation in a voxel at a given time point (t) in GLM2 is given by

$$s_t = \beta_{\text{GLM2}} * \cos\left(6*(a_t - \varphi)\right)$$

Finally, the magnitude of grid-cell-like representations in an ROI can then be quantified by averaging the parametric modulation regressor's parameter estimates (β_{GLM2}) over all constituent

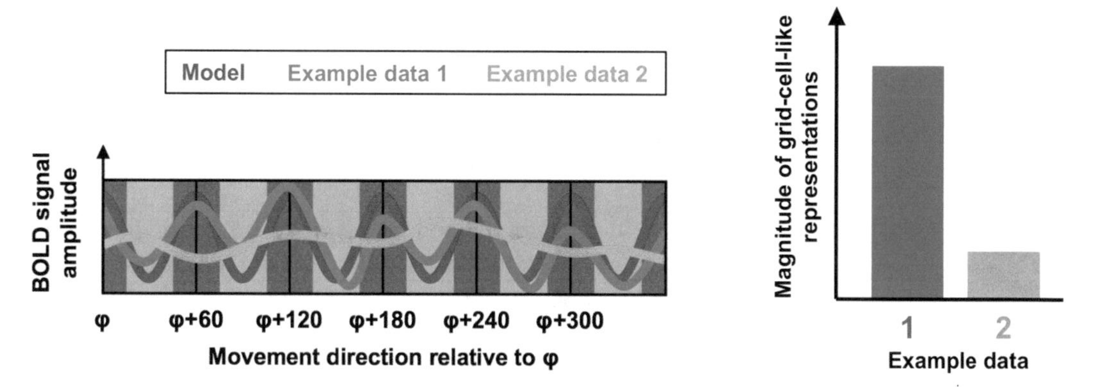

Fig. 8 Estimating the magnitude of grid-cell-like representations. Left: The red curve shows the expected pattern of the BOLD signal amplitude modulated by movement direction relative to φ ("model"). High signal peaks are expected for movements aligned with φ or a 60° multiple of φ (green sectors). The blue curve shows BOLD signal change for an example dataset 1, in which the data is highly similar to the model (i.e., strong modulation of BOLD signal by movement direction). The orange line shows BOLD signal change for an example dataset 2, in which the data is largely independent from the model (i.e., weak modulation of BOLD signal by movement direction). Right: Comparing both example datasets shows that the dataset with the stronger modulation of the BOLD signal by movement direction (example dataset 1) represents a higher magnitude of grid-cell-like representations, compared to the model with weaker modulation

voxels. Based on this model, magnitudes of grid-cell-like representations in a single voxel are expected to be positive for changes in mean grid orientation of less than ±15° between estimation and test datasets or negative for changes of more than ±15°. This is due to the grid orientations ranging from 0° to 60°, and therefore the maximally detectable change in grid orientation is 30°.

Importantly, the stability of grid orientations over time as well as their coherence across voxels within a brain region relates to the magnitude of the grid-cell-like representations. First, if the grid orientations calculated in GLM1 fluctuate over time, the parametric modulation regressor in GLM2 will result in a small parameter estimate (indicative of a low grid-cell-like representation magnitude). This is because GLM2 tests whether the BOLD signal follows a sinusoidal model defined by a specific mean grid orientation that is assumed to be constant over time. Second, if all voxels within an ROI provide a different orientation value (i.e., low orientation coherence across voxels), then the resulting mean grid orientation within this ROI would be random, and the coding of translation events in GLM2 would be arbitrary, again resulting in a low grid-cell-like representation magnitude. An inability to detect grid-cell-like representations, therefore, can result either from a lack of grid orientation stability over time, a lack of orientation coherence across voxels within a brain region of interest, or both.

3 Different Variants, Adaptations, and Alternatives to the General Analysis Method

Having detailed the general analysis method to identify grid-cell-like representations using fMRI, we now turn to different versions of this analysis that have been used in the literature.

3.1 Control Analyses

Given that grid cells are defined by a sixfold symmetric pattern of their firing fields, all published studies investigating grid-cell-like representations aimed to confirm this hexadirectional signal in the BOLD response. This can be tested by examining whether the sixfold sinusoidal model (as shown in Fig. 8) provides a better model fit compared to control models with different periodicities (e.g., fivefold or sevenfold). Analogous to calculating the model fit for the sixfold symmetric model, the fit for a control model with different periodicity can be calculated by replacing the multiplication factor 6 with an alternative value (x). Specifically, the two parametric modulation regressors included in GLM1 are then defined by

$$\sin\left(a_t * x\right) \text{ and } \cos\left(a_t * x\right)$$

and the resulting grid orientations can be calculated by

$$\arctan(\beta_{\sin} / \beta_{\cos}) / x$$

whereas the magnitude of symmetric representations with periodicity x can be assessed in GLM2 by defining the parametric modulation regressor with

$$\cos\left(x * \left(a_t - \varphi\right)\right)$$

where x defines the control model's periodicity (e.g., $x = 7$ when testing for a sevenfold symmetric model).

Several previous studies have also carried out further control analyses to demonstrate that the hexadirectional modulation of the fMRI signal with movement direction is neither a general physiological phenomenon evident in all brain regions nor an artifact of the fMRI signal per se but is specific to anatomically plausible brain regions such as the entorhinal cortex [12]. This is normally achieved by selecting a separate ROI in which one would not expect to see grid-cell-like representations. It is unclear, however, where exactly one should expect this pattern of data in the human brain, given that human grid-cell-like representations have been found in regions outside of the entorhinal cortex [12, 14].

3.2 Comparing Activity Associated with Aligned Versus Misaligned Orientations

In the general analysis method, the magnitude of grid-cell-like representations within an ROI is estimated in GLM2 by modeling the difference between a grid event's orientation and the mean grid orientation using a parametric modulation regressor. Alternatively, it is also possible to categorize grid events into a number of discrete

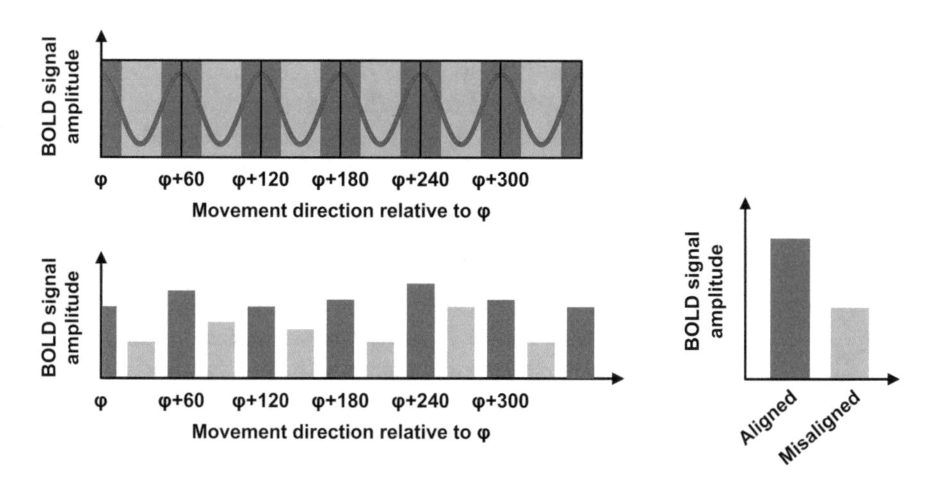

Fig. 9 Analyzing signal amplitudes for aligned versus misaligned orientation bins. After estimating the putative grid orientation within an ROI in GLM1, grid events can be classified in GLM2 as either "aligned" or "misaligned" with this mean grid orientation. In this scheme, unlike the general analysis method, GLM2 does not test for a continuous sinusoidal fit of the model (top panel). Instead, the signal amplitude is estimated for multiple directional bins, as illustrated in an example dataset (bottom panel). Higher signal amplitudes are expected for orientation bins aligned as compared to those misaligned with the mean grid orientation (φ)

categories based on the alignment of their orientation with the mean grid orientation (Fig. 9). For example, several studies have separated grid events into two regressors, one for "aligned" and another for "misaligned" events (e.g., [16]), depending on whether the difference between a grid event's angle and the grid axis was smaller ("aligned" regressor) or larger ("misaligned" regressor) than $\pm 15°$. Parameter estimates can then be generated and compared for each category, in order to test whether, as would be predicted, the signal amplitude is higher for aligned versus misaligned events.

The same approach can be used to categorize grid events into more than two categories. For example, studies have used 12 different regressors, each representing a 30° bin [12, 14]. Six regressors comprised aligned trials—those events within $\pm 15°$ of the mean grid orientation (or a 60° multiple of it)—while the remaining six regressors comprised misaligned trials, that is events offset from the mean grid orientation by 30° (plus a 60° multiple of this value) $\pm 15°$. This approach not only allows for comparisons between aligned and misaligned events but also between specific orientations (e.g., one could compare signal amplitudes during movements in "north" versus "south" direction).

3.3 Different Cross-Validation Methods

Partitioning the data into estimation and test datasets results in a significant amount of data being lost in which the orientation of the grid-cell-like representation is initially calculated. A variation of the analysis, in which a cross-validation scheme was used so that all of the data served as the estimation and test datasets, was demonstrated by Nau et al. [21]. This method was novel also in two other respects. First, different scan runs were averaged in an effort to boost the signal-to-noise ratio; second, the test of the model fit was carried out voxel-wise, meaning that a mean grid orientation was not calculated across the entire entorhinal cortex ROI. Consistent with the general analysis method, in the first level analysis, different trajectories were modeled for each of the nine separate fMRI scan runs. The data were then partitioned into thirds, with two-thirds averaged to increase the signal-to-noise ratio for each trajectory after which the standard sine and cosine regressors were fit to these averaged data. The remaining third of the data were then averaged to form the test dataset and individual trajectories categorized as aligned or misaligned based upon the voxel-wise putative grid orientation derived from the estimation data. This procedure was repeated three times until all partitions of the data had served once as the test dataset and twice as the estimation dataset. To determine the voxel-wise model fit, trials in the test dataset were coded as aligned versus misaligned based upon the individual voxel's putative orientation and contrasted, with the resulting beta coefficients transformed into *z*-scores via random shuffling of the data. Specifically, within each test fold, the direction labels were randomly permuted 1,000 times and the contrast repeated with the shuffled labels to generate a null distribution. The mean and standard deviation of this null distribution was then used to determine the *z*-score for each voxel, and the *z*-scores per voxel averaged over the threefold of analysis. Finally, an average *z*-score describing the difference in activity associated with aligned versus misaligned trajectories was determined by averaging over all voxel-wise *z*-scores in an ROI.

3.4 Multivariate Analysis

The studies described thus far have used univariate analysis methods to assess grid-cell-like representations in the human brain. Multivariate methods, in which the pattern of responses over a number of voxels is extracted, have also demonstrated hexadirectional modulation in the human entorhinal cortex. In Bellmund et al. [22], prior to scanning participants were required to overlearn a virtual environment containing a number of different buildings that served as goal locations, arranged in two regularly spaced hexagons. In the fMRI scanner, the names of two buildings were presented simultaneously, and the participant had to imagine standing at the first building while facing the second building. The angular disparity between the two buildings was then used to categorize trials as either aligned or misaligned with the grid symmetry. Specifically, trials could be binned according to whether the angular disparity comprised a phase 0° modulo 60 (i.e., the same hexadirectional symmetry as reported in grid cells) or 30°

modulo 60 (i.e., offset 30° from the grid's putative axis). The responses of individual voxels in the entorhinal cortex associated with these trials were extracted, Pearson correlations carried out between these values, and the difference in pattern similarity calculated for 0° modulo 60 and 30° modulo 60 trials. This so-called representational similarity analysis [23] revealed that in left entorhinal cortex, the hexadirectional symmetrical trials were more similar to one another than those offset by 30°.

3.5 Method Application to Electrophysiological Recordings

The original method for analyzing population activity of grid cell firing in humans has been developed and applied in fMRI research. Most recently, however, several studies adapted this approach in order to analyze grid cell signals also in the local field potential (LFP) using electrophysiological recordings in humans [24–26]. Similar to human single-cell recordings, LFP recordings are mainly obtained from patients with intractable epilepsy during seizure monitoring to localize seizure foci and guide potential surgical treatment. LFP recordings reflect the electrophysiological signal generated by the summed electric current of multiple neurons within a small volume of nerve tissue near an implanted electrode. Given that LFP recordings represent activity of a population of neurons (as is the case with fMRI), researchers hypothesized that the putative firing of grid cell populations could be detected also in the LFP signal by applying the same analysis approach used to detect grid-cell-representations in the fMRI BOLD signal.

While the general analysis logic and approach remained unchanged, there are several differences between the analysis approaches for fMRI versus LFP recordings. In fMRI, the analysis method is applied to individual voxels, whereas the same analysis pipeline is applied to single electrodes or electrode channels for LFP recordings. The total number of fMRI voxels within the brain, however, is very large (the exact number of voxels depends upon the fMRI sequence parameters but can number around one million), as compared to the much lower number of implanted electrodes for LFP recordings. fMRI, therefore, allows for the sampling of many different voxels simultaneously, while analyses using LFP recordings are typically limited to a small number of brain regions. The LFP signal has a much higher temporal resolution (in the millisecond range), compared to fMRI recordings where the usual duration of one datapoint is around 1 s or more. This high temporal resolution of the LFP signal allows for measurement of the temporal dynamics of neuronal activity that is not possible with fMRI.

The high temporal resolution of the LFP signal also allows for frequency analysis of the grid-cell-like representations. While fMRI analyses typically use the BOLD signal intensity (one value per time point) as an indicator of neuronal activity, one way in which the measured electrophysiological LFP signal can be analyzed is via the examination of different oscillatory frequencies. Previous studies have filtered the net electrophysiological signals into different frequency bands, such as "low theta" or "delta" (1 to 4 Hz), "high theta" (5–8 Hz), "alpha" (8–12 Hz), "beta" (12–30 Hz), "low gamma" (30–80 Hz), and "high gamma" (80–150 Hz) oscillations. The precise nomenclature and cutoff frequencies for each frequency band, however, differed across studies. Analyses of grid-cell-like representations were then applied for each frequency band separately and using the oscillatory power within the specific frequency boundaries.

Using this method, both Maidenbaum et al. [25] and Chen et al. [24] reported converging evidence for entorhinal grid-cell-like representations in the LFP activity of the high theta frequency band, as shown in Fig. 10 for one example electrode.

Lastly, one study applied the same analysis to intracranial LFP recordings in an epilepsy patient and extended the approach to magnetoencephalography (MEG) data in healthy participants [26]. MEG is a noninvasive recording technique used to measure magnetic fields in the brain that are induced by synchronous activity of neuronal populations. As both MEG and LFP recordings are electrophysiological signals with millisecond accuracy, the analysis method described above for analyzing grid-cell-like signals in different frequency bands of LFP recordings can also be applied to MEG data. Using this method, Staudigl et al. [26] found evidence for grid-cell-like signals in broadband high-frequency activity (60–120 Hz) consistently in both LFP and MEG recordings.

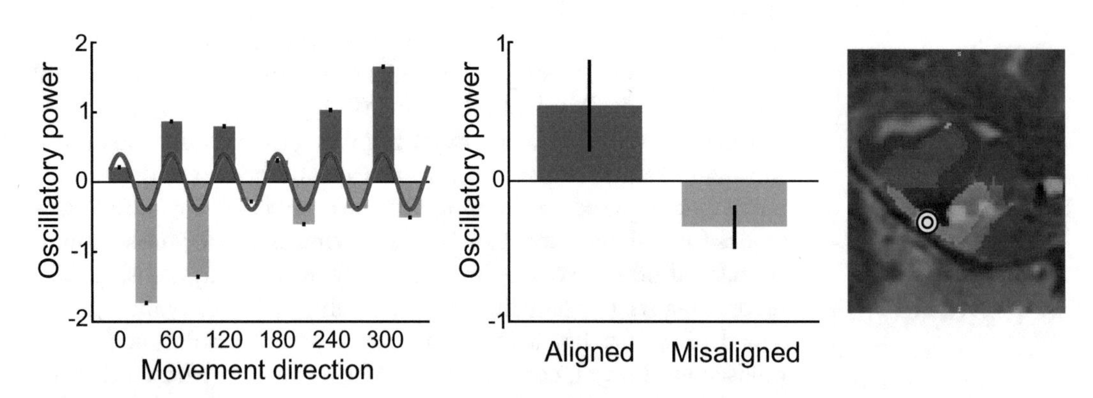

Fig. 10 An example electrode from Maidenbaum et al. [25] exhibiting significant hexadirectional modulation of the LFP signal by movement direction. Left: Higher oscillatory power (*z*-scored) in the "high theta" frequency band (5–8 Hz) is shown during movements aligned with one of the main axes of the grid, and the oscillatory power change for different movement direction bins follows the expected sinusoidal model (illustrated in red). Middle: Significantly higher theta power for movements aligned versus misaligned with the grid axis, averaged across all directional bins. Right: Illustration of the implanted electrode's location (circular marker) in the entorhinal cortex, shown on an fMRI-CT overlay in coronal view. Figure adapted from Maidenbaum et al. [25]

Given that MEG does not allow for the precise localization of signals in subcortical structures such as the entorhinal cortex or the hippocampus (for a review see [27]), the results reported in Staudigl et al. [26] were source localized to the anterior medial temporal lobe, but not the entorhinal cortex per se.

3.6 Circular–Linear Correlation Method

As described in the previous section, Maidenbaum et al. [25] adapted the general analysis method to detect grid-cell-like representations using LFP data and found hexadirectional modulation of the signal in the human entorhinal cortex, mirroring earlier results in the fMRI BOLD signal. To validate their findings, they also applied an alternative analysis technique comprising a circular–linear model [28, 29]. In this model, they correlated the circular distribution of movement directions with a matching linear distribution of oscillatory power. This method has the advantage that the dataset does not need to be split into separate estimation and test datasets and may therefore allow for increased statistical power in the estimation of grid orientations. Moreover, it avoids random biases arising from the selection of different events for estimation and test dataset, respectively. Using this method, they were able to validate the results obtained with the general analysis method, as the results were largely similar across the two different analysis approaches. While this method has only been applied in the study of Maidenbaum et al. [25] to intracranial LFP signals, the same analysis logic could readily be applied also to fMRI recordings.

4 Current State of Knowledge and Future Directions

In this final section, we provide a brief overview as to the current findings regarding grid cell coding in humans, before highlighting future directions for the field. Consistent with the rodent literature, grid-cell-like representations in humans were first demonstrated using spatial navigation paradigms [12, 16, 17]. In contrast to rodents, nonhuman primates and humans have foveal vision meaning that they are better able to visually explore the world around them. Consequently, the same neural mechanisms underlying spatial navigation in rodents may also underpin the visual exploration of stimuli in primates. Evidence of grid cell activation during visual exploration was first demonstrated in monkeys [7], where a population of neurons in the entorhinal cortex showed a firing pattern characteristic of grid cells while the animal visually explored pictures of scenes. These findings were replicated in humans using both MEG [26] and fMRI [21, 30]. In summary, grid-cell-like representations are not limited to one's physical location in the environment, but appear to extend also to the visual exploration of the world around us [31].

There is evidence also that grid-cell-like representations support more abstract spatial processing. For example, Horner et al. [15] demonstrated grid-cell-like representations when participants were required to imagine a trajectory in a virtual environment, even in the absence of visual input. Similarly, Bellmund et al. [22] found that simply imagining the direction between two locations in a virtual environment was associated with a sixfold symmetry. Furthermore, grid cells may also provide the cognitive framework to encode more abstract properties. Aronov et al. [32] found that grid cells in rodents also mapped the relationship between different auditory frequencies. Consistent with these data, in human fMRI Constantinescu et al. [14] manipulated the visual features of a pictorial stimulus along two dimensions and found evidence of grid-cell-like representations when participants were required to mentally navigate this conceptual space. These data support recent accounts that have suggested spatially tuned neural populations may underpin the organization of knowledge in humans [2].

The behavioral relevance of grid-cell-like representations has also been demonstrated. For example, inter-individual differences in properties of grid-cell-like representations are associated with behavioral differences in navigation abilities, such as spatial memory and path integration performance [12, 17]. Given that changes in navigational abilities are among the earliest symptoms of Alzheimer's disease ([33–35], for a review see [36]), these findings also highlight the potential clinical relevance of grid-cell-like representations to explain normative or pathological age-related decline in navigational abilities. Consistent with this notion, reductions in the magnitude of grid-cell-like representations have been found in healthy older adults, as well as in healthy young adults at increased genetic risk for Alzheimer's disease [16, 17]. A promising future research approach, therefore, will be to explore whether grid-cell-like representations can provide a biomarker for the integrity of the grid cell system and aid early detection of Alzheimer's disease by identifying the earliest neuropathological changes before the onset of cognitive deficits.

Several questions remain also regarding the anatomical location, and organization, of the grid cell system in humans. Given that grid cells were first discovered in the entorhinal cortex, both human and animal studies have focused on this region when analyzing data. In humans, however, grid-cell-like representations have been reported in cortical regions outside of the entorhinal cortex [12, 14]. Moreover, in rodents there is evidence that grid cells form a number of separate modules, with cells within a module sharing more similar properties (e.g., grid orientations) relative to cells in different modules [37]. Lastly, rodent studies also suggest that grid cells are not uniformly distributed across different cortical layers, but that they are most populous in more superficial layers (e.g., layer II of the entorhinal cortex in rats). Recent advances in analyzing layer-specific differences in human fMRI data, as well as fMRI imaging with high field strengths (e.g., 7T) that allow for an increased spatial resolution, may, therefore, help elucidate the precise anatomical location and the modular nature of the grid cell system in humans.

In sum, although we have seen rapid progress in the characterization of grid-cell-like representations in humans, and their behavioral relevance, there remain a number of key questions, which are outlined in Box 1. With the help of this chapter, and the availability of open-source software for the analysis of fMRI data [13], we hope to provide researchers around the world with the tools to address these fundamental questions regarding the grid cell system in humans.

Box 1

- How do grid cells support higher-order cognitive functions beyond pure spatial navigation?
- Do distinct populations of grid cells support different cognitive functions, such as spatial navigation versus navigation in visual or conceptual space?
- How might simultaneous representations of different spaces be coordinated, for example, during movements in the environment while visually exploring the space around us?
- Can grid cell changes due to neuropathological processes explain symptoms of disorientation and navigational decline, as seen, for example, in Alzheimer's disease?
- Can measures of grid cell activation serve as a biomarker to facilitate early detection of neurodegenerative diseases such as Alzheimer's disease?
- Does the grid cell system operate in a network-like fashion spanning different brain regions and different modules, and what are the precise anatomical locations of this network's components?
- Are locations of grid cells in the human brain limited to specific cortical layers?

References

1. Moser EI, Kropff E, Moser M-B (2008) Place cells, grid cells, and the brain's spatial representation system. Annu Rev Neurosci 31:69–89
2. Bellmund JLS, Gärdenfors P, Moser EI, Doeller CF (2018) Navigating cognition: spatial codes for human thinking. Science 362. pii: eaat6766:eaat6766
3. O'Keefe J, Dostrovsky J (1971) The hippocampus as a spatial map. Preliminary evidence from unit activity in the freely-moving rat. Brain Res 34:171–175
4. Hafting T, Fyhn M, Molden S et al (2005) Microstructure of a spatial map in the entorhinal cortex. Nature 436:801–806
5. Boccara CN, Sargolini F, Thoresen VH et al (2010) Grid cells in pre- and parasubiculum. Nat Neurosci 13:987–994
6. Yartsev MM, Witter MP, Ulanovsky N (2011) Grid cells without theta oscillations in the entorhinal cortex of bats. Nature 479:103–107
7. Killian NJ, Jutras MJ, Buffalo EA (2012) A map of visual space in the primate entorhinal cortex. Nature 491:761–764
8. Jacobs J, Weidemann CT, Miller JF et al (2013) Direct recordings of grid-like neuronal activity in human spatial navigation. Nat Neurosci 16:1188–1190
9. Barry C, Hayman R, Burgess N, Jeffery KJ (2007) Experience-dependent rescaling of entorhinal grids. Nat Neurosci 10:682–684
10. Sargolini F, Fyhn M, Hafting T et al (2006) Conjunctive representation of position, direction, and velocity in entorhinal cortex. Science 312:758–762
11. Kropff E, Carmichael JE, Moser M-B, Moser EI (2015) Speed cells in the medial entorhinal cortex. Nature 523:419–424
12. Doeller CF, Barry C, Burgess N (2010) Evidence for grid cells in a human memory network. Nature 463:657–661
13. Stangl M, Shine J, Wolbers T (2017) The GridCAT: a toolbox for automated analysis of human grid cell codes in fMRI. Front Neuroinform 11:47
14. Constantinescu AO, O'Reilly JX, Behrens TEJ (2016) Organizing conceptual knowledge in humans with a gridlike code. Science 352:1464–1468
15. Horner AJ, Bisby JA, Zotow E et al (2016) Grid-like processing of imagined navigation. Curr Biol 26:842–847
16. Kunz L, Schröder TN, Lee H et al (2015) Reduced grid-cell-like representations in adults at genetic risk for Alzheimer's disease. Science 350:430–433
17. Stangl M, Achtzehn J, Huber K et al (2018) Compromised grid-cell-like representations in old age as a key mechanism to explain age-related navigational deficits. Curr Biol 28:1108–1115
18. Berron D, Vieweg P, Hochkeppler A et al (2017) A protocol for manual segmentation of medial temporal lobe subregions in 7Tesla MRI. Neuroimage Clin 15:466–482
19. Yushkevich PA, Piven J, Hazlett HC et al (2006) User-guided 3D active contour segmentation of anatomical structures: significantly improved efficiency and reliability. NeuroImage 31:1116–1128
20. Yushkevich PA, Pluta JB, Wang H et al (2015) Automated volumetry and regional thickness analysis of hippocampal subfields and medial temporal cortical structures in mild cognitive impairment. Hum Brain Mapp 36:258–287
21. Nau M, Navarro Schröder T, Bellmund JLS, Doeller CF (2018) Hexadirectional coding of visual space in human entorhinal cortex. Nat Neurosci 21:188–190
22. Bellmund JL, Deuker L, Navarro Schröder T, Doeller CF (2016) Grid-cell representations in mental simulation. elife 5:12897–12901
23. Kriegeskorte N, Mur M, Bandettini P (2008) Representational similarity analysis—connecting the branches of systems neuroscience. Front Syst Neurosci 2:4
24. Chen D, Kunz L, Wang W et al (2018) Hexadirectional modulation of theta power in human entorhinal cortex during spatial navigation. Curr Biol 28:3310–3315.e4
25. Maidenbaum S, Miller J, Stein JM, Jacobs J (2018) Grid-like hexadirectional modulation of human entorhinal theta oscillations. Proc Natl Acad Sci 115:10798–10803
26. Staudigl T, Leszczynski M, Jacobs J et al (2018) Hexadirectional modulation of high-frequency electrophysiological activity in the human anterior medial temporal lobe maps visual space. Curr Biol 28:3325–3329.e4
27. Pu Y, Cheyne DO, Cornwell BR, Johnson BW (2018) Non-invasive investigation of human hippocampal rhythms using magnetoencephalography: a review. Front Neurosci 12:273
28. Fisher NI (1995) Statistical analysis of circular data. Cambridge University Press, Cambridge
29. Berens P (2009) CircStat: a MATLAB toolbox for circular statistics. J Stat Softw 31:1–21

30. Julian JB, Keinath AT, Frazzetta G, Epstein RA (2018) Human entorhinal cortex represents visual space using a boundary-anchored grid. Nat Neurosci 21:191–194
31. Nau M, Julian JB, Doeller CF (2018) How the brain's navigation system shapes our visual experience. Trends Cogn Sci 22:810–825
32. Aronov D, Nevers R, Tank DW (2017) Mapping of a non-spatial dimension by the hippocampal-entorhinal circuit. Nature 543:719–722
33. Hort J, Laczó J, Vyhnálek M et al (2007) Spatial navigation deficit in amnestic mild cognitive impairment. Proc Natl Acad Sci 104:4042–4047
34. Laczó J, Andel R, Vyhnalek M et al (2010) Human analogue of the Morris water maze for testing subjects at risk of Alzheimer's disease. Neurodegener Dis 7:148–152
35. Mokrisova I, Laczo J, Andel R et al (2016) Real-space path integration is impaired in Alzheimer's disease and mild cognitive impairment. Behav Brain Res 307:150–158
36. Coughlan G, Laczó J, Hort J et al (2018) Spatial navigation deficits—overlooked cognitive marker for preclinical Alzheimer disease? Nat Rev Neurol 14:496–506
37. Stensola H, Stensola T, Solstad T et al (2012) The entorhinal grid map is discretized. Nature 492:72–78
38. Kesner R (2013) Neurobiological foundations of an attribute model of memory. Comp Cogn Behav Rev 8:29–59
39. Grieves RM, Jeffery KJ (2017) The representation of space in the brain. Behav Process 135:113–131

Neuromethods (2020) 151: 281–289
DOI 10.1007/7657_2019_25
© Springer Science+Business Media, LLC 2019
Published online: 29 August 2019

Hyperaligning Neural Representational Spaces

J. Swaroop Guntupalli

Abstract

Neural populations in different brain regions represent different domains of information, but accounting for how population responses in homologous regions in different brains encode the same fine distinctions has been elusive. Common models of cortical functional architectures based on anatomy account for coarse regional topography that encode coarse-scale information such as visual versus auditory stimulation or perception of animate versus inanimate entities but fail to account for fine-scale information that captures distinctions between two songs or two insects. We proposed a method of functional alignment called hyperalignment that aligned high-dimensional neural representational spaces to derive a new common model of cortical functional architecture. This model is based on a common representational space rather than a common cortical topography. By modeling functional topographies as weighted sums of overlapping topographic basis functions, our model also accounts for coarse-scale regional topography and goes further to capture fine-scale topographies that coexist with coarse topographies and carry finer distinctions. In this chapter we present steps for an experimenter to use hyperalignment in their own study to derive a common model representational space and perform analyses in that space.

Keywords Hyperalignment, Neural decoding, MVPA, RSA, Neuroimaging, Functional alignment, Representational spaces

1 Introduction

Brain represents information in perceptions, thoughts, and knowledge as activity in populations of neurons. Technologies such as functional magnetic resonance imaging (fMRI), electrophysiology, electroencephalography (EEG), magnetoencephalography (MEG) let us measure the neural activity, and multivariate pattern analysis (MVPA) lets us decode the information represented in the response patterns [1, 2]. MVPA models the cortical representation of information as response vectors in a high-dimensional space in which each dimension is a local measure of neural activity, such as a voxel in fMRI or a channel in EEG and MEG or a unit in electrophysiology. Univariate methods that typically relate the average response in a brain region (a single feature) to behavior do not account for the fine-scale distributed information stored in the response patterns within that region. This approach provides a reasonable correspondence across brains when aligned based on anatomical features. However, anatomy-based alignment is inadequate for one-to-one correspondence across brains of individual features such as voxels (in contrast to the whole region). This is evident in the discrepancy between classification accuracies of within-subject and between-subject classification analyses [3–5]. Representational similarity analysis (RSA) [6] provides a clever work around to feature correspondence by comparing the similarities among two sets of vectors in these spaces instead of directly comparing vectors. However, this does not generalize to new stimulus sets or allow for building common decoding or encoding models.

Methods for aligning two-dimensional cortical surfaces based on functional responses in addition to sulcal curvature have been proposed to address this discrepancy but fell short [7–8]. We proposed a novel functional alignment method—hyperalignment—that aligned high-dimensional feature spaces to a common space and showed that between-subject classification performance can be as good as within-subject classification performance and that our method aligned the fine-scale representations embedded in the regions [3–5]. Hyperalignment allows for building models of representation that are common across subjects and capture fine-grained distinctions in the information represented at a fine spatial scale that facilitates building common decoding and encoding models [9–11]. In contrast, areal models of the functional representation are on a spatial scale that captures only a coarser topography of dissociable, more global functions, such as visual versus auditory stimulation or perception of animate versus inanimate entities.

Hyperalignment builds a common high-dimensional model of neural representational spaces that is based on a large set of response-tuning basis functions that are shared across brains (Fig. 1). Each shared basis function is associated with individual-specific topographic basis functions. Once individuals' data are transformed into the model dimensions with common response-tuning basis functions, patterns of response across dimensions afford between-subject multivariate pattern classification with high accuracy. This model is a radical departure from previous models of cortical functional architecture as it is based on a common representational space rather than a common cortical topography. By modeling functional topographies as weighted sums of overlapping topographic basis functions, our model also accounts for coarse areal topography and goes further to capture fine-scale topographies that coexist with coarse topographies and carry finer distinctions. In our previous studies, we derived a common model space based on neural responses to complex, dynamic stimuli, movies that provide a broad sampling of visual, auditory, and social percepts. Results show that the common model space accounts for population codes in occipital, temporal, parietal, and prefrontal cortices, indicating that the principles underlying the model are valid for neural representations of widely divergent domains of information [4, 5]. In addition, we showed that a common model

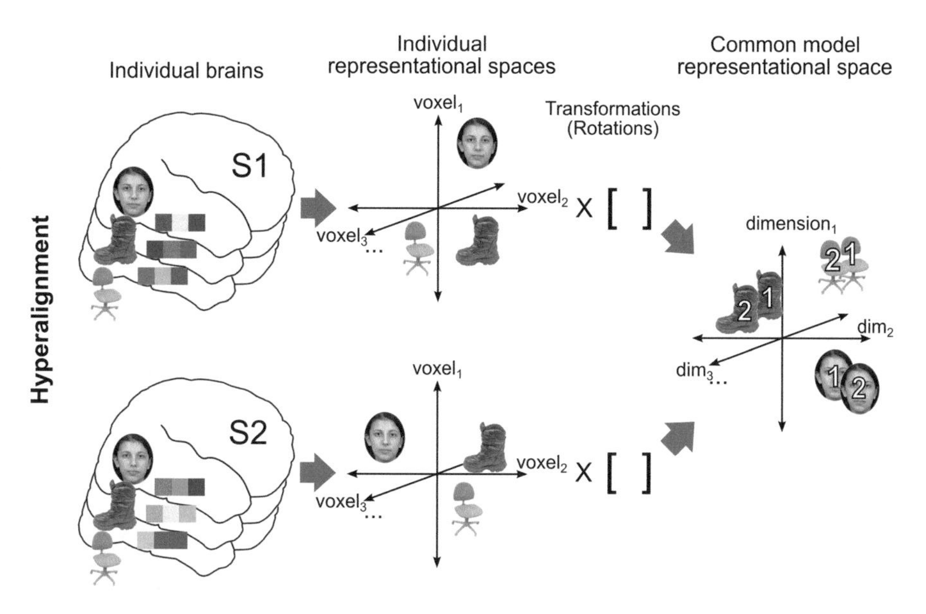

Fig. 1 Schematic of hyperalignment. Neural response patterns encoding information about stimuli such as faces, shoes, and chair can be decoded using MVPA in different subjects in their own representational spaces. However, between-subject classification is not possible due to inadequate alignment of individual features across subjects. Hyperalignment aligns individual subject representational spaces into a common space where features are aligned facilitating common models of encoding and decoding. Transformations for alignment are estimated from a separate dataset such as a movie that provides a broad sampling of neural response patterns to optimizing this alignment

space based on neural responses to a limited set of stimuli such as a few items from a specific category (e.g., six different species of animal) generalizes well to that subcategory but not to a diverse stimulus set such as a movie [3].

In this chapter, we present steps to use hyperalignment in any experimental setup detailing recommendations and requirements for different steps in the pipeline. Our end goal is for the experimenter to be able to analyze multi-subject data in a common representational space that preserves fine-scale information needed for MVPA.

2 Methods

Hyperalignment jointly estimates a common model of neural representational space and the transformations that align each individual subject's neural representational space to that common model. In this section, we will detail the steps involved in running hyperalignment pipeline in an fMRI experiment (*see* **Note 1**) from data acquisition, preprocessing, parameter estimation, and validation. The specifics listed are recommendations, and the experimenter has freedom to choose their preferred variations that do not violate

Table 1
Standard pipeline for hyperalignment-based analyses

Hyperalignment pipeline
1. Anatomical and functional data acquisition
2. Data preprocessing
(a) Preprocess anatomical data
(b) Extract surfaces
(c) Align surfaces across all subjects
(d) Preprocess functional data
(e) Align all functional data to a common anatomical template that correspond to surfaces
(f) If applicable, extract function ROIs and align them to the same anatomical template and extract the functional data only from these ROIs
(g) Split data into training and testing sets
3. Hyperalignment parameter estimation
(a) Zero-center and scale functional data by z-scoring each feature
4. Hyperalignment of new data
5. Analyses in the common space

the assumptions detailed in respective sections. See Table 1 for the pipeline overview. Following are typically required for data acquisition and analysis: 3T or higher-capacity MRI scanner for data acquisition; Linux or Mac workstation for processing and analysis; neuroimaging software such as AFNI, FSL, FreeSurfer, fMRIPrep, etc. [12–15]; and PyMVPA for hyperalignment [16].

2.1 Data Acquisition

As discussed before, MVPA treats each voxel or feature as an independent variable, which means the dimensionality of an individual's representational space is in the order of tens of thousands in a typical fMRI dataset (restricting to a region of interest or a dimensionality reduction step can reduce this number, but it is typically still in the order of hundreds to thousands). This necessitates a need for adequate high-quality fMRI data, preferably of at least 30 min duration during which participants performed an attentive task that engages multiple brain systems or the areas of interest to the study.

2.1.1 MRI Parameters

Functional and anatomical data should be acquired in a scanner with 3T or higher field strength and a head coil system that provides high-quality functional data with as few artifacts as possible. We and others have successfully performed hyperalignment on fMRI data from a variety of scanner systems and between two different systems [3, 5], so there is no hard and fast requirement on the specifics but a good quality dataset is always preferred.

Typical dataset should contain a high-resolution T1-weighted anatomical scan for each subject. Voxel resolution for anatomical scan should be 1mm or smaller. Functional imaging should be acquired at a resolution of 3mm or smaller. Time repetition (TR) for fMRI should be on the order of 1–3 s that provides good signal-to-noise ratio depending on the scanner system.

2.1.2 Task

Functional imaging data should be acquired while the subjects perform an attentive task that engages as many brain systems as possible or system of interest to the study. For response-based hyperalignment, it is critical for all participants to have same number of samples corresponding to the same stimulus or behavior. An engaging naturalistic movie viewing is a good example of a task that engages many brain systems while synchronizing this experience across participants.

For data acquired during a task in block design or event-related design, with each individual experiencing the stimuli in randomized order, make sure the processed data is reordered such that each temporal sample corresponds to same event across participants. Connectivity-based hyperalignment relaxes this requirement. However, the assumption of similar brain systems being engaged during data acquisition across participants should still be honored. If a functional localization task is available for the function of interest to the experiment, it can be added to the task for defining regions of interest (ROIs).

For a rich, naturalistic, dynamic stimulation like movie viewing, 30 min has been shown to be adequate in our previous studies, and for task specific stimulation, we and others have used leave-one-run-out cross-validation scheme to use as much data as possible to estimate hyperalignment parameters and use the rest for cross-validation (*see* **Note 2**). For connectivity-based hyperalignment, we have successfully used ~15 min of resting state data from the human connectome project (HCP) which is considered to be of high quality.

2.2 Preprocessing

2.2.1 Anatomical Data

Anatomical scans should be corrected for field bias, and cortical surfaces should be extracted using FreeSurfer and aligned to a standard surface such as fsaverage in FreeSurfer. Surface meshes should be resampled to match a standard mesh with same number of nodes, and a given node should correspond to the same anatomical location across subjects as best as possible based on the cortical folding, sulcal curvature, and any other markers available in the study (see HCP data for more details on different markers and their usage in alignment) [17]. Better alignment of anatomy gives a good start for parameter estimation. Aligning to a standard dataset meshes such as those available in FreeSurfer or HCP will give the user access to published parcellations of brain into regions of interest [17].

2.2.2 Functional Data

Functional data should be preprocessed using any software of choice such as AFNI, FSL, FreeSurfer, fMRIPrep, etc. Preprocessing should include correcting for head motion and aligning the data to the same reference template brain of that individual, the same template brain that is aligned to the surface mesh in the previous step. Goal of this alignment step is to facilitate the use of surface searchlight method which needs correspondence between surface and volume to provide voxels within a specified distance on the cortical surface (instead of volume) to a surface node [18–20] for whole-cortex hyperalignment [4]. If any addition experiments such as functional localization are included, those should be aligned to the same template brain of the individual. If the experimenter only wishes to perform hyperalignment within a functionally defined ROI, then aligning all the data of an individual to one brain template for that individual is sufficient, even though we recommend aligning to a common reference template such as MNI. It is critical for the same voxel in different experiments to correspond to the same location as best as possible across studies within an individual brain. Removing contributions from signals of no interest such as head movement, breathing, heart beats, linear and quadratic trends, and average response from ventricles, white matter, or regions outside the brain etc. is recommended.

For data from experiments with a known design matrix (unlike resting state or uncoded movie), fitting a GLM and removing signal unrelated to the design are also recommended. For movie-like stimuli, the residual time series of each surface vertex in each run was normalized to zero mean and unit variance.

2.3 Parameter Estimation

Hyperalignment can be performed over the whole brain or within ROIs to the study.

Whole brain hyperalignment performs parameter estimation within individual searchlights evenly spaced across the cortex [18–20] and covering all voxels and aggregates the parameters across searchlights to create a single transformation. It takes more computational resources, and the functional data used for parameter estimation is expected to be collected during tasks that engage multiple brain systems.

Response-based hyperalignment uses the data provided after preprocessing step by treating each voxel's responses as its fingerprint. Connectivity-based hyperalignment first computes the connectivities from every voxel or feature to regions spread across the brain and uses this connectivity as the voxel's fingerprint [5].

Core of hyperalignment algorithm is to estimate a transformation between two sets of features. In classic response-based hyperalignment, we constrain this transformation to be a rotation and use Procrustes transformation [21] to rotate each individual's representation space into a common space constructed from the average of all individuals in an iterative procedure. Data are centered and scaled by independently removing the mean and scaling to unit variance for each feature prior to estimating rotation matrices.

Preprocessed data should be loaded from ROIs in each individual into a data matrix Di of shape $Ns \times Nv$, where Ns is the number of samples and Nv is the number of voxels or features. If the ROIs are of different sizes across individuals, we suggest using a feature selection method to select an equal number of top Nv features based on their SNR or between-subject correlation metric proposed in Haxby et al. [3].

We recommend using PyMVPA for running hyperalignment. Estimating hyperalignment transformations and applying the transformations to a new dataset can be performed in PyMVPA once the data is loaded in PyMVPA [22].

Python Code

```
from mvpa2.suite import *
# Load your data for estimating hyperalignment parameters
# into datasets_est, and for validation analyses into datasets_val.
hyperalignment = Hyperalignment()
mappers = hyperalignment(datasets_est)
datasets_est_in_common_space = [m.forward(ds_) for m, ds_ in
    zip(mappers, datasets_est)]
datasets_val_in_common_space = [m.forward(ds_) for m, ds_ in
    zip(mappers, datasets_val)]
# Save the data in common space and run your analyses of choice in
    common space.
```

2.4 Analyses in Common Space

Once the transformation matrices are estimated, new data from the same individuals can be mapped into a common representational space. Validation data should be preprocessed and aligned to that individual's brain template used for estimating the parameters in the previous step. Data is also expected to be centered and scaled in a similar way resulting in a data matrix Di of shape $Ns \times Nv$. This data is then aligned by applying the transformation Ri for that subject resulting in aligned data Ci ($Ci = Di \times Ri$).

Data in common space can be treated as if each feature in common space has correspondence across all subjects unlike each voxel in original individual's brain space.

Data in common space can be used for between-subject MVPA, such as between-subject classification, between-subject encoding models, inter-subject correlation of response patterns, inter-subject correlation of connectivity patterns, or building a common encoding model. Having access to more data across multiple subjects should lead to better encoding models overall. Decoding individual differences in behavior has also been shown to improve in common space [9].

3 Notes

1. Even though we focus on fMRI data, preliminary work on hyperalignment has been shown to work on other modalities such as MEG and EEG.
2. Hyperalignment procedure has many parameters to optimize to align the data, so it is prone to overfitting to that dataset. It is therefore very important to not use the validation dataset in estimating the parameters.

References

1. Haxby JV, Gobbini MI, Furey ML et al (2001) Distributed and overlapping representations of faces and objects in ventral temporal cortex. Science 293:2425–2430. https://doi.org/10.1126/science.1063736
2. Haxby JV, Connolly AC, Guntupalli JS (2014) Decoding neural representational spaces using multivariate pattern analysis. Annu Rev Neurosci 37:435–456. https://doi.org/10.1146/annurev-neuro-062012-170325
3. Haxby JV, Guntupalli JS, Connolly AC et al (2011) A common, high-dimensional model of the representational space in human ventral temporal cortex. Neuron 72:404–416. https://doi.org/10.1016/j.neuron.2011.08.026
4. Guntupalli JS, Hanke M, Halchenko YO et al (2016) A model of representational spaces in human cortex. Cereb Cortex 26:2919–2934. https://doi.org/10.1093/cercor/bhw068
5. Guntupalli JS, Feilong M, Haxby JV (2018) A computational model of shared fine-scale structure in the human connectome. PLoS Comput Biol 14:e1006120. https://doi.org/10.1371/journal.pcbi.1006120
6. Kriegeskorte N, Mur M, Bandettini P (2008) Representational similarity analysis—connecting the branches of systems neuroscience. Front Syst Neurosci 2. https://doi.org/10.3389/neuro.06.004.2008
7. Sabuncu MR, Singer BD, Conroy B et al (2010) Function-based intersubject alignment of human cortical anatomy. Cereb Cortex 20:130–140. https://doi.org/10.1093/cercor/bhp085
8. Conroy BR, Singer BD, Guntupalli JS et al (2013) Inter-subject alignment of human cortical anatomy using functional connectivity. Neuroimage 81:400–411. https://doi.org/10.1016/j.neuroimage.2013.05.009
9. Feilong M, Nastase SA, Guntupalli JS, Haxby JV (2018) Reliable individual differences in fine-grained cortical functional architecture. Neuroimage 183:375–386. https://doi.org/10.1016/j.neuroimage.2018.08.029
10. Taschereau-Dumouchel V, Cortese A, Chiba T et al (2018) Towards an unconscious neural reinforcement intervention for common fears. Proc Natl Acad Sci U S A 115:3470–3475. https://doi.org/10.1073/pnas.1721572115
11. Naselaris T, Kay KN, Nishimoto S, Gallant JL (2011) Encoding and decoding in fMRI. Neuroimage 56:400–410. https://doi.org/10.1016/j.neuroimage.2010.07.073
12. Cox RW (1996) AFNI: software for analysis and visualization of functional magnetic resonance neuroimages. Comput Biomed Res 29:162–173
13. Smith SM, Jenkinson M, Woolrich MW et al (2004) Advances in functional and structural MR image analysis and implementation as FSL. Neuroimage 23:S208–S219. https://doi.org/10.1016/j.neuroimage.2004.07.051
14. Fischl B, Sereno MI, Tootell RBH, Dale AM (1999) High-resolution intersubject averaging and a coordinate system for the cortical surface. Hum Brain Mapp 8:272–284
15. Esteban O, Markiewicz CJ, Blair RW et al (2019) fMRIPrep: a robust preprocessing pipeline for functional MRI. Nat Methods 16:111. https://doi.org/10.1038/s41592-018-0235-4
16. Hanke M, Halchenko YO, Sederberg PB et al (2009) PyMVPA: a python toolbox for multivariate pattern analysis of fMRI data. Neuroinformatics 7:37–53. https://doi.org/10.1007/s12021-008-9041-y
17. Glasser MF, Coalson TS, Robinson EC et al (2016) A multi-modal parcellation of human cerebral cortex. Nature 536:171–178. https://doi.org/10.1038/nature18933
18. Kriegeskorte N, Goebel R, Bandettini P (2006) Information-based functional brain

mapping. Proc Natl Acad Sci U S A 103:3863–3868. https://doi.org/10.1073/pnas.0600244103

19. Chen Y, Namburi P, Elliott LT et al (2011) Cortical surface-based searchlight decoding. Neuroimage 56:582–592. https://doi.org/10.1016/j.neuroimage.2010.07.035
20. Oosterhof NN, Wiestler T, Downing PE, Diedrichsen J (2011) A comparison of volume-based and surface-based multi-voxel pattern analysis. Neuroimage 56:593–600. https://doi.org/10.1016/j.neuroimage.2010.04.270
21. Schönemann PH (1966) A generalized solution of the orthogonal procrustes problem. Psychometrika 31:1–10. https://doi.org/10.1007/BF02289451
22. Hyperalignment for between-subject analysis. PyMVPA User Manual. http://www.pymvpa.org/examples/hyperalignment.html. Accessed 31 Mar 2019

Neuromethods (2020) 151: 291–305
DOI 10.1007/7657_2019_31
© Springer Science+Business Media, LLC 2019
Published online: 2 October 2019

A Practical Guide to Functional Magnetic Resonance Imaging with Simultaneous Eye Tracking for Cognitive Neuroimaging Research

Michael Hanke, Sebastiaan Mathôt, Eduard Ort, Norman Peitek, Jörg Stadler, and Adina Wagner

Abstract

The simultaneous acquisition of functional magnetic resonance imaging (fMRI) with in-scanner eye tracking promises to combine the advantages of full-brain coverage of brain activity measurements with a fast and unobtrusive capture of eye movement behavior and attentional deployment. Despite its applicability to a wide variety of research questions, ranging from investigations of gaze control and attention guidance to the use of eye movement events as a response modality for gaze-contingent fMRI experiments, only few studies employ this kind of data acquisition. In this chapter we identify technical challenges, describe all necessary components and procedures for conducting such a study, and give practical advice on how these can be integrated in a common MRI laboratory setup. The chapter concludes with notes on the analysis of such datasets and summarizes key data properties and their implications of a joint analysis of fMRI and eye tracking data.

Keywords Functional magnetic resonance imaging, Eye tracking, Hardware, Gaze-contingent stimulation, Eye movement event detection, Pupillometry

1 Introduction

Both fMRI and eye tracking have been employed extensively in cognitive science research. However, usually these techniques are used in isolation, and their combination only occurs relatively infrequently in the literature. Possibly, practical difficulties counter the conceptual benefits of this combination, resulting in its underutilization. With its high temporal resolution of up to 2000 Hz, eye tracking is able to measure rapid movements of the eye and thus represents a fast behavioral measure that can complement the low temporal sampling rate of fMRI data acquisition.

The combination of eye tracking and fMRI can serve several purposes. It can be used to monitor eye movements and validate task compliance by verifying whether participants fixate a fixation point, correctly execute instructed eye movements, or stay awake throughout an experiment. In passive viewing scenarios, such as movie watching, when performance measures like task accuracy or reaction times are unsuitable, eye tracking is a candidate technique to assess attentiveness or fatigue. Online measurement of gaze location can also be used for gaze-contingent stimulation. For example, to prevent saccade-related brain activity from contaminating brain activity specific to a condition of interest, conventional methods rely on passively recording gaze and exclude problematic trials during data analysis, which can lead to a significant loss of data. By using a gaze-contingent manipulation, on the other hand, trials can be stalled or aborted online once the eyes moved and restarted if necessary, so that, ideally, more data can be retained (see Ort et al. [1] for an implementation of such a procedure). Gaze-contingent stimulation can also be used for simulating particular types of visual impairments, such as foveal vision loss [2].

More generally, eye tracking complements fMRI data with insights into higher cognitive processes. To name just a few applications, gaze position and its trajectories during natural viewing can serve as an indication for attentional deployment [3–5], pupil dilation or blink frequency has been associated with cognitive load and memory processes [6–8], and gaze path characteristics are a measure of language comprehension [9–11]. Finally, gaze and fMRI data can be used to investigate the neural basis of gaze control itself. Gaze-contingent paradigms even enable measurement of brain activity under conditions that would otherwise be unattainable, for example, studying neural processes of transsaccadic memory, which requires precise temporal manipulation of visual stimulation either before, during, or after a saccade (e.g., [12, 13]).

This chapter provides a practical guide for conducting studies with simultaneous fMRI and eye tracking acquisition. It focuses on (1) hardware setup, (2) implications for experiment paradigms and procedures, and (3) data preprocessing and analysis considerations, with the aim to provide the reader with a clear idea of what is needed to translate the theoretical advantages of this data-modality combination into practical assets that can be employed for cognitive science. This chapter does not provide information on the individual acquisition methods and assumes that the reader is familiar with design and procedures of fMRI studies and eye tracking experiments in isolation.

2 Methods

Compared to fMRI only, combined fMRI and eye tracking imposes additional constraints and requirements on setup and procedures which are outlined in the following sections.

2.1 Hardware Setup

In-scanner eye tracking requires additional MR-compatible hardware. Several vendors of eye tracking equipment offer add-ons to their products for this purpose. These typically comprise of an MR-compatible camera with a telephoto lens, an MR-compatible infrared (IR) light source, an IR-reflective replacement mirror for visual stimulation, and (optical) wiring to connect this hardware to the remaining standard components of an eye tracker system outside the shielded cabin of an MRI scanner. However, despite the availability of such kits, a fair bit of customization is generally required in order to maximize data quality and procedural simplicity. Key challenges are compatibility with existing equipment for visual stimulation and noninterference with the MRI scanner components. In many cases, required customization to a given scanner environment will involve a custom-built camera mount and a stimulation mirror contraption.

Positioning of the eye tracking camera and IR light source is arguably the most crucial aspect of the hardware setup. If the targeted installation can only be temporary, i.e., has to be moved between scans, reliable repositioning has to be possible and practiced. One solution is to build a dedicated frame structure that the eye tracker components attach to, and to only move the entire structure, while leaving any fine adjustments of camera focus and angle fixed. The frame's target position can be marked on the floor, and repositioning the camera setup is then achievable by non-experts. Figure 1a shows an example of an MR-compatible telephoto lens and illumination kit (SR Research Ltd., Mississauga, Ontario, Canada) mounted on a wooden frame structure standing at the back of a Philips Achieva dStream 3T MRI scanner.

Camera and IR spotlight must be placed as close to the light path of the visual stimulation as possible, while not interfering with it. Reliable eye tracking requires a clearly and constantly visible corneal reflection. The spatial constraints of the scanner environment (bore diameter, head coil shape) can make this challenging. The situation is compounded by the tendency of participants to partially close their eyes while lying on their back in the scanner bore. Given the variability of equipment, no general recommendation can be given other than to minimize the distance between camera, light source, and eye. In the example setup shown in Fig. 1a, the only viable positioning option was to use the small gap between the projection screen and the scanner bore to emit IR light and capture its reflection on the cornea. Figure 1b illustrates the resulting camera field of view. Due to the view from the above stimulation, the corneal reflection is barely visible below the eyelid. Increasing distance between IR light source and eye, as well as shadows cast by the head coil structure, can prevent a corneal reflection or reduce its image contrast. This would result in signal loss or reduced gaze coordinate precision, particularly for gaze locations at lower screen coordinates, as the eyelid tends to close

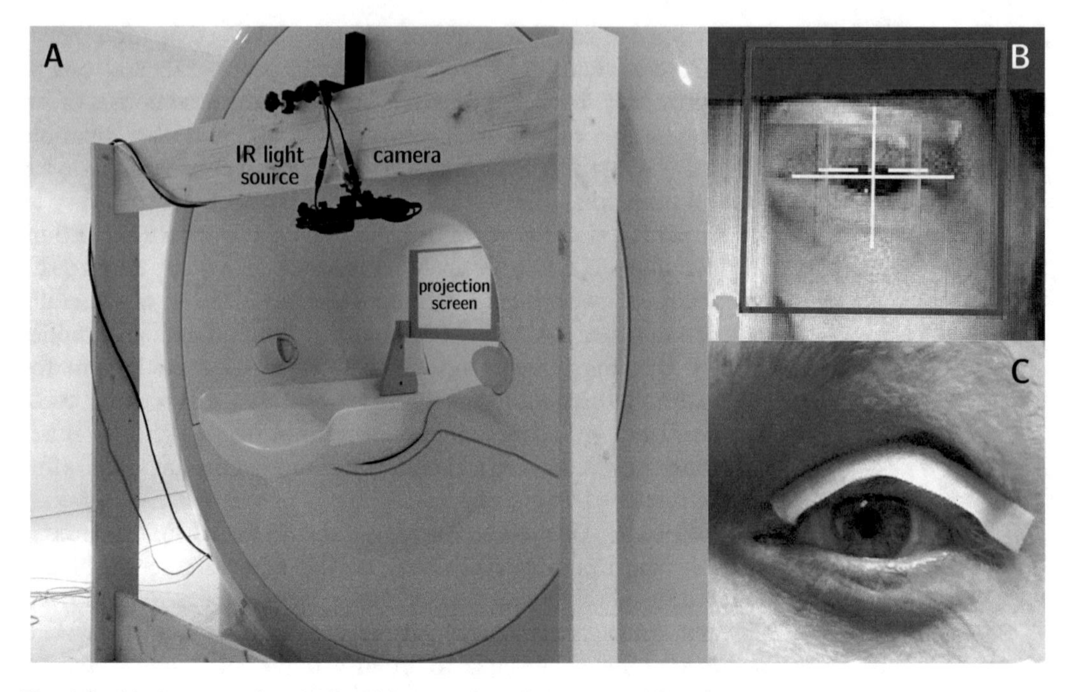

Fig. 1 Positioning examples. (**a**) Possible mounting of camera and light source in a back-projection configuration, outside the light path of the projector. Camera and spotlight utilize a small gap between the top of the screen and the scanner bore, with a camera-to-eye distance of about 1 m. (**b**) Camera angle and field of view of a participant's eye inside an MR head coil. Insufficient lighting and partial pupil occlusion by lid or eyelashes represent major challenges. (**c**) Taping eyelashes to the lid with medical tape can increase the usable measurement range

slightly with downward movement of the eyeball. In such cases, participants' long eyelashes can further reduce the usable measurement range. Taping the eyelashes to the lid with medical tape, as depicted in Fig. 1c, can be a pain-free, though slightly uncomfortable, option to mitigate this problem.

MR eye tracking kits typically include a replacement for the mirror through which participants view the visual stimulation. Standard mirrors cannot be used for two reasons: (1) they generally do not reflect (enough) IR light at the wavelengths used by the light sources, and (2) the reflective surface is at the back of the glass, leading to ghost images. Both aspects prevent or impair high-precision eye tracking, and consequently, a first-surface, IR-reflective mirror is required. It is possible that the provided mirror mount does not fit a particular head coil. In such cases it may be possible to use the provided mirror with an existing spare mount. However, in some cases the physical dimensions of the provided mirror will not afford proper positioning, and an alternative customized mirror will have to be obtained (see the notes on Sect. 3.5 for additional information).

2.2 Integration with Stimulation and Response Logging Equipment

Outside the scanner cabin, the eye tracker has to be connected to the existing hardware for stimulation and experiment control. A common fMRI setup includes a dedicated computer for delivering stimulation, and recording behavior responses, in addition to the MR scanner console machine. This additional computer typically has access to the trigger signal emitted by the MR scanner at the start of each volume acquisition, and experiment implementations use it to synchronize stimulation timing with the fMRI data acquisition. Simultaneous eye tracking requires the synchronization of a third time series, the continuously measured gaze coordinates, in relation to the MR signal acquisition and the stimulation and response timing. While pretty much all eye trackers offer the possibility to programmatically start a recording session, so that it can be initiated by the stimulus computer upon receiving the MR trigger signal, this may not be sufficient for achieving optimal synchronicity.

Two major sources of temporal uncertainty and instability exist in such a setup: (1) video presentation delay, i.e., the time from triggering a stimulation on the machine running an experiment implementation and its appearance in the MR scanner cabin, and (2) eye tracker response latency after the initial "start" signal. It is advisable to measure the video presentation delay using a photodiode over a longer period of time prior to a study or ideally during any acquisition. A photodiode can be placed on a corner of the screen inside the MR cabin, and a visual marker can be shown at this location for any significant event in an experiment. The diode is covered, such that this additional visual signal is not perceived by a participant. A reliable estimate of the visual presentation delay is crucial for synchronizing gaze coordinate time series and stimulation protocol. It is possible that there is no constant delay for a particular stimulation setup, especially when devices for video stream splitting or transcoding for optical signal transmission are used. In such cases, it may be that the video card in the computer running the experiment has no access to the real vertical synchronization signal of the projector or screen used for visual stimulation inside the MR cabin. This can lead to seemingly random timing fluctuations of a magnitude that can be relevant in experiments with critical timing requirements (see [14], Fig. 4a for an example).

If an eye tracker supports logging of custom events, it is recommended to use this functionality to create a duplicate protocol of stimulation, response, and acquisition events within the eye tracker log. This should include the information when the experiment control computer has received each MR acquisition trigger signal, as the most reliable source of timing information. Temporal variations of this signal in the eye tracker log can be used as an upper-bound estimate of the temporal precision of the acquired data in the context of a joint analysis. Experiment control computer and eye tracker should use a dedicated connection for message

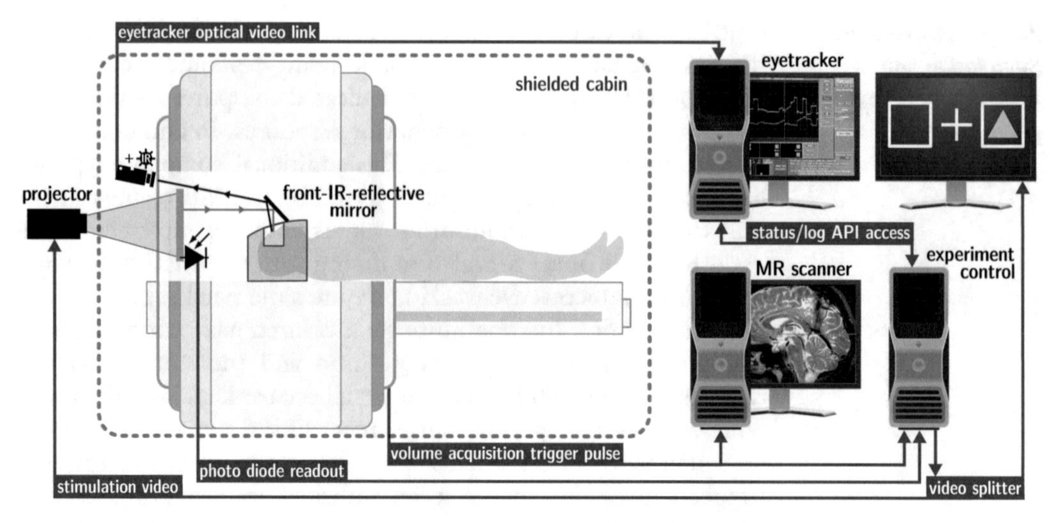

Fig. 2 Schematic depiction of hardware components and wiring setup for simultaneous fMRI and eye gaze recording

passing, such as a direct loopback network cable. A connection over a shared medium, such as a standard LAN, has the potential to introduce additional, load-dependent timing inaccuracies. Figure 2 depicts all involved hardware components and their interconnections.

2.3 Procedural Considerations

The addition of eye tracking to an fMRI study requires procedures to be adjusted to accommodate two additional steps: (1) participant positioning for optimal camera visibility and (2) the eye tracker calibration for converting corneal reflection geometry into screen coordinates.

Given the machine-aided, identical positioning of the head coil within the scanner bore, camera angle and focus adjustment should not be necessary after the initial setup. Consequently, physical stimulus dimensions and viewing distance only have to be measured once for a configuration of the eye tracker measurement parameters. However, each participant has to be carefully positioned within the head coil for optimal data quality. This positioning is complicated in comparison to a normal fMRI study, because the optimal configuration may only be assessable while the participant is inside the scanner, as opposed to the normal positioning routine that takes place while the patient table is moved all the way outside the scanner. Lighting and camera visibility of the eye can only be confirmed when the head is in its final measurement position, and a participant may need to be instructed to slightly move in or out of the head coil or tilt their head in order to achieve a good compromise of screen visibility for the participant and eye visibility for the tracker camera (see Fig. 1b and the note on Sect. 3.6). For participants that need MR-compatible glasses, and no contact lenses are available, the positioning procedure is further complicated, as the angle of the glasses has to be adjusted to avoid reflection of the IR spotlight on the lens surface in the vicinity of the tracked eye. This need for in-scanner adjustments directly counteracts the desire for immobility of a participant's head for optimal MR image quality. It is important that participants achieve a stable and relaxed head position, as otherwise movement during the experiment is a likely consequence, with implied negative consequences for MR image quality and eye gaze measurement precision. This initial positioning can take a considerable amount of time, should be practiced, and accounted for in the scan time allotment.

Before measurement can start, the eye tracker has to be calibrated. This is generally done by a short sequence during which participants have to fixate a target marker in various locations of the screen. Especially in the case of a suboptimal positioning, calibration can fail. Depending on the specific procedure, participants might be motivated to adjust their head position as a result. Therefore, this initial calibration should take place before the start of any MRI scan.

It is advisable to validate the quality of an eye tracker calibration, often using a target dot sequence identical to the one used for calibration, in order to compare the measure coordinates with the target coordinates. A validation immediately after calibration can be used to confirm a suitable setup, and validation at the end of the measurement can be used to assess degradation of measurements over time. In long-running experiments, it can be beneficial to use breaks for repeated reassessment of data quality and on-demand recalibration. If only a subset of the screen area is required for a particular experiment (e.g., a paradigm with a center fixation task), calibration can be limited to the relevant locations, in order to facilitate this step and avoid calibration failure due to measurement problems in irrelevant parts of the screen.

2.4 Gaze-Contingent Stimulation

With a hardware setup as depicted in Fig. 2 and an eye tracking system that supports online reporting of gaze coordinates or eye movement events, it is possible to conduct experiments with gaze-contingent stimulation with no further adjustments other than programming the experimental procedure. Gaze-contingent stimulation uses information on gaze location or eye movement events (see note on Sect. 3.1) to control the visual display, either to compensate for eye movements (e.g., gaze-centered visual stimulation) or as a dedicated behavioral response modality (e.g., response delivery via saccades to target locations). To guarantee a seamless experience, it is critical that the temporal latency of the combined system for eye tracking and stimulation, including all necessary processing for gaze position sampling, event parsing, and stimulation updating, is low and stable. For the same reason, the eye tracker has to be able to sample at a high frequency. Otherwise changes to the visual display might not be applied with sufficient temporal precision, and participants would be affected by that delay [15, 16].

Likewise, the precision of gaze position measurement should be sufficiently high given the requirements of an experiment. Therefore, experimenters should try to reduce the calibration error as much as possible (ideally below ~0.5°) and rerun the calibration whenever the precision starts to deteriorate. If the spatial precision of a measurement setup is insufficient, the spatial layout of the visual stimulation should be adjusted to avoid undesirable response artifacts. For example, the area for saccade targets can be increased to avoid repeated, failed response attempts that disrupt the flow of an experiment and distract from an actual task. It is advisable to test the effectiveness and reliability of gaze-contingent stimulation paradigm implementations upfront on a range of participants, as interindividual measurement quality differences may introduce undesirable confounds in error rates or reaction time measurements.

An advantage of in-scanner, gaze-contingent stimulation is that relatively minimal head motion occurs and a calibration may remain remarkably stable throughout an entire scanning session. However, a downside of the participants' lying position is that looking straight ahead can be strenuous (the natural fixation when lying down is slightly shifted toward the lower half of the visual field). Therefore, saccades to the top half of the visual field are more effortful and feel less natural and are thus less accurate than when sitting up. Similarly, and as mentioned before, gaze positions in the lower parts of the visual field can be more prone to signal loss, because the eyelid or eyelashes may obstruct parts of the pupil. These constraints have to be taken into account when developing gaze-contingent stimulation paradigms for combined eye tracking and fMRI studies.

3 Notes

3.1 Detection of Eye Movement Events

The raw eye gaze coordinates provided by an eye tracker are rarely used "as is." Instead, the main focus often lies on particular eye movement events, such as saccades, smooth pursuit-type tracking movements, or fixations, that are made in response to a particular stimulation or experimental condition. In principle, such events can be manually annotated based on the gaze coordinate time series after the conclusion of a measurement. In practice, however, event detection is commonly carried out using an abundance of computer algorithms (see Holmqvist et al. [15] for an overview and Andersson et al. [17] for a recent benchmark of available solutions). For gaze-contingent paradigms, any suitable algorithm and its actual implementation have to have real-time processing capabilities, i.e., be able to detect and report events based on a single-pass analysis of a gaze coordinate stream. It can be a substantial technical challenge to utilize third-party tools within a given software framework for conducting behavioral experiments. Therefore, the selection of a detection algorithm implementation generally involves considerations of compatibility, in addition to optimal algorithmic performance. One software solution for real-time event detection is *PyGaze*, a *Python*-based toolbox for eye tracking [18] that can either be run as a stand-alone software or as plugin in *OpenSesame* (a general experiment builder) [19]. Many eye tracking systems also support the online detection of eye movement events. However, these capabilities are of limited use for scientific studies, as they typically rely on closed source, proprietary algorithms and implementations that cannot be properly verified or exhaustively described.

As mentioned before, optimal data quality for both fMRI and eye tracking may not be achievable for any given participant and experiment. In the context of such combined studies, any necessary compromise will typically result in suboptimal eye tracking results. Compared to dedicated laboratory setups, these can lead to reduced spatial precision or even complete signal loss. Figure 3 illustrates this difference, showing increased noise in the gaze coordinate measurement (possibly due to lack of image contrast as a result of suboptimal lighting) that translate to noisier eye movement velocity estimates, which are the primary input for a particularly effective class of event detection algorithms [17].

Failure of algorithmic eye movement event detection due to suboptimal data quality can incur substantial costs in terms of additional data post-processing or manual annotation. Consequently, care should be taken to tune an acquisition setup before

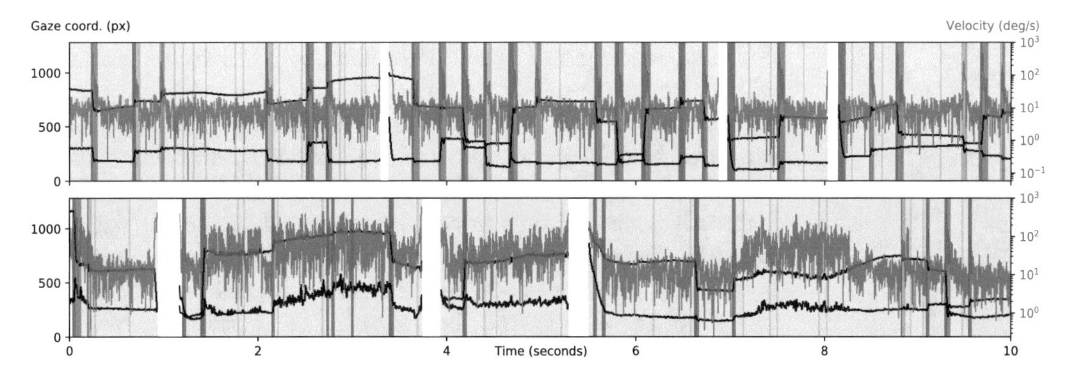

Fig. 3 Exemplary data quality comparison for a high-quality lab sample (top row) and a lower-quality sample from an MRI acquisition setup (bottom row). MRI sample data may exhibit more frequent signal loss (white) and higher and more variable noise in recorded gaze coordinates (black) that translates to noisier velocity estimates (gray). Reduced data quality represents a challenge for manual or automated eye movement event detection. Exemplary event detection results are depicted by colored time series segments: periods of fixation (green), pursuit (beige), saccades (blue), and high-/low-velocity post-saccadic oscillations (dark/light purple)

the start of a study to yield the maximum attainable data quality. Additionally, event detection algorithms that are capable of processing lower-quality data, such as REMoDNaV [20], can be employed when optimal data quality could not be achieved. However, such solutions may not be universally applicable. In the case of REMoDNaV, the algorithm is designed for offline data processing and is not suitable for event detection in the context of gaze-contingent stimulation.

Regardless of a specific choice of event detection algorithm, it is advisable to always check detection results for plausibility. Candidate strategies for verification are to (1) compute and inspect key eye movement properties, such as movement velocities, saccade frequency, or the log-linear relationship of saccade distance and peak velocity [21], and confirm sensible event statistics and (2) visualization of raw and preprocessed data together with event detection results, as shown in Fig. 3, to screen any acquisition for detection artifacts.

3.2 Analysis of Pupillometry

The size of the human pupil ranges from about 2 to 8 mm in diameter (reviewed in Mathot [22]). Most of the variance within that range is explained by light and distance to fixation: the pupil constricts in response to brightness increases (the pupil light response); and the pupil constricts when gaze shifts from far to near (the pupil near response). A third factor that affects pupil size is mental effort, arousal, or more generally any psychological process that "activates the mind" [23]; this always results in pupil dilation, although the effect of psychological processes on pupil size is far smaller than that of light or fixation distance.

3.3 Preprocessing of Pupil-Size Data

Pupil-size data requires preprocessing. Detailed recommendations can be found elsewhere [24, 25], but the following steps are generally performed.

First, missing and invalid data need to be identified and dealt with. Identifying missing data is easy, because most eye trackers report missing data as a pupil size of 0. Identifying invalid data is less easy, but, in general, pupil-size data is invalid when it falls outside of the physiological range (<2 mm or >8 mm) or when it is directly preceded or followed by missing data (before data goes missing, there is often a pronounced-but-artifactual drop in pupil size). Once periods of missing or invalid data have been identified, these can either be discarded by marking them as *nan* (not-a-number) values or be reconstructed with a linear or cubic-spline interpolation [26, 27].

Second, for most eye trackers, measured pupil size depends on eye position. In small part, this effect is real: pupil size really does change as a function of (at least vertical) eye position [28]. But in large part, this effect is artifactual and results from changes in the angle from which the eye tracker records the pupil. When viewed from the side, the pupil is recorded as an ellipse; when viewed from the front, the pupil is recorded as a circle, and with most eye trackers, this difference in viewing angle translates into a difference in measured pupil size. There are several ways to correct for these position artifacts (e.g., [29–31]). But in general, it is important to be aware that eye position and pupil size are interdependent measures.

Third, when using an event-related design, baseline correction is generally applied to pupil size [22]. That is, pupil size is measured at a suitable moment (the baseline) just before the event of interest (e.g., the appearance of a stimulus), and this baseline value is then subtracted from the pupil-size values during the epoch of interest (e.g., after the appearance of a stimulus).

3.4 Combining Pupil-Size and fMRI Data

When using an event-related design, deconvolution can be applied to pupil-size time series in a similar way as it can be applied to BOLD time series [32]. That is, pupil size is assumed to result from the combination of an impulse at time 0 and a hypothetical response function. The impulse strength is then deduced from the observed pupil response. The main difference between deconvolution of pupil-size and BOLD time series is the hypothetical response function (see Fig. 4). For BOLD time series, the hemodynamic response function is slower and peaks after about 5 s. For pupil time series, the response function as estimated by Hoeks and Levelt [33] for psychological events ("attention pulses") is much faster and peaks after about 1 s.

When not using an event-related design, the pupil-size time series needs to be downsampled to match the temporal resolution of the BOLD time series (e.g., from 1,000 to 0.5 Hz). For most research questions, the appropriate form of downsampling is to take the average or median pupil size during each bold sample, but depending on the research question, more advanced techniques can be used for downsampling as well. For example, pupil-size variability or the number of "pupil dilation events" [34] can be determined for each BOLD sample.

Also, when not using event-related design, the difference in response functions between BOLD and pupil-size time series (Fig. 4) should be taken into account. In its simplest form, this can be done by temporally offsetting the pupil-size time series such that the pupil-size signal at time 0 s is matched with the BOLD signal at time 4 s.

3.5 Mirror Alternative

Whenever an eye tracker vendor's MRI kit does not provide an IR-reflective mirror that fits the dimensions of the available space within the scanner bore, it is possible to obtain an affordable alternative from suppliers of consumer electronic parts. For example, mirrors for consumer back-projection TVs or other home cinema products are often suitable IR-reflective, first-surface

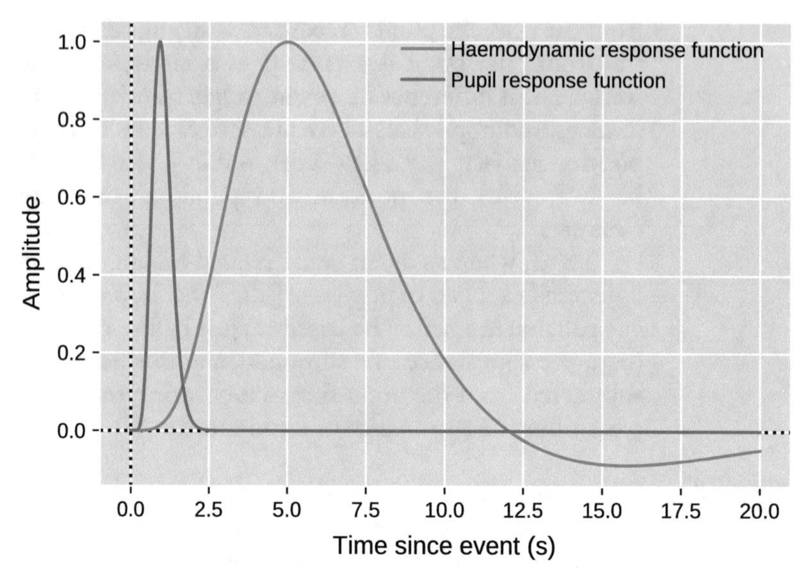

Fig. 4 The pupil response function (orange line) peaks about 1 s after the triggering event. The hemodynamic response function (blue line) peaks about 5 s after the triggering event

mirrors. Some manufacturers also provide trapezoid shapes in addition to rectangular mirrors which can be instrumental in configurations with very little available space. Typically, such mirrors can be ordered with thicker glass layers (~5 mm) for improved handling robustness, in contrast to high-precision mirrors for optical physics (<1 mm thickness). A typical reflectivity of ~95% is sufficient for visual stimulation and eye tracking.

3.6 Participant Positioning Aid

The positioning of a participant for optimal eye tracking data quality can be facilitated, if a live video feed of the eye tracking camera is presented on the stimulation screen inside the scanner during positioning. Using this visual guide, participants can typically find the optimal compromise between screen visibility and visibility of their eye, quickly. If the eye tracker is capable of showing a visual indicator for a detected corneal reflection, this can be used to quickly assess whether all relevant gaze coordinates can be measured or whether eyelid or eyelashes occlude the light for some eye positions. Participants only need to be instructed to sequentially move their gaze to the extreme locations in the visual stimulus for an assessment.

3.7 Developing and Conducting Gaze-Contingent Experiments

As the experimental procedure relies on correctly extracting information on the current gaze position, it is critical that the experiment is sufficiently piloted and robust against a variety of potential issues: *Invalid gaze coordinate samples* can occur at any time when no measurement is possible (e.g., when the eyes are closed). To make sure that an experiment does not crash halfway in a session, a fallback procedure must be in place that is executed when an invalid gaze sample is encountered (e.g., assume the actual coordinate is unchanged and keep querying the eye tracker for a valid update).

Lost Calibration: Even though participants move much less in an MR scanner than in a sitting position in a laboratory eye tracking experiment, it is still likely that a calibration is invalidated during a session. If the progression of the experiment depends on precise gaze positions, this would be detrimental. To avoid having to abort and restart a run, it makes sense to incorporate break points in such an experiment, during which the user interface for (re-)calibration can be accessed, in order to fix the issue. It is important to appropriately instruct participants regarding this possibility beforehand, so that they will know what to do, in case their task is suddenly interrupted by a calibration screen.

"Atypical" Eyes and Eye Movements: The eyes of some individuals can be notoriously hard to track, without any upfront indication. Piloting prospective participants will reduce the chance for having to abort an experiment session, because eye tracking quality is insufficient. However, even if an individual initially seemed to be a suitable participant, the in-scanner environment (lighting conditions, half-closed eyelids) can substantially impact tracking reliability. To some extent, these issues can be mitigated by instructing participants to open their eyes wider than usual, but this can cause undesired side effects on the task performance and brain activity. These problems are frequently restricted to a subset of the visual field and to large eccentricities. Therefore, they might be mitigated by avoiding these regions of the visual display.

However, even if the eye tracking itself works sufficiently well, it is still possible that the employed algorithms for event detection are not appropriate for the eye movements of a particular participant. For example, saccade detection is usually based on velocity and acceleration of an eye movement. If individuals make unusually slow saccades, some algorithms might not be able to capture these eye movements. Training individuals can mitigate the issue, but the benefit is limited. Running behavior-only pilots outside the scanner will help to select suitable individuals. If custom-made algorithms are used, they have to be validated on a large enough sample.

Even though it generally makes sense to use event detection algorithms that accurately capture all types of events, they are vulnerable to noisy data and will occasionally fail during an experiment. Therefore, it is oftentimes possible to approximate events with some heuristic, making the experimental procedure more robust without sacrificing eye tracking precision (too much). For example, if the visual display is supposed to change during a saccade from one stimulus to another and these two stimuli are far apart

(>8°), instead of using a saccade detection algorithm, one can simply use the raw gaze position and update the display once the midline between the two points is crossed. Similarly, if the trial progression depends on which stimulus is fixated and stimuli are far enough apart, it might be viable to increase the size of the region around each stimulus that triggers a fixation. Even though using such heuristics can greatly improve the robustness and smoothness of an experiment, one always has to consider whether the loss in precision is acceptable in light of the specific research question.

References

1. Ort E et al (2019) The role of proactive and reactive cognitive control for target selection in multiple-target search. BioRxiv:559500. https://doi.org/10.1101/559500
2. Geringswald F, Baumgartner F, Pollmann S (2012) Simulated loss of foveal vision eliminates visual search advantage in repeated displays. Front Hum Neurosci 6:134. https://doi.org/10.3389/fnhum.2012.00134
3. Hansen DW, Ji Q (2010) In the eye of the beholder: a survey of models for eyes and gaze. IEEE Trans Pattern Anal Mach Intell 32:478–500
4. Van der Stigchel S, Meeter M, Theeuwes J (2006) Eye movement trajectories and what they tell us. Neurosci Biobehav Rev 30 (5):666–679
5. Theeuwes J, Belopolsky A, Olivers CN (2009) Interactions between working memory, attention and eye movements. Acta Psychol 132 (2):106–114
6. Van Gerven M et al (2004) Memory load and the cognitive pupillary response in aging. Psychophysiology 41:167–174
7. Mathôt S et al (2015) Large pupils predict goal-driven eye movements. J Exp Psychol Gen 144(3):513–521. https://doi.org/10.1037/a0039168
8. Lean Y, Shan F (2011) Brief review on physiological and biochemical evaluations of human mental workload. Hum Factors Man 22 (3):177–187
9. Gordon PC et al (2006) Similarity-based interference during language comprehension: evidence from eye tracking during reading. Exp Psychol Learn Mem Cogn 32(6):1304–1321
10. Rayner K (2009) Eye movements and attention in reading, scene perception, and visual search. Q J Exp Psychol (Hove) 62(8):1457–1506
11. Peitek N et al (2018) Simultaneous measurement of program comprehension with fMRI and eye tracking: a case study. In: Proceedings of the international symposium of empirical software engineering and measurement (ESEM). ACM, New York, pp 24:1–24:10
12. Henderson JM (1997) Transsaccadic memory and integration during real-world object perception. Psychol Sci 8(1):51–55
13. Mathôt S, Theeuwes J (2011) Visual attention and stability. Philos Trans R Soc Lond Ser B Biol Sci 366(1564):516–527. https://doi.org/10.1098/rstb.2010.0187
14. Hanke M et al (2016) A studyforrest extension, simultaneous fMRI and eye gaze recordings during prolonged natural stimulation. Sci Data 3:160092
15. Holmqvist K et al (2011) Eye tracking: a comprehensive guide to methods and measures. OUP, Oxford
16. Richlan F et al (2013) A new high-speed visual stimulation method for gaze-contingent eye movement and brain activity studies. Front Syst Neurosci 7:24. https://doi.org/10.3389/fnsys.2013.00024
17. Andersson R et al (2017) One algorithm to rule them all? An evaluation and discussion of ten eye movement event-detection algorithms. Behav Res Methods 49(2):616–637. https://doi.org/10.3758/s13428-016-0738-9
18. Dalmaijer ES, Mathôt S, Van der Stigchel S (2013) PyGaze: an open-source, crossplatform toolbox for minimal-effort programming of eyetracking experiments. Behav Res Methods 46:913–921. https://doi.org/10.3758/s13428-013-0422-2
19. Mathôt S, Schreij D, Theeuwes J (2012) OpenSesame: an open-source, graphical experiment builder for the social sciences. Behav Res Methods 44:314–324. https://doi.org/10.3758/s13428-011-0168-7
20. Dar AH, Wagner AS, Hanke M (2019) REMoDNaV: robust eye movement detection for natural viewing. bioRxiv:619254. https://doi.org/10.1101/619254

21. Bahill AT, Clark MR, Stark L (1975) The main sequence, a tool for studying human eye movements. Math Biosci 24(3–4):191–204. https://doi.org/10.1016/0025-5564(75)90075-9
22. Mathôt S (2018) Pupillometry: psychology, physiology, and function. J Cogn 1(1):1–16. https://doi.org/10.5334/joc.18
23. Goldwater BC (1972) Psychological significance of pupillary movements. Psychol Bull 77(5):340–355. https://doi.org/10.1037/h0032456
24. Kret ME, Sjak-Shie EE (2018) Preprocessing pupil size data: guidelines and code. Behav Res Methods:1–7. https://doi.org/10.3758/s13428-018-1075-y
25. Mathôt S et al (2018) Safe and sensible preprocessing and baseline correction of pupil-size data. Behav Res Methods:1–13. https://doi.org/10.3758/s13428-017-1007-2
26. Hershman R, Henik A, Cohen N (2018) A novel blink detection method based on pupillometry noise. Behav Res Methods 50 (1):107–114. https://doi.org/10.3758/s13428-017-1008-1
27. Mathôt S (2013) A simple way to reconstruct pupil size during eye blinks. https://doi.org/10.6084/m9.figshare.688001
28. Mathôt S, Melmi JB, Castet E (2015) Intrasaccadic perception triggers pupillary constriction. PeerJ 3(e1150):1–16. https://doi.org/10.7717/peerj.1150
29. Brisson J et al (2013) Pupil diameter measurement errors as a function of gaze direction in corneal reflection eyetrackers. Behav Res Methods 45(4):1322–1331. https://doi.org/10.3758/s13428-013-0327-0
30. Gagl B, Hawelka S, Hutzler F (2011) Systematic influence of gaze position on pupil size measurement: analysis and correction. Behav Res Methods 43(4):1171–1181. https://doi.org/10.3758/s13428-011-0109-5
31. Hayes TR, Petrov AA (2016) Mapping and correcting the influence of gaze position on pupil size measurements. Behav Res Methods 48(2):510–527. https://doi.org/10.3758/s13428-015-0588-x
32. Wierda SM, van Rijn H, Taatgen NA, Martens S (2012) Pupil dilation deconvolution reveals the dynamics of attention at high temporal resolution. Proc Natl Acad Sci 109 (22):8456–8460. https://doi.org/10.1073/pnas.1201858109
33. Hoeks B, Levelt WJ (1993) Pupillary dilation as a measure of attention: a quantitative system analysis. Behav Res Methods Instrum Comput 25(1):16–26. https://doi.org/10.3758/BF03204445
34. Joshi S et al (2016) Relationships between pupil diameter and neuronal activity in the locus Coeruleus, Colliculi, and cingulate cortex. Neuron 89(1):221–234. https://doi.org/10.1016/j.neuron.2015.11.028

Neuromethods (2020) 151: 307–310
DOI 10.1007/7657_2019_29
© Springer Science+Business Media, LLC 2019
Published online: 23 July 2019

Correction to: Topographic Mapping of Parietal Cortex

Summer Sheremata

Correction to:
Chapter "Topographic Mapping of Parietal Cortex" in: S. Pollmann, Neuromethods,
https://doi.org/10.1007/7657_2019_23

This chapter was inadvertently published with incorrect Tables 1 and 2. The correct presentation is given here.

The updated online version of this chapter can be found at
https://doi.org/10.1007/7657_2019_23

Table 1

```
# Create anatomy file and copy to output directory

mri_convert ${subj_anat_dir}/brain.mgz ${subj_anat_dir}/mri/brain.nii

3dcopy ${subj_anat_dir}/mri/brain.nii ${subj}_anat_stripped+orig
```

```
# Motion correct

touch out.pre_ss_warn.txt

foreach run($runs)

   3dvolreg -verbose -zpad 1 -base pb00.${subj}.r01.tcat+orig.'[2]' \
       -1Dfile dfile.r${run}.1D -prefix pb01.${subj}.r${run}.volreg \
       -cubic pb00.$subj.r$run.tcat+orig

end
```

```
# Detrend

foreach run($runs)

   3dTstat -prefix tempMean pb01.${subj}.run${run}.volreg

   3dDetrend -prefix tempDetrend -vector dfile.r${run}.1D \
       -polorrt 2 pb01.${subj}.r${run}.volreg+orig

   3dcalc -a tempDetrend+orig -b tempMean+orig -expr "(a+b)" -float \
       -prefix pb02.${subj}.r${run}.detrended

   rm temp*

end
```

```
# Smoothing

foreach run($runs)

  3dmerge -1blur_fwhm 4.0 -doall -prefix pb03.${subj}.r${run}.smoothing \
       pb02.${subj}.r${run}.detrend+orig

end
```

(continued)

Table 1 (continued)

```
# Average

3dcalc -a pb03.retina2.r01.smoothing+orig.HEAD \
       -b pb03.retina2.r02.smoothing+orig.HEAD \
       -c pb03.retina2.r03.smoothing+orig.HEAD \
       -d pb03.retina2.r04.smoothing+orig.HEAD \
       -expr '(a+b+c+d)/4' -prefix average.ret
```

```
# Automask and remove mean

3dcopy average.ret  AFNI_pRF+orig.BRIK

3dTstat -mean -prefix AFNI_pRFmean AFNI_pRF+orig.BRIK

3dAutomask -prefix automask AFNI_pRF+orig.BRIK

3dcalc -a AFNI_pRF+orig. -b AFNI_pRFmean+orig. -c automask+orig. \
       -expr '100*c*(a-b)/b'  -prefix e.scale.demean
```

```
# Run model

3dNLfim -input e.scale.demean+orig -mask automask+orig -noise Zero \
        -signal Conv_PRF
        -sconstr 0 -10.0 10.0  \
        -sconstr 1 -1.0 1.0    \
        -sconstr 2 -1.0 1.0    \
        -sconstr 3 0.0 1.0     \
        -BOTH -nrand 10000 -nbest 5 -bucket 0 Buckslow.PRF -snfit snfitslow.PRF

3dcalc -a Buckslow.PRF+orig'[1]' -b Buckslow.PRF+orig'[2]' -expr 'sqrt(a^2+b^2)' \
        -prefix polarslow.m

3dcalc -a Buckslow.PRF+orig.'[1]' -b Buckslow.PRF+orig.'[2]' -expr 'atan2(a,-b)' \
        -prefix polarslow.ph

3dbucket -glueto Buckslow.PRF+orig.  polarslow.m+orig.  polarslow.ph+orig.
```

Table 2

```
# Deoblique and align to epi

3dWarp -prefix Buckslow.PRF_warped+orig -deoblique Buckslow.PRF+orig

3dWarp -prefix PRF_mc_warped+orig -deoblique pb01.${subj}.r01.volreg+orig

align_epi_anat.py -anat ${subj}_anat_stripped+orig -anat_has_skull no \
    -epi PRF_mc_warped+orig -epi_strip 3dAutomask -cost lpc+ZZ \
     -multi_cost nmi lpa -suffix _al2epi -epi_base 48 -prep_off

# Align to SUMA

@SUMA_Make_Spec_FS -NIFTI -fspath ${subj_anat_dir}/ -sid ${anat_subj}

@SUMA_AlignToExperiment -wd -exp_anat anat_stripped_al2epi+orig.BRIK     \
 -surf_anat ${subj_anat_dir}/SUMA/${anat_subj}_SurfVol+orig.     \
 -align_centers -strip_skull surf_anat \
 -prefix ${anat_subj}_SurfVol_ns_centre_AInd_Exp


foreach hemi (lh rh)

3dVol2Surf -spec ${subj_anat_dir}/SUMA/std.141.${anat_subj}_${hemi}.spec   \
          -surf_A smoothwm -surf_B pial  \
          -sv ${anat_subj}_SurfVol_ns_centre_AInd_Exp+orig      \
          -grid_parent $output_dir/Buckslow.PRF_warped+orig      \
          -oob_value 0-map_func ave -f_steps 10 -f_index nodes.  \
          -out_niml std_${hemi}_Buckslow.PRF_warped+orig.BRIK.niml.dset

end
```

Also, the chapter has been published without updating the following correction:

p. 12, First paragraph last sentence has been amended to ‘Here I have inverted the left hemisphere so that the colors in both hemispheres represent the same locations vertically.’

INDEX

Stefan Pollmann (ed.), *Spatial Learning and Attention Guidance*, Neuromethods, vol. 151,
https://doi.org/10.1007/978-1-4939-9948-4, © Springer Science+Business Media, LLC, part of Springer Nature 2020

S

T

U

V

W

Z

Printed in the United States
by Baker & Taylor Publisher Services